PENGUIN REFERENCE BOOKS

A DICTIONARY OF ELECTRONICS

S. Handel, who was born in London, was educated at the Regent Street Polytechnic and University College. He spent over thirty years as a professional engineer in various branches of electronics and was a Member of the Institution of Electrical Engineers. For many years he practised as an independent consultant, and during that time contributed many articles on the subject to technical journals.

He believed that whatever applied electronics has achieved is quite insignificant compared to its potentialities. It makes technically possible a civilization in which physical and mental drudgery will be an anachronism and everyone will have free access to the whole of human culture.

He also published *The Electronic Revolution* in Pelicans.

He died in 1971.

D1332652

S. HANDEL

A DICTIONARY OF ELECTRONICS

THIRD EDITION

PENGUIN BOOKS

Penguin Books Ltd, Harmondsworth, Middlesex, England
Penguin Books Inc., 7110 Ambassador Road, Baltimore, Maryland 21207, U.S.A.
Penguin Books Australia Ltd, Ringwood, Victoria, Australia
Penguin Books Canada Ltd, 41 Steelcase Road West, Markham, Ontario, Canada
Penguin Books (N.Z.) Ltd, 182–190 Wairau Road, Auckland 10, New Zealand

—

First published 1962
Reprinted 1964
Second edition 1966
Reprinted 1967, 1968
Third edition 1971
Reprinted 1972, 1973, 1974

—

Copyright © the Estate of S. Handel, 1962, 1966, 1971

—

Made and printed in Great Britain
by Cox & Wyman Ltd,
London, Reading and Fakenham
Set in Monotype Times

PREFACE

OF the 5,000 or so entries in this dictionary about one third did not exist before 1950; yet many of the new words are already part of our common language. Could anything show more clearly the startling growth of electronics and its impact on our lives?

The hardest part of writing this book was to decide which words to leave out. Those omitted could fill another book, and by the time that was written there would be many more. Acronyms, following a war-time fashion, run riot. The author feels that what he needs in the role of lexicographer (defined in Dr Johnson's dictionary as 'a harmless drudge') is a RED PENCIL: a Reliable Electronic Device for Printing Every Name Composed of Initial Letters. The difficulty is that we cannot dismiss these ingenious creations as amusing but ephemeral – words such as 'radar' and 'maser' show the fallacy in that assumption. On the other hand a useful word such as 'gubbins' – universally understood by electronic engineers to mean 'a highly technical piece of equipment which *can* be made to work' – can hardly be included in a serious work of reference. Again, as a keen collector of names ending in '-tron', the author was amazed to find in a recent issue of *British Communications and Electronics* a list of 561 members of the TRON family which included dozens of names he had never even heard of. Some items, such as 'patron' and 'matron', have no claim to be included here, but others have been excluded regretfully for lack of space. So that, if through ignorance, prejudice, or timidity the author has omitted any word which a reader feels strongly should be included, he would be grateful to be informed of it.

The aim of the book is to include the maximum amount of authentic information which is economically feasible. Thus the definitions have been made as terse as is consistent with accuracy and clarity. This should prove acceptable to technical readers, many of whom may be suffering from mental indigestion due to a surfeit of words in technical journals, proceedings of learned societies, research reports, academic texts, etc. For non-technical readers every effort has been made to explain any words not to be found in any English dictionary in terms of words which will be found there. This may entail hunting through a series of entries, but those who stick to it will be rewarded by gaining some insight into the fascinating world of electronics.

S.H.

London
June 1962

ACKNOWLEDGEMENTS

THE list of British and American sources, references, journals, colleagues, societies, institutions, and electronic firms consulted in the preparation of this book is far too long to enumerate, but the author would like to mention some of them.

Wherever possible definitions have been based on the recommendations of the British Standards Institution, the American Standards Association, the British and American Institutions of Electrical Engineers and Institutions of Radio Engineers, the International Electrotechnical Vocabulary, and the American National Television Standards Committee.

Books which have been particularly valuable include: Dr. R. Sarbacher's monumental *Encyclopaedic Dictionary of Electronics and Nuclear Engineering*; Van Nostrand's *International Dictionary of Physics and Electronics*; *Electronic Engineer's Reference Book* edited by Hughes; and Kaye and Laby's *Physical and Chemical Constants*.

The libraries of the I.E.E., Science Museum and Patent Office were unfailingly helpful.

Firms providing information and illustrations of their products include: A.E.I. Ltd, Decca Ltd, English Electric Ltd, E.M.I. Electronics Ltd, Ericsson Telephones Ltd, Ferranti Ltd, Hughes International Ltd, Motorola Semiconductor Products Inc., Mullard Ltd, and Standard Telephones and Cables Ltd. Trade names are indicated by the suffix mark ® after the name itself.

The author also thanks many friends and colleagues for their helpful suggestions and wishes to put on record his special thanks to Mr Wilfred Hardy, and Mr Cecil Misstear, who drew the illustrations.

NOTE TO THIRD EDITION

IT is almost ten years since the first edition of this book was set up and this is a long time in the life of a technology which is growing as fast as electronics. Some of the achievements of applied electronics in this period which come to mind are:

A world-wide satellite system for television and telephone communication is now in commercial operation.

The first man was landed on the moon and the event was televised to nearly a third of the world's population.

The first robot exploration vehicle, Lunokhod I, is transmitting detailed information about the composition of the moon. There has been an enormous rate of growth in the trunk telephone system: e.g. for industrial usage in the U.K. it is 116%. The rapid development of microcircuits and in particular the technique of large-scale integration has transformed not only the electronics industry but also many of the industries which it serves.

The low-priced computer has become a reality and the mini-computer is on the horizon.

The electronics industry is the second largest in the U.S.A. and it is probable that it will be the largest before the end of this century.

It is hoped that these revolutionary developments have been faithfully reflected in this dictionary and every effort had been made to keep it up to date. A number of valuable suggestions from readers have been incorporated and comments and suggestions will continue to be welcome.

What is quite certain is that in the next ten years we shall feel the full impact of the electronic revolution.

January 1971   S.H.

Note Re Cross-References

Words in italics in the definitions refer
the reader to separate entries.

A

A BATTERY. The battery used to supply power for the *heaters* of electronic valves (U.S.A.). Heater battery (G.B.).

AB PACK (U.S.A.). A package of A and B *batteries* providing a complete source of power for battery-operated valves.

ABAC. A graphical system of lines or curves, usually on a chart, which can be used to solve equations without calculation.

ABAMPERE (ABAMP). The unit of current in the *Electromagnetic System of Units*. It is that current which, if flowing in straight parallel wires 1 cm. apart in free space, produces a force of 2 dynes per cm. length of each wire. 1 abampere = 10 *amperes*. See *Units*.

ABC. Abbreviation for *automatic brightness control* in television receivers.

ABERRATION. An image defect in the electronic lens system of a *cathode-ray tube*.

ABNORMAL GLOW DISCHARGE. A *glow discharge* in which the working voltage increases as the current increases.

ABOHM. The unit of resistance in the *electromagnetic system*; equal to 10^{-9} *ohms*. See *Units*.

ABSOLUTE ADDRESS. In *computers*, the code description of a specific register of storage element allocated by the designer.

ABSOLUTE ALTIMETER. An electronic instrument which measures the distance from an aircraft to the surface of the ground or sea.

ABSOLUTE AMPERE. One tenth of an *abampere*; since 1950, the legal standard of electricity.

ABSOLUTE UNITS. Units defined in terms of the fundamental units of length, mass, time, and charge.

ABSORPTION COEFFICIENT, ELECTRONIC. The fractional decrease in the intensity of a beam of radiation per number of electrons per unit area.

ABSORPTION CURRENT. That part of the current in a *dielectric* which is proportional to the rate of accumulation of electric charge within the dielectric.

ABSORPTION WAVEMETER. A device for measuring *wavelength* by tuning a resonant circuit so that it absorbs the maximum energy from the source to be measured.

AC. Abbreviation for *alternating current*.

AC BRIDGE. Abbreviation for alternating current *bridge*.

AC/DC RECEIVER. A radio receiver designed to work from either alternating or direct current supplies.

AC INTERRUPTION. The opening of a circuit at the input side of a *rectifier*.

11

ACCELERATING ANODE. In a *cathode-ray tube*, the electrode which is given a high positive voltage to increase the velocity of electrons in the electron beam.

ACCELERATING ELECTRODE. Any electrode in an electronic valve or tube which serves to increase the velocity of electrons in the *electron beam*.

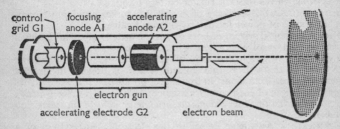

Fig. 1. Accelerating electrodes in a typical cathode-ray tube (diagrammatic)

ACCELERATION SPACE. The region following the *electron gun* of a valve or tube in which the emitted electrons are accelerated.

ACCELERATION VOLTAGE. The voltage between anode and cathode of a *velocity-modulated valve* such as a *klystron*.

ACCELERATOR, HIGH VOLTAGE. A machine, or part of a machine, used to accelerate charged particles, such as ions or electrons, to great speeds by the application of high voltages. See *Betatron*, *Cyclotron*, *Synchro-cyclotron*.

ACCELEROMETER. An instrument which measures the acceleration to which it is subjected. Electronic kinds use some form of *transducer* which usually produces a voltage proportional to the acceleration.

ACCENTUATION. A method of emphasizing selected bands of frequencies in an *audio amplifier*.

ACCEPTANCE ANGLE. Of a *phototube*: the solid angle inside which all incident light reaches the *photocathode*.

ACCEPTOR CIRCUIT. A circuit adjusted so that it presents the minimum impedance to a desired frequency.

ACCEPTOR IMPURITY. In a crystalline *semiconductor*, a minute trace of an added substance whose atoms attract electrons from the atoms in the crystal lattice and so create *holes* in the main structure. This gives rise to *P-type conductivity*. For example one part in ten thousand of boron can increase the conductivity of silicon one million times.

ACCEPTOR LEVEL. The name given to a band of energy in the *energy level diagram* of a *semiconductor* crystal lattice to which *acceptor impurities* have been added. It is slightly above the normal *valence*

ACTUATOR

band, and at all temperatures above absolute zero some electrons are thermally transferred to this level, leaving holes in the lower energy band.

ACCESS TIME. In a *memory* or *storage device*, the time interval between the instant at which the information is demanded from the store and the instant when it is available in a useful form.

ACCUMULATOR. (1) A storage battery, usually with a liquid electrolyte. (2) In a computer, a device which stores one or more numbers and progressively adds numbers to this store.

ACE. An early electronic computer designed and studied at the National Physical Laboratory, Teddington.

ACORN VALVE (TUBE). A very small electronic valve, sometimes shaped like an acorn, with very small, closely spaced electrodes to reduce electron *transit time* to a minimum.

ACOUSTIC DELAY LINE. A device which transmits and delays sound pulses by re-circulating them in a liquid or solid medium.

ACOUSTIC FEEDBACK. In an *audio amplifier*: *feedback* of a sound wave from a loudspeaker to some preceding part of the system so as to augment the output. Beyond a critical point, howling or self-sustained oscillation results.

ACOUSTICS. The science of sound, including its production, transmission, and effects.

ACTINIUM (Ac). Radioactive element. Atomic No. 89.

ACTION CURRENT POTENTIAL. The short, very small current or potential wave following a stimulus.

ACTIVATION. The treatment applied to an emitting surface to establish or increase its *emission* of electrons.

ACTIVATION, THERMAL. The role played by the thermal energy of an *intrinsic semiconductor* in maintaining the concentration of free electrons and holes in the crystal structure.

ACTIVATOR (SENSITIZER). An impurity in a solid which produces, or greatly increases, the *luminescence* of the solid. Examples are thallium in potassium chloride and copper in zinc sulphide.

ACTIVE. Of a device, component, or part of an electric circuit: containing a source of energy.

ACTIVE AREA. The area of a metallic *rectifier* which acts as the rectifying junction and conducts current in the forward direction.

ACTIVITY. (1) In *piezoelectric* crystals, the magnitude of oscillation relative to the exciting voltage. (2) In radioactivity, the intensity or strength of a radioactive source.

ACTUATING TRANSFER FUNCTION. In a *feedback control loop*: the mathematical relation between the loop input signal and the actuating signal.

ACTUATOR. A device for providing remotely controlled mechanical movement, usually in a straight line. Linear actuators consist of motor, speed-reducing gear, jack-screw, and motion-limiting switches.

13

ADALINE. An ADAptive LINEar pattern recognition machine built at Stanford University, U.S.A.

ADAPTIVE CONTROL. A form of control in which the controls adapt themselves automatically to give optimum results from the controlled process.

ADDER. In a *computer*, a device which can form the sum of two or more numbers or quantities impressed on it.

ADDITRON. A Canadian radial double-beam valve with electrostatic focusing, used as a binary adder in high-speed digital computers. It has three control grids, two current-collecting electrodes, a screen grid, and a central cathode.

ADDRESS. Information, usually in the form of a number, which identifies a particular location in a memory or storage device.

ADDRESS, MULTIPLE. Used in describing *digital computers*; e.g. two, three, or four address machines.

ADDRESS, ONE (or SINGLE). A system of *computer* instruction in which each complete instruction explicitly describes one operation and only one storage location is involved.

ADMITTANCE. The reciprocal of *impedance*. Symbol: Y. Thus, if Z = impedance, R = resistance, X = reactance:

$$Y = \frac{1}{Z} = \frac{1}{R + jX} = \frac{R}{IZI^2} - j\frac{X}{IZI^2} = G + jB, \text{ where}$$

$G = \dfrac{R}{IZI^2}$ is the *conductance*, and $B = -\dfrac{X}{IZI^2}$ is the *susceptance*.

The practical unit of admittance is the *mho*.

ADP. Abbreviation for automatic *data processing*.

AEOLIGHT (U.S.A.). A *glow discharge* lamp with a cold cathode filled with a mixture of permanent gases, in which the intensity of illumination varies with the applied signal. It is used as a modulating light for motion-picture sound recording.

AERIAL. That part of a radio system from which energy is radiated into, or received from, space. Antenna (U.S.A.).

AERIAL ARRAY. An assembly of simple aerials arranged in space so that transmitted or received energy is concentrated in one or more directions.

AF. Abbreviation for *audio frequency*.

AF AMPLIFIER AND TUNING INDICATOR. A valve comprising a *pentode* and an electronic indicator.

AF TRIODE. A *triode* valve designed primarily for low frequencies.

AFC. Abbreviation for *automatic frequency control*.

AFTERGLOW (PERSISTENCE). A *phosphorescent* material continues to glow after excitation and the effect may last from a fraction of a second to several years, according to the *phosphor*.

AGC. Abbreviation for *automatic gain control*.

AIR CAPACITOR. A *capacitor* in which the main dielectric is air.

ALCOMAX®. An alloy of iron, nickel, aluminium, cobalt, and copper which has exceptionally high *coercivity* and is used as a material for permanent magnets.

ALGOL. A symbolic *language* specifically designed to *programme* mathematical problems for *computers* used in scientific and engineering applications, without reference to the common *code* of a particular computer.

ALGORITHM. A set of well-defined rules, or process, for solving a problem in a finite number of steps.

ALIGNED GRID VALVE. See *Beam power valve*.

ALIGNMENT. In electronic equipment, adjustment of tuned circuits so that they respond in a desired way at a given frequency.

ALLOWED BAND. In solid-state physics, a range of levels of energy which electrons can attain in a material.

ALLOY DIODE. A *semiconductor diode* made from a wafer of semiconductor (silicon) in which the rectifying junction is made by an alloying process.

ALLOYED TRANSISTOR. A *transistor* made by alloying a *semiconductor* such as germanium or silicon with a metal such as indium.

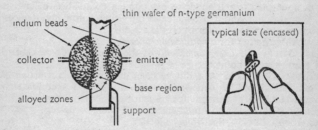

Fig. 2. Alloy junction transistor

ALNICO®. An alloy of nickel, iron, aluminium, cobalt, and copper which provides a high energy permanent magnet material.

ALPHA COUNTER. An electronic system for counting alpha particles which includes a special tube or counter chamber.

ALPHA CURRENT-GAIN FACTOR. Of a *transistor* with common base connection: the ratio of the small current change in the collector to the small current change in the emitter which produces it. Alpha is always less than 1.

ALPHA CUT-OFF. The frequency at which the current amplification – alpha – of a *transistor* has fallen to 0·7 of its low frequency value.

ALPHA PARTICLE. A positively charged particle emitted from an atom nucleus, composed of two protons and two neutrons. It is identical with a Helium nucleus.

ALPHA RAY. A stream of *alpha particles*.

ALPHANUMERIC. In *computer* terminology: by letters and numbers.

ALTERNATING CURRENT. A flow of electrons in which the direction of movement periodically reverses and the average value over a period is zero. The number of complete cycles in a second is called the *frequency*.

ALTERNATOR. A device designed to supply *alternating current*.

ALTIMETER, CAPACITANCE. An altimeter which measures the height of an aircraft in terms of its capacitance to earth.

ALUMINA. Aluminium Oxide, Al_2O_3. Used as an insulator in thermionic valves because of its excellent electrical and thermal resistance at high temperatures.

ALUMINIUM (Al). Metallic element, Atomic No. 13. As it is a good conductor, ductile, malleable, not easily corroded, light in weight, and in good supply, it is extensively used in electronic equipment.

ALUMINIUM ANTIMONIDE. A *semiconductor* having useful properties for *transistors* operating at temperatures up to 500°C.

ALUMINIZED SCREEN. The screen of a *cathode-ray tube* on the inside of which a thin layer of aluminium has been deposited. The coating is connected to the anode to overcome accumulation of charge on the phosphor, and it also increases the visible output and reduces the effect of *ion* bombardment on the screen.

AM. Abbreviation for *amplitude modulation*.

AMBIGUITY. Of a *servomechanism*: the condition of seeking more than one null point.

AM/FM RECEIVER. A radio receiver capable of converting both amplitude- and frequency-modulated waves into perceptible signals.

AMMETER. An instrument for measuring electric current in *amperes*. See also *Milliammeter* and *Microammeter*.

AMMONIA CLOCK. See *Atomic* clock.

AMPERE (A). The ampere is that current which, if maintained in two straight parallel conductors of infinite length, of negligible circular cross-section and placed 1 metre apart in vacuum, would produce between these conductors a force equal to 2×10^{-7} newton per metre of length.

A

symbol

AMPERE-HOUR (Ah). A unit of electric charge equivalent to the amount of electricity in a current of 1 *ampere* flowing for 1 hour. 1 amp-hour = 3,600 *coulombs*.

AMPERE-TURN. A measure of *magnetomotive force*. The product of the number of turns of a coil and the current in amperes which flows through it.

AMPERE-TURNS, CONTROL. A name expressing the magnitude and polarity of the control magnetomotive force required for the operation of a *magnetic amplifier* at a specified output.

AMPERE-TURN GAIN. In *magnetic amplifiers*, the ratio of the product of the load current and gate turns to the product of control current and control turns.

AMPERE-TURNS PER METRE. A measure of *magnetizing force, H.* The magnitude of *H* at any point in a magnetic circuit depends on the current in the circuit, the number of turns, and the geometry of the circuit.

AMPLIDYNE. A rotary *magnetic amplifier*, based on a quick response dc generator, which needs only a small energizing power. In combination with an electronic amplifier it allows accurate control of position, speed, acceleration, voltage, and current. See also *Servomechanism.*

AMPLIFICATION. A general expression applied to the increase of the magnitude of any kind of signal by an electronic device.

AMPLIFICATION, MAGNETIC. Amplification achieved by the use of the non-linear properties of *saturable reactors.*

AMPLIFICATION, POWER. (1) In an *amplifier*, the ratio of the power level at the output terminals to that at the input terminals. (2) In a *magnetic amplifier*, the product of voltage amplification and current amplification using a specified control circuit. (3) In a *transducer*, the ratio of the power delivered to the load under specified operating conditions, to the power absorbed by the input circuit.

AMPLIFICATION FACTOR (OF A VALVE). The ratio of the small change of voltage on the *anode* required to counteract a very small change on the *control grid* to keep the anode current constant.

AMPLIFIER. A device whose output is a magnified function of its input and which draws its power from sources other than the input signal. The following types of amplifier are defined in this dictionary: *audio, balanced (push-pull), booster, bootstrap, buffer, cascade, cascode, cathode follower, chopper, class A, class A B, class B, class C, class D, direct-coupled, direct-current, distributed, feedback, grounded-anode, grounded-cathode, grounded-grid, head, intermediate frequency, linear, magnetic, operational, parametric, paraphase, power, push-pull, push-push, quiescent*

symbol

push-pull, radio-frequency, regenerative, servo, single-ended, stagger-tuned, transistor, tuned, video frequency, voltage, wide-band.

AMPLISTAT®. A self-saturating *magnetic amplifier* circuit.

AMPLITRON®. A *microwave* amplifying tube made by the Raytheon Mfg Co., U.S.A.

AMPLITUDE. The maximum departure of the value of an *alternating current* or wave from the average value.

AMPLITUDE, INITIAL. The maximum displacement from a reference point of a periodic function.

AMPLITUDE, PEAK TO PEAK. Of an oscillating quantity, the difference between extreme values in a complete cycle.

AMPLITUDE COMPARISON. The process of indicating the time at which two (or more) waveforms reach the same amplitude.

AMPLITUDE DISCRIMINATOR. A circuit which performs an *amplitude comparison.*

AMPLITUDE DISTORTION. The type of distortion which may occur in an *amplifier* or other device in which the output amplitude is not an exactly linear function of the input amplitude.

AMPLITUDE LIMITER. See *Limiter.*

AMPLITUDE MODULATION (AM). The *modulation* of a simple *sine wave* by a fluctuating signal so that its instantaneous amplitude varies in sympathy with the signal. It is one of the main methods of transmission of information by radio waves: the constant amplitude radiofrequency sine wave – the *carrier* – being modulated by sound vibrations or picture signals.

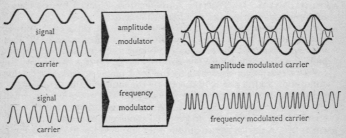

Fig. 3. Amplitude and frequency modulation compared

AMPLITUDE MODULATION RECEIVER (AM RECEIVER). A receiver for converting amplitude-modulated waves into perceptible signals, usually of *audio-frequency.*

AMPLITUDE OF OSCILLATION. The peak value of a *sine wave.*

AMPLITUDE SEPARATION. The process of separating all values of a periodic function greater or less than a given amplitude, or those values lying between two given amplitudes. See also *Clipper.*

AMPLITUDE VS FREQUENCY RESPONSE CHARACTERISTIC. The variation with frequency of gain or loss in any device or system.

ANALOGUE COMPUTER. A *computer* which can perform operations on numbers, where the numbers are represented by some physical quantity. In most modern computers electrical quantities such as voltage and current are used.

ANALOGUE-TO-DIGITAL CONVERTER. A device for translating analogue information into digital form.

ANDERSON BRIDGE. A modification of the *Maxwell bridge* with two independent balance conditions, used for measuring *inductance* in terms of *capacitance* and *resistance.*

18

AND GATE. A *gating* circuit, with two or more input wires, in which the output wire gives a signal if, and only if, all input wires receive coincident signals.

symbol

ANGLE MODULATION. See *Frequency modulation*.

ANGLE OF DEFLECTION. In a *cathode-ray tube*, the angle through which the electron beam is deflected.

ANGLE OF FLOW. That proportion, expressed as an angle, of the cycle of an alternating voltage during which current flows.

ANGSTROM. A very small unit of length equal to 10^{-10} metres or 1/10,000th of a *micron*.

ANHARMONIC OSCILLATOR. An *oscillator* in which the restoring force varies linearly with the displacement of the system from its equilibrium position, and therefore produces no harmonics.

ANION. An *ion* which deposits on the *anode*. Anions form that part of an *electrolyte* which carries the negative charge and travels against the conventional direction of the current. In a *battery* the deposition of anions makes the anode negative.

ANISEIKON. An electronic flow detector.

ANODE. (1) The *electrode* via which current enters a device. It is the positive terminal of an electroplating cell but the negative terminal of a battery. (2) The electrode in a vacuum tube into which the principal electron stream flows. *Plate* (U.S.A.).

symbol

ANODE, HOLDING. A small auxiliary anode in a *mercury pool rectifier* supplied with direct current to keep the *cathode spot* energized during intervals when the main anode current is zero.

ANODE BEND DETECTOR. A *detector* in which a triode valve is operated over the curved portion of its anode-current/grid-voltage *characteristic*.

ANODE BREAKDOWN VOLTAGE. Of a *cold-cathode glow discharge tube* the anode voltage required to cause conduction across the main gap, with the starter gap not conducting and with all other electrodes at cathode potential before breakdown.

ANODE CHARACTERISTIC. A graph of the relation between the *anode current* of a valve and the anode voltage.

ANODE CURRENT. Usually, the current flowing from the *anode* to the *cathode* of a *thermionic valve*. The current inside the valve is due to the flow of negative electrons in the opposite direction to the conventional current, i.e. from cathode to anode.

ANODE DARK-SPACE. A narrow dark zone observed near the surface of the *anode* in a *glow discharge tube*.

ANODE DISSIPATION. Power dissipated as heat owing to bombardment of an anode by electrons and ions.

ANODE EFFICIENCY. Of a valve or tube: the ratio of ac power in the anode load circuit to the dc power input to the anode. Plate efficiency (U.S.A.).

ANODE FALL. In a *gas discharge tube*, the fall of potential between the *positive column* and the *anode*.

ANODE GLOW. A narrow bright zone on the anode side of the *positive column* in a *gas tube*.

ANODE RAYS. In a vacuum valve or tube, ions which are sometimes formed from impurities in the metallic *anode*. If the anode is treated with certain alkali-earth salts or oxides, copious emission of positive rays may be produced. See *Canal rays*.

ANODE RESISTANCE. Of a valve or tube, the ratio between a small change in anode voltage and the resulting small change in anode current.

ANODE SATURATION. Of a valve or tube, the condition in which electrons are no longer attracted by the *anode*.

ANODE SHEATH. In a *gas discharge tube*, the electron boundary between the *plasma* and the anode which forms when the current demand by the anode circuit is more than the random electron current at the anode surface.

ANODE SHIELD. In a *mercury-arc rectifier*: a metal shield round the anode to protect it from massive ionization or radiation.

ANODE VOLTAGE, PEAK FORWARD. Of a valve, the maximum instantaneous anode voltage in the direction in which the valve is designed to pass current.

ANODE VOLTAGE, PEAK INVERSE. Of a valve, the maximum instantaneous anode voltage in the direction opposed to that in which the valve is designed to pass current.

ANODIGE. An analogue-to-digital converter in which the input voltage is compared to a staircase waveform, in which the appropriate number of 'steps' are counted as pulses.

ANODIZE. Place a protective film on a metal surface by electrolytic or chemical action.

ANOTRON. A *cold-cathode glow discharge diode* in which the *cathode* is usually made of sodium and the *anode* of calcium.

ANTENNA. See *Aerial*.

ANTI-CAPACITANCE SWITCH. A switch designed to present the smallest practical series *capacity* in the open position.

ANTI-COINCIDENCE CIRCUIT. A circuit with two input terminals which produces an output pulse if one input terminal receives a pulse but not if both terminals receive pulses simultaneously or within a specified time interval. See *Coincidence gate*.

ANTI-COINCIDENCE COUNTER. An electronic *counter* using an *anti-coincidence* circuit to record specifically events which do not occur together or nearly together. See *Coincidence counter*.

ANTI-CYCLOTRON. A form of *travelling wave tube*.

ANTI-HUNTING CIRCUIT. A stabilizing circuit often inserted in a *feedback* system of control to prevent self-oscillation.

ANTIMONY (Sb). Metallic element, Atomic No. 51.

ANTI-REGENERATION DEVICE. See *Neutralization*.

ANTI-RESONANCE (PARALLEL IMPEDANCE). The condition of maximum *impedance* for two or more *impedors* connected in *parallel*.

ANTI-RESONANT FREQUENCY. The frequency at which the *impedance* of a given system is very high, as opposed to the condition of resonant frequency when the impedance is low.

ANTI-TRANSMIT–RECEIVE SWITCH (ATR). See *Transmit–receive switch*.

APEX. A *digital computer* made and studied at Birkbeck College, University of London.

APPLE TUBE. A special *cathode-ray tube* developed by the Philco Corp. U.S.A., for *colour television*, whose screen consists of an ordered array of thin vertical stripes of red, green, and blue *phosphors*. These stripes are excited independently in response to the amounts of red, green, or blue light called for by the transmitted signal. *Compatibility* with black-and-white receivers is achieved by exciting the three colours in coordination to produce white light.

APPLEGATE DIAGRAM. A graphical representation of *bunching* in *velocity-modulated valves*.

APPLETON LAYER. A conducting layer in the *ionosphere* about 180 km. from the earth's surface, discovered by Sir Edward Appleton.

APPLICATORS. In *dielectric heating*, appropriately shaped conducting surfaces enclosing the article to be heated, between which an alternating field of high frequency is established.

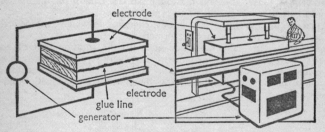

Fig. 4. Applicators in a typical dielectric heating installation

APT. Acronym for Automatically Programmed Tools: a universal computer-assisted program system for automatic multi-axis machining work.

AQUADAG® COATING. A film of graphite applied to the inside glass surfaces of some valves and tubes – particularly cathode-ray tubes – which acts as a source of *secondary electrons* and as a post-deflection *accelerating anode*.

ARC. A low voltage high current electrical discharge. Cf. *Spark*.

ARC, MERCURY. An electric *arc* which passes through mercury vapour.

ARC BACK (BACKFIRE). The *arc* from anode to cathode in a gaseous rectifier valve.

ARC BAFFLE. An obstruction placed inside *mercury-arc rectifiers* to prevent mercury splashing on to the anode.

ARC CONVERTER. An oscillatory device which employs an electric *arc* to generate alternating current.

ARC DISCHARGE. A discharge between electrodes in a gas characterized by relatively low voltage drop and high current density.

ARC DROP. The voltage drop between the anode and cathode of a *gas-tube rectifier*.

ARC KEEP ALIVE. See *Anode, holding*.

ARC LAMP. The brilliant light which accompanies an electric *arc* struck between suitable electrodes can be used as a source of illumination. Most of the light comes from the incandescent crater which forms at the positive electrode.

ARC THROUGH. In a gas-filled valve or tube, conductance via the anode in the normal or forward direction, but at a time when the anode should not be conducting.

ARCH®. Articulated Computing Hierarchy: a system of *computer* control designed by Elliott Automation Ltd, in which a number of computers are interlinked and each computer is controlled by one preceding it.

ARGON (A). Gaseous element, Atomic No. 18. One of the inert gases used extensively in gas-filled valves and tubes.

ARITHMETIC UNIT. The section of a *computer* in which arithmetical and logical operations are performed.

ARMATURE. Originally the name given to the rotating part of an electrical machine, but in modern usage any structure in which a voltage is induced by a magnetic field as in motors, generators, relays, buzzers, gramophone pick-ups, and electro-magnetic loudspeakers.

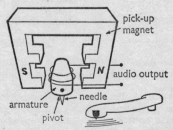

Fig. 5. Armature in one type of magnetic pick-up

ARMSTRONG OSCILLATOR. An *oscillator* circuit, invented by E. H. Armstrong, in which both anode and grid circuits are tuned.

ARSENIC (As). Semi-metallic element, Atomic No. 33. Used as a *donor impurity* and in intermetallic semiconductor compounds.

ARTICULATION. The amount of speech transmitted by a complete communication system, including the speaker, which is understood correctly by the listener. Per cent articulation is the amount expressed as a percentage of the total.

ARTIFICIAL EAR. A device for testing and calibrating earphones which is designed to simulate the electrical characteristics of the average human ear.

ARTIFICIAL LARYNX. A device to assist the speech of a person who has undergone the operation of tracheotomy. The larynx is replaced by a vibrating reed which is actuated by the air breathed in through the opening in the throat.

ARTIFICIAL LINE. A network which simulates the electrical characteristics of a *transmission line*.

ARTIFICIAL LOAD. A substitute for an *aerial, transmission line,* or similar device, which dissipates the correct amount of power without radiating it.

ARTIFICIAL VOICE. A device for testing and calibrating close-speaking microphones, consisting of a small loudspeaker in a baffle designed to simulate the acoustical properties of the human head.

ASDIC. Originally Allied Submarine Detection Investigation Committee. Now used generally for submarine detection devices based on the work of the committee.

ASPECT RATIO. In television, the ratio of the frame width to the frame height of the picture.

aspect ratio = $\frac{W}{H}$

British and American standard: $\frac{4}{3}$

Fig. 6. Aspect ratio

ASSEMBLER (OR ASSEMBLY PROGRAMME). A *digital computer* programme in which a machine language is constructed from a symbolic language.

ASTABLE CIRCUIT. A circuit with two quasi-stable states which can generate a continuous waveform with no *trigger*.

ASTATINE (At). Radioactive element, Atomic No. 85.

ASTON DARK-SPACE. In a *glow discharge* device, a narrow dark band close to the cathode.

ASYNCHRONOUS COMPUTER. A *computer* in which the signal to start an operation is the completion of the previous operation.

ATLAS®. A large *digital computer* made by Ferranti Ltd, capable of accepting about one million completed instructions per second. Thermionic valves are replaced entirely by *transistors* in the computer circuits and any number of programmes can be handled concurrently while the machine's internal supervision ensures maximum efficiency of operation. At the end of 1961 it was claimed to be the most powerful computer in the world.

ATMITE®. A non-linear resistor of silicon carbide.

ATOM. The smallest part of an *element* which can enter into chemical combination.

ATOMIC CHARGE. The product of the number of electrons an atom has gained or lost in *ionization* and the charge on one electron.

ATOMIC CLOCK. A very accurate clock in which the frequency is controlled by the vibrations of molecules or atoms. In the ammonia clock, for example, a signal is sent through a waveguide containing ammonia gas and the vibrations of excited ammonia molecules absorb the energy of the signal at a very constant frequency – about 23,870 megacycles – which is the basis of the time scale. See *Caesium clock*.

ATOMIC DISTANCE. The average distance between the centres of two atoms.

ATOMIC FREQUENCY. The natural frequency of vibration of atoms in a molecule, especially in the solid state.

ATOMIC KERNEL. An atom which has lost the electrons in its outer shell, called the *valence electrons*.

ATOMIC NUMBER. The number of *protons* in an atomic nucleus and hence the positive charge of the nucleus. It is equal to the number of *electrons* surrounding the nucleus.

ATOMIC SPECTRUM. The band of radiations which accompanies internal changes in atoms; in particular, the transition of *electrons* from one shell to another.

ATOMIC STRUCTURE. It is now thought that an atom consists of a number of *electrons* moving in the electric field created by the positively charged massive *nucleus*. Although it is useful to designate shells or orbits which may be occupied by electrons, modern theory does not ascribe exact orbits in the ordinary sense of the word.

ATOMICHRON. An *atomic clock* built in the U.S.A. which embodies automatic correction for drifts in frequency.

ATR TUBE. See *Transmit-receive switch*.

ATTENUATION. Decrease in the magnitude of any kind of signal by an electronic active or passive device.

ATTENUATION, SIDEBAND. Attenuation of certain transmitted components of a *modulated signal*, excluding the *carrier*, relative to the amplitude of those components produced by *modulation* before transmission.

ATTENUATION BAND. See *Rejection band.*

ATTENUATION CONSTANT. For a travelling plane wave at a given frequency, the rate of exponential decrease of amplitude of voltage, current, or field-component in the direction of propagation. Expressed in *nepers* or *decibels* per unit length.

ATTENUATION DISTORTION. See *Amplitude distortion.*

ATTENUATION EQUALIZER. An *equalizer* designed to compensate for *amplitude distortion* in a system.

ATTENUATION VOLTAGE. Of a *transducer*, the ratio of the voltage at the input of the transducer to the voltage across a specified load at its output. Expressed in *decibels.*

ATTENUATOR. A *network* or *transducer* designed to reduce the amplitude of a wave without distorting it. It may be fixed or variable: a fixed attenuator is sometimes called a *pad.* Usually calibrated in *decibels.*

ATTENUATOR VALVE (TUBE). A gas-filled switching valve in which a gas discharge is used to control *radio-frequency* power by reflection or absorption.

ATTOAMPERE. An extremely small unit of electric current equal to 10^{-18} *ampere*, or about 6 electrons per second.

ATTRACTION, ELECTRICAL. The force between electric charges of opposite sign expressed in *Coulomb's law.*

ATTRACTION, MAGNETIC. The kind of force which an iron magnet exerts on a piece of iron.

AUDIBILITY. The intensity of a received signal usually expressed in decibels referred to one milliwatt (*dbm*).

AUDIBILITY THRESHOLD. The intensity of a sound which can just be heard by a normal human ear in a quiet place. It varies with the frequency of the sound and with other factors (including subjective factors) so that measurements are usually based on a number of trials of a given subject.

AUDIO (FREQUENCY). Any frequency at which a sound wave is normally audible: in practice from about 15 cycles per second to 20,000 cycles per second.

AUDIO (-FREQUENCY) AMPLIFIER. Any device designed to amplify audio frequencies.

AUDIO-FREQUENCY CHOKE. An *inductor*, usually wound on an iron core, designed to impede audio-frequency currents.

AUDIO-FREQUENCY HARMONIC DISTORTION. The change in the content of *harmonics* in an audible signal after transmission.

AUDIO-FREQUENCY PEAK LIMITER. A device in an audio-frequency system to cut off amplitudes above a specified maximum.

AUDIO (-FREQUENCY) TRANSFORMER. A *transformer* having an iron core used for coupling *audio-frequency* circuits.

AUDIOGRAM. A graph showing hearing loss as a function of frequency.

AUDIOMETER. An instrument for measuring acuteness of hearing.

AUDIOMETRY. The use of *audiometers* for the study of hearing.

AUDION. The original three-electrode *vacuum valve* invented by De Forest in the U.S.A.

AUGER EFFECT. A change is an atom from an excited to a lower energy state, accompanied by the emission of an *electron* but without radiation.

AUGER ELECTRON. An electron ejected from an excited atom in the *Auger effect*.

AUTOCODE. A program *language* which a *computer* can accept and which can be changed into a *machine code* by a *compiler*.

AUTODYNE. A circuit in which the same elements and valves are used as *oscillator* and *detector*.

AUTODYNE RECEPTION. A system of *heterodyne* reception using an *autodyne* circuit.

AUTOELECTRIC EFFECT. The emission of electrons from a *cold cathode* subjected to a strong electric field.

AUTOLECTOR®. A British-designed machine which reads computer printed or handwritten forms straight into a computer store at high speeds.

AUTOMATIC BIAS. See *Cathode bias, Grid bias*.

AUTOMATIC BRIGHTNESS CONTROL. A *television receiver* circuit designed to maintain the average brightness of the reproduced image at a predetermined level.

AUTOMATIC CHECK. In a *computer*, a check performed by built-in equipment each time a particular operation is performed.

AUTOMATIC COMPUTER. A *computer* which automatically starts, continues, and completes sequences of operations on information supplied to it or already contained in its storage elements.

AUTOMATIC CONTRAST CONTROL. In *television receivers*, the automatic control of the gain of the vision *intermediate-frequency* circuits to ensure constant contrast in the picture for a given setting of the manual contrast control.

AUTOMATIC CONTROL. In electrical switching, the opening and closing of switching devices in an automatic sequence to maintain a desired performance.

AUTOMATIC CONTROL SYSTEM. A chain of electrical, magnetic, mechanical, hydraulic, pneumatic, and electronic devices, or combinations of these, interconnected so that one or more elements in the chain are controlled. See *Feedback, Servomechanism*.

AUTOMATIC CONTROLLER. A device which measures any variable quantity or condition and acts to correct or limit deviation of the measured value from a selected reference.

AUTOMATIC FOCUSING. In a *television tube*: *electrostatic focusing* in which the anode of the tube is internally connected through a resistor to the cathode, so that no external focusing voltage is needed.

AUTOMATIC FREQUENCY CONTROL (AFC). A circuit which automatically maintains the frequency of any source of alternating voltage within specified limits.

AUTOMATIC GAIN CONTROL (AGC). See *Automatic volume control*.

AUTOMATIC PILOT (ELECTRONIC). An equipment used to provide

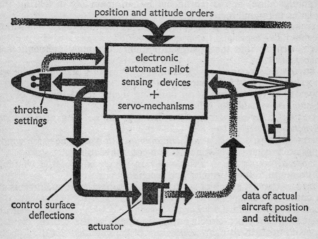

position and attitude orders

electronic
automatic pilot
sensing devices
+
servo-mechanisms

throttle
settings

control surface
deflections

actuator

data of actual
aircraft position
and attitude

Fig. 7. Principle of automatic pilot

automatic signals from an electronic computer to maintain an aircraft in a given flight path. It includes *amplifiers*, gyroscopes, and *servo* systems.

AUTOMATIC PROGRAMMING. For a *computer*, any system in which the computer is used to transform *programming* instructions from a form readily prepared by a human being to a form more suitable for a computer.

AUTOMATIC SEQUENCE-CONTROLLED CALCULATOR. A *computer* in which the sequence of operations is automatically controlled by the computer itself.

AUTOMATIC TRACKING. In radar, the process of keeping the radar beam automatically on the target while automatically determining the range of the target.

AUTOMATIC VOLUME COMPRESSION. An *audio-frequency* circuit which decreases the gain for large signals and increases the gain for small signals, thus reducing the range of variation in volume of the output.

AUTOMATIC VOLUME CONTROL (AVC). A method of obtaining automatically a substantially constant output volume over a specified range of variation in input volume. Also the device used to accomplish this.

AUTOMATIC VOLUME EXPANSION. An *audio-frequency* circuit which increases the *dynamic range* of recorded or transmitted signals (usually music) at the reproducer by automatically increasing the gain as the signal increases.

AUTOMATION. A generalized description of the application of various techniques – especially those using electronic devices – to any process or system so that the minimum amount of human intervention is necessary.

AUTONOMOUS WORKING. A form of concurrent working in which a portion of an *ADP* system performs a substantial sequence of operations in response to an initiating signal, while the rest of the system is available for other operations.

AUTOSYN®. An angular position control device. See also *Selsyn*, *Synchro*.

AUTOTRANSDUCTOR. A *magnetic amplifier* in which the same windings are used as power windings and control windings.

AUTOTRANSFORMER. A *transformer* in which part of the winding is common to both primary and secondary circuits.

AUXOCHROME. A group of atoms which absorbs radiation and acts as a colour carrier.

AVAILABLE POWER. Of a *linear* source of electrical energy, the mean square of the open-circuit terminal voltage of the source, divided by four times the *resistive component* of the *impedance* of the source.

AVAILABLE POWER GAIN. Of a linear *transducer* at a specified frequency, the ratio of the available signal power from the output to the available signal power from the input source.

AVALANCHE. A cascade multiplication of *ions*. See *Townsend discharge*.

AVALANCHE BREAKDOWN. Of a *semiconductor diode*, a breakdown caused by the cumulative multiplication of *carriers* resulting from field-induced *impact ionization*. See *Zener breakdown*.

AVC. Abbreviation for *automatic volume control*.

AXIOTRON. A high-vacuum *thermionic diode* in which the anode current is controlled by a magnetic field due to the heater current.

AZIMUTH MARKER. A bright radial line on the *cathode-ray oscilloscope* of a radar set, usually adjusted to indicate true North on the *Plan Position Indicator*.

B

B BATTERY. The power source for the anode circuit of a battery-operated electronic device (U.S.A.). H.T. Battery (G.B.).

B POWER SUPPLY (U.S.A.). A device which supplies direct current power to the *anode* of a valve.

BACK EMISSION. Reverse *emission*.

BACK HEATING. In a *magnetron*, the overheating of the cathode or filament as a result of bombardment by returning high velocity electrons out of phase with the oscillations.

BACK LIGHTING. A technique used to increase the sensitivity of a television camera *iconoscope* which depends on illuminating the tube and rear surface of the mosaic without increasing the direct illumination of the mosaic itself.

BACK PORCH. In a *television signal*, the period of time immediately after a *synchronizing pulse* during which the signal is held at *black level*.

BACK SCATTERING. The scattering or reflecting into the sensitive part of a measuring instrument of radioactive radiations which originally had no component of motion towards that part.

BACKFIRE. See *Arc back*.

BACKGROUND CONTROLS. In *colour television*, the three individual *brightness controls*.

BACKGROUND COUNTS. In a *counter tube*, counts caused by radiation from sources other than the one being measured, or by radioactive contamination of the tube itself.

BACKGROUND PROJECTION. In a television studio, the optical projection of a background appropriate to the studio scenes.

BACKLASH. The incomplete rectification of an alternating current in a gas rectifier valve.

BACKPLATE. In television *camera tubes*, the electrode to which the stored charge image is capacitatively coupled.

BACK-PORCH EFFECT. In *transistors*, the continuation of collector current for a short time after the input signal has fallen to zero, due to storage of *minority carriers* in the base region.

BACK-TO-BACK CONNECTION. In *thyratron* or *ignition* rectifiers, a method of operation to provide control of alternating current to a load.

BACKWARD WAVE. In a *travelling wave tube* (*valve*), a wave whose group velocity is opposed to the direction of motion of the electron stream.

BACKWARD WAVE TUBE (VALVE). A *travelling wave tube* in which the direction of wave propagation is opposed to the direction of the electron stream. See Figure 8.

symbol

29

BACKWASH DIODE (OVERSWING DIODE)

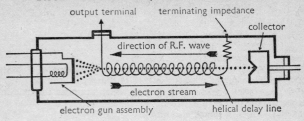

Fig. 8. Principle of backward wave valve

BACKWASH DIODE (OVERSWING DIODE). A diode valve connected across the line of a *pulse modulator* so that it absorbs energy during reversals of the voltage pulses which might otherwise cause failure of the modulator.

BAFFLE. (1) In a *gas tube*, an auxiliary structure placed in the arc path and having no connection to the outside. It is used to control the flow of charged particles or mercury vapour and may be of conducting or insulating material. (2) In acoustics, a structure or partition used to increase the effective length of the external transmission path between two points.

BAKELITE. A synthetic resin compound which is a good insulator and has some desirable mechanical characteristics.

BALANCE COIL. A coil for supplying a three-wire circuit from a two-wire circuit.

BALANCE CONTROLS. Variable circuit components used to obtain conditions of electrical equilibrium in the operation of electrical instruments and *bridges.*

BALANCED (PUSH-PULL) AMPLIFIER. An amplifier in which there are two identical signal branches connected to operate in opposite phase and with input and output connections each balanced to ground.

BALANCED WIRE CIRCUIT. One in which the two sides are electrically alike and symmetrical with respect to ground and other conductors.

BALLAST LAMP (VALVE, TUBE). A resistance element sealed in a glass or metal envelope, which may be evacuated or filled with hydrogen, used as a series component in a circuit to maintain nearly constant current.

BALLAST RESISTOR. A resistor in which the resistance increases as the current through it increases. Thus, if the voltage across it varies within prescribed limits it tends to maintain the current constant. See *Barretter, Thermistor.*

BALLISTIC GALVANOMETER. A *galvanometer* whose deflection is proportional to the transient charge passing through it.

BALUN (BALancing UNit). A *resonant line* designed to work between a balanced system which is symmetrical to ground, and an unbalanced system in which one side only is grounded.

BANANA JACK. A jack which accepts a *banana plug*.

BANANA PLUG. A single conductor plug with a spring metal tip which somewhat resembles a banana in shape.

BANANA TUBE. A special *cathode-ray tube* developed by Mullard Ltd as the basis of a system of *colour television*. A single repetitive line *scan* is produced inside the tube and is then shifted to produce the *field scan* by cylindrical lenses which are mounted inside a drum rotating around the tube. The tube proper takes the form of a long cylinder with phosphor stripes, in the three primary colours, parallel to its axis, and having its *electron gun* in a narrower tube joined to it at one end. The name apparently follows the fashion set by the American *apple tube*.

BAND. (1) In communications, a range of frequencies within two prescribed limits used for a defined purpose. (2) In electronic computers, a group of tracks on a magnetic drum. (3) A closely spaced group of energy levels in atoms. (4) The set of closely spaced spectral lines produced by transitions in the vibrations of electrons inside atoms.

BAND EDGE ENERGY. In a *semiconductor*, the band of energy between two defined limits. The upper limit is the highest permissible energy of a freed electron and the lower limit is the lowest energy which an electron can possess and still remain free.

BAND EXPANSION FACTOR. In *FM* radio, the ratio of the channel band width to modulating frequency. It is equal to twice the *deviation ratio*.

BAND SPREADING. (1) A method of *double-sideband* transmission in which the sidebands are widely separated in frequency from the carrier. (2) In a radio receiver, the use of a trimmer capacitor as an adjunct to the main tuner to extend the tuning in a selected band over the whole tuning scale.

BAND SWITCH. A switch used to select any of the frequency bands in which any electronic equipment is designed to operate.

BAND WIDTH. Of a *tuned amplifier*, the width of the band of frequencies over which the *power amplification* does not drop to less than an assigned fraction – usually one-half – of the power amplification at resonance.

BAND WIDTH COMPRESSION. In *television*, a technique based on the use of two channels for the picture information to improve the definition of the picture.

BAND-ELIMINATION (EXCLUSION) FILTER. A filter designed to attenuate a band of frequencies between two specified values.

BAND-IGNITER VALVE (TUBE). A *mercury-pool valve* in which the control electrode is in the form of a metallic band outside the glass envelope, or as a coating around an electrode dipped in the mercury. See *Capacitron*. See Figure 9.

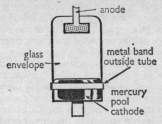

Fig. 9. Band-igniter valve

BAND-PASS FILTER. A filter network made by coupling two resonant circuits tuned to the same (or nearly the same) frequency, which passes a defined band in the neighbourhood of this frequency and attenuates all others.

BAND-PASS TUNING. A method of tuning radio receivers by varying simultaneously the capacitors or the inductors in the resonant circuits of a *band-pass filter*.

BANG-BANG CONTROL (SERVO). In a control system, a method of operating the controls in which the correcting signals are always applied to the full extent of the *servomechanism*.

BANK WINDING. A method of winding wire in coils in which single turns are wound successively in each of two or more layers and the winding is completed without return. The main object is to reduce the *self-capacitance* of the coil.

BANTAM. A very small vacuum valve or tube.

BAR GENERATOR. See *Bar pattern*.

BAR MAGNET. Any *ferromagnetic* material in the form of a bar with a north-seeking pole at one end and a south-seeking pole at the other.

BAR PATTERN. A pattern of repeating bars or lines on the screen of a *television receiver*, produced by a *pulsed signal generator* and used to measure the linearity of the *scanning systems*.

BARIUM (Ba). Silvery-white metallic element, Atomic No. 56. Because of its low *work function*, widely used as an electron emitter in the cathodes of thermionic valves.

BARIUM TITANATE (BaTiO$_3$). A crystalline ceramic which has pronounced *ferroelectric* and *piezoelectric* properties and is used in compact capacitors and certain *transducers*.

BARKHAUSEN CRITERION. The condition for sustained oscillations in a self-excited single valve *oscillator*, worked out by the German scientist Barkhausen, that the voltage multiplication of the amplifier must exactly counterbalance the voltage-dividing ratio of the *feedback* circuit.

BARKHAUSEN EFFECT. If a ferromagnetic material is enclosed by a coil connected to an amplifier and loudspeaker, magnetic changes in the material caused, e.g. by approximating a bar magnet to it or by increasing the current in a field coil around it, are accompanied by rustling sounds in the loudspeaker. The effect, attributed to discontinuities in the magnetization, was first demonstrated by Barkhausen.

BARKHAUSEN VALVE (TUBE). See *Positive-grid oscillator valve*.

BARKHAUSEN-KURZ OSCILLATOR. A *retarding-field* type of *oscillator* in which the frequency of oscillation depends solely on the *transit time* of electrons in a valve or tube.

BARNETT EFFECT. The magnetization produced in a *ferromagnetic* material solely by rotating it in the absence of an external magnetic field.

BARRAGE RECEPTION. A system of radio reception in which interference from any particular direction is minimized by selecting from a system of directional aerials those which provide maximum signal to interference ratio.

BARREL DISTORTION. A defect in the lens system of a *cathode-ray tube* which causes the image of a square to appear barrel-shaped.

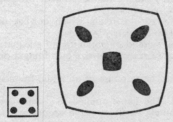

Fig. 10. Barrel distortion

BARRETTER. A sensitive metallic resistor, usually enclosed in a glass container, whose resistance increase with temperature is used as the basis of some measuring instruments.

BARRIER. A junction between two solids – usually *semiconductors* – across which the movement of electric charges is not symmetrical.

BARRIER-GRID STORAGE. See *Radechon*.

BARRIER-LAYER PHOTO-CELL. See *Photovoltaic cell*.

BARRIER-LAYER RECTIFIER. A solid *rectifier* consisting of a *semiconductor* between rectifying and non-rectifying metal electrodes. The barrier layer at the rectifying contact is a very thin layer of semiconductor immediately adjacent to the metal contact, whose *resistivity* is much greater than for the bulk material and is dependent on the direction and magnitude of the voltage applied across the contact.

BARYE. The absolute unit of pressure in the c.g.s. system, equal to a pressure of one dyne per sq. cm.

BASE. The part of a *junction transistor* which separates the *emitter* and *collector* regions.

BASE ELECTRODE. The electrode attached to the *base* of a *junction transistor*.

BASE LINE. The glowing line on a *radar* screen which represents the track followed by a radar scanner.

BASE REGION. In a *semiconductor* device, the region which surrounds the *base electrode*.

BASEBAND. In the process of *modulation*, the frequency band occupied by all the transmitted signals which modulate the carrier.

BASIC FREQUENCY. In any oscillatory movement which can be analysed as the sum of a number of sinusoidal components of different frequencies, the basic frequency is the frequency of the component considered to be the most important.

BASKET WINDING. A method of winding a coil so as to reduce the *distributed capacitance*, in which the adjacent turns are separated except at crossing points, as in a basket weave.

BASS. The low frequency part of the *audio-frequency* spectrum.

BASS BOOST. Increase in the amplification of low frequencies compared to that of high frequencies.

BASS COMPENSATION. Circuits which correct the low frequency response of audio-frequency systems.

BASS REFLEX. A method of obtaining an adequate *baffle* in a loud-speaker cabinet of reasonable size by adding a port opening at the front.

BASS RESPONSE. The response of an audio-frequency system to low frequencies.

BATHOCHROME. The name applied to certain groups of atoms in organic compounds which lower the frequency of radiation absorbed by those compounds.

BATTERY. A combination of *cells* connected together so as to produce useful electrical energy.

BAUD. In telecommunications: a unit of measure of data flow, interchangeable with bits per second (see *bit*). The name is derived from the *Baudot Code*.

BAUDOT CODE. The standard five-channel typewriter code which has been used for over a century by the telegraph industry.

BAYONET BASE. The base of any lamp, tube, or valve which has a pin on opposite sides and which can be inserted into a suitable socket and rotated so that the pins are retained in slots in the socket.

BEACON. (1) A signal station. (2) The signal produced to serve as a landmark or location for ships or aircraft.

BEACON, CODE. A *beacon* which transmits a characteristic series of dots and dashes by which it may be identified.

BEACON, HOMING. A device which guides an aircraft to an airport or any other desired objective.

BEACON, RADAR. A device which receives signals from a radar transmitter and transmits signals to a radar receiver, which can then determine its position.

BEACON, RADIO. A non-directional radio transmitter at a fixed location which emits a characteristic signal whose main purpose is to allow mobile stations to determine their bearings.

BEACON RECEIVER. A radio receiver for converting waves from a *radio beacon* into perceptible signals.

BEAD. An insulator shaped like a bead; especially as an insulating support for the inner wire of a *coaxial line*.

BEAD MEMORY. See *Ferrite bead memory*.

BEAD THERMISTOR. A *thermistor* made from a small piece of semi-conducting material into which two terminal wires are inserted.

BEAGLE. An automatic search jammer.

BEAM. An essentially unidirectional flow of particles or of electro-magnetic radiation.

BEAM, HOLDING. A diffuse beam of electrons for regenerating the charges retained on the surface of a *storage tube*.

BEAM ALIGNMENT. In *camera tubes*, an adjustment of the electron beam so that it is perpendicular to the surface of the target.

BEAM ANGLE. The solid angle of the cone of electrons emerging from the *cross-over area*.

BEAM BENDING. In *camera tubes*, deflection of the scanning beam by the electrostatic field of the stored charges on the target.

BEAM CONVERGENCE. In a *colour television* picture tube, the focusing of the three electron beams at a *shadow-mask* opening.

BEAM COUPLING. The production of an alternating current in a circuit between two electrodes through which a *density-modulated* electron beam is passed.

BEAM COUPLING COEFFICIENT. The ratio of the alternating current produced to the beam current in *beam coupling*.

BEAM CURRENT. Of a *cathode-ray tube*: the electron current of the beam arriving at the screen.

BEAM DEFLECTION TUBE. An electronic tube in which the output current is controlled by the transverse movement of the *electron beam*.

symbol

BEAM LOADING. The production of an *admittance* between two electrodes when a constant-density electron beam is directed across the gap between them.

BEAM POWER VALVE (TUBE). A valve in which use is made of directed electron beams to increase the power-handling capacity of the valve and in which control grid and screen grid are essentially aligned. In *tetrode* and *pentode* beam power valves, the beam is formed by two pencil-beam-forming plates. See Figure 11.

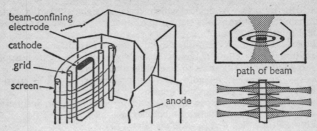

beam-confining electrode

cathode

grid

screen

anode

path of beam

Fig. 11. Beam power valve

BEAM-CURRENT MODULATION. The process of varying with time the magnitude of the *beam current* passing through a surface.

BEAM-FORMING ELECTRONIC VALVES (TUBES). The following types are defined in this dictionary: *beam deflection mixer, beam power, camera, cathode-ray, dissector, electron microscope, iconoscope, image converter, image orthicon, indicator, induction output, monoscope, orthicon, velocity-modulated.*

symbol for beam forming electrodes

BEAM-SWITCHING TUBE. See *Cyclophon.*

BEAT. Periodic pulsation resulting from the interference of two wave trains of different frequency.

BEAT FREQUENCY. The difference frequency resulting from the interaction between two signals of different frequency applied to a *non-linear* circuit (or to the ear).

BEAT FREQUENCY OSCILLATOR. An *oscillator* designed to mix a variable frequency with a fixed frequency and thus obtain a *beat frequency.*

BEAT NOTE. The wave of difference frequency created when two *sinusoidal* waves differing in frequency are applied to a *non-linear* device.

BEATING. The combination of two or more frequencies to produce *beats.*

BECQUEREL RAYS. Penetrating radiation emitted by uranium salts made up of *alpha, beta,* and *gamma rays.*

BEL. A unit which expresses the ratio of two values of power, the number of bels being the logarithm, to the base 10, of that ratio. See *Decibel.*

BEND, E-PLANE (E-TYPE). A bend in a *waveguide* in the plane of the electric field.

BEND, H-PLANE (H-TYPE). A bend in a *waveguide* in the plane of the magnetic field.

BERYLLIA. Beryllium oxide, a material which insulates electrically like a ceramic but conducts heat like a metal.

BERYLLIUM (Be). Hard, white metallic element, Atomic No. 4. Similar in appearance and chemical properties to magnesium.

BETA CIRCUIT. In a *feedback amplifier*, the circuit which transmits a proportion of the output back to the input.

BETA CURRENT GAIN FACTOR. Of a transistor with common-base connection: the ratio of the change in the collector current to the small change in the base current required to produce it. Beta is always greater than 1.

BETA DECAY (DISINTEGRATION). A radioactive transformation of a nucleus which results in a change in the *atomic number* by $+$ or -1 without change in the *mass number*.

BETA PARTICLE. A negative electron (*negatron*) or positive electron (*positron*) emitted from a nucleus in *beta decay*.

BETA RAY. Stream of *beta particles*.

BETA RAY SPECTROMETER. An instrument for determining the distribution in energy or momentum of *beta rays*, *photoelectrons*, and *conversion electrons*. It consists of an evacuated enclosure containing a radioactive source, a detector, and a means of deflecting the rays travelling between source and detector.

BETA THICKNESS GAUGE. See *Beta-absorption gauge*.

BETA-ABSORPTION GAUGE. An instrument which measures the thickness or density of a material by measuring the absorption of *beta rays* in a sample of the material.

BETATRON. An electron *accelerator* capable of producing electron beams of high energy and *X-rays* of very high penetrating power, in which the electrons are accelerated by a rapidly changing electric field in a circular orbit of constant radius. Contrast *Cyclotron*.

BEV. Abbreviation for Billion (10^9) *Electron volts*.

BEVATRON. A six thousand million volt *accelerator* of protons and other atomic particles at Berkeley University, California.

BFO. Abbreviation for *beat frequency oscillator*.

B-H CURVE. See *Magnetization curve*.

BIAS. In electronics, a voltage whose main function is to determine the operating point on the characteristic of an electronic device.

BIAS CELL. A small electric *cell* which can supply about $1\frac{1}{2}$ volts for long periods.

BIAS LIGHTING. See *Back lighting*.

BIAS METER. A device used in *teletypewriters* for measuring signal bias directly in per cent.

BIAS RESISTOR. A *resistor* across which a *bias* voltage is developed.

BIAS WINDING. In a *saturable reactor*, a control winding through which a biasing *magnetomotive force* is applied.

BIDIRECTIONAL CURRENT. A current which may flow in either direction through a conductor.

BIDIRECTIONAL PULSES. *Pulses* some of which rise in one direction and the rest in the other direction.

BIDIRECTIONAL READ-OUT. A display panel which shows an actual plus or minus figure.

BIFILAR WINDING. A technique of winding *non-inductive resistors* in which the resistance wire is doubled on itself and wound double from the looped end.

Fig. 12. Bifilar winding

BIMETAL (LIC) STRIP. A composite strip of two metals welded together, used to open and close contacts in *thermostats* and *time-delay relays*.

BIMORPH CELL. An assembly of two plates of *piezoelectric* material designed to bend in proportion to an applied voltage; or, conversely, a single cell producing twice the normal voltage by the application of pressure.

BIN. A name sometimes given to *magnetic tape memory* units in which the tapes are stored in a single cabinet or bin.

BINAC. The first American high speed electronic *digital computer* operating serially on *binary numbers*.

BINARY CELL. In a *computer* storage system, an element which can have one or other of two stable states and can therefore store information.

BINARY CODE. A *code* in which each element may have one or other of two distinct states, values, or numbers.

BINARY CODED DECIMAL SYSTEM. A system of representing numbers in which the decimal digits of the number are expressed in *binary code*.

BINARY NUMBER SYSTEM. A system which has only two different digits, usually 0 and 1, and has two as its base. Thus the number 10 in the decimal system, which has ten different digits and a base or radix of ten, is represented as 1010 in the binary system.

BINARY POINT. The radix point in a binary system analogous to the decimal point in the decimal system.

BINAURAL INTENSITY EFFECT. The relationship which enables human ears to locate a sound which reaches both ears at the same frequency and in the same phase, but with different intensities.

BINAURAL PHASE EFFECT. The relationship which enables human ears to locate a sound which is incident on both ears at the same frequency and intensity but out of phase.

BINDING ENERGY, ELECTRON. The energy necessary to remove an electron from an atom. Identical with *ionization potential*.

BINISTOR®. A *semiconductor* device with negative resistance characteristics which depend on an external voltage supply.

BIONICS. The characteristics, functions, and phenomena of living systems, especially in their relations to electronic systems.

BIPOLAR. Having two *poles* or containing *charges* of opposite polarity.

BIPOLAR TRANSISTOR. A *transistor* in which both negative and positive charge carriers play an essential part. See *Majority carrier, Minority carrier, Transistor*.

BISMUTH (Bi). Metallic element, Atomic No. 83. Its electrical *resistance* increases with increasing strength of a magnetic field in which it is placed.

BISTABLE. A circuit with two stable states.

BIT. (1) A unit of information content corresponding to the decision between one of two possible states. (2) A unit of capacity in a *storage* device. The capacity, in bits, of a storage device is the logarithm to base two of the number of possible states of the device. (3) A binary digit.

BITTER PATTERN. The pattern displayed in a *ferromagnetic* powder lying on the surface of a magnetized crystal.

BLACK BODY. An ideally perfect absorber and emitter of radiation. The nearest practical form is a cavity with opaque walls and a small opening, maintained at uniform temperature.

BLACK BOX. Any self-contained unit or part of an electronic equipment which can be treated as a single package. Such units are often housed in black boxes for military or industrial use.

BLACK COMPRESSION (SATURATION). In television, the reduction in gain applied to a *picture signal* at those levels corresponding to dark areas in the picture, with respect to the gain at that level which corresponds to the mid-range light value.

BLACK LEVEL. The instantaneous amplitude of the television signal which produces a black area in the received picture.

BLACK SCREEN. A plastic filter placed between a television picture tube and the viewer to reduce reflected light and so increase contrast.

BLACK-AFTER-WHITE. A defect in a television picture which causes a black line to follow a white object.

BLACKER-THAN-BLACK LEVEL. In television, a level of greater instantaneous amplitude than the *black level*, used for synchronizing and control signals.

BLACK-FACE TUBE. A television *picture tube* which makes use of unilluminated dark-grey *phosphors* to reduce reflection from external light sources.

BLACK-OUT. A temporary loss of sensitivity in an electronic valve or other device after a strong short pulse.

BLANK. Make a device or channel ineffective for a desired interval. In cathode-ray and picture tubes blanking is used to eliminate the return trace. See *Blanking signal*.

BLANKING LEVEL. In television, the level of a composite picture signal which separates the range containing picture information from that containing synchronizing information.

BLANKING SIGNAL. In television, a train of recurrent pulses, related in time to the scanning process, used to effect *blanking*.

BLASTING. In *audio* and *radio amplifiers*, a form of *distortion* due to heavy overloading.

BLATTNERPHONE. A type of *magnetic sound recorder* and reproducer which uses *ferromagnetic* wire.

BLEEDER RESISTOR. A *resistor* permanently connected across the output of a power supply.

BLEMISH. In *charge-storage* tubes: an imperfection in the storage surface which may produce a spurious output.

BLISTER. See *Radome*.

BLOCH WALL. The transition layer between two oppositely magnetized regions in a *ferromagnetic* material.

BLOCK. In *data processing*: a set of associated words or characters treated as a unit.

BLOCK DIAGRAM. A diagram in which the important units of any system are drawn in the form of blocks, usually rectangular.

BLOCK READER. A *tape reader* which will read a complete *block* of information at the same time. This type of reader enables a machine controller to operate without a memory.

BLOCKED IMPEDANCE. The *impedance* of an electromechanical or electro-acoustical *transducer* measured when the mechanical motion is blocked. Contrast *Motional impedance*.

BLOCKING. In valves, the interruption of anode current by a high negative bias on the grid.

BLOCKING CAPACITOR. A *capacitor* which effectively limits the flow of low frequency alternating current (or direct current), without materially affecting the flow of high frequency current.

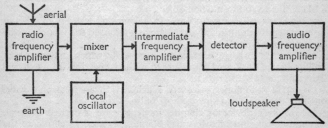

Fig. 13. Block diagram of a superheterodyne radio receiver

BLOCKING OSCILLATOR. A transformer-coupled feedback oscillator in which anode current flows for only one half of the cycle before the oscillation is stopped by *blocking*.

BLOODHOUND. A British ground-to-air missile.

BLOOMING. In cathode-ray and picture tubes, a mushrooming effect on the electron beam caused by setting the brightness control too high, which produces expansion, defocusing, and consequent distortion of the image.

BLUMLEIN INTEGRATOR. See *Miller Integrator*.

BMEWS. Ballistic Missile Early Warning System.

BODE DIAGRAM. A diagram showing the relation between the gain or phase shift and the frequency of an amplifier or servomechanism.

BODY CAPACITANCE. The *capacitance* introduced into an electric circuit by the proximity of the human body.

BOLOMETER. (1) In electronics, a small resistive element whose resistance changes with the heat produced when it dissipates *microwave* power and can so be used as a detector or indicator. (2) A very sensitive metallic resistance thermometer used to measure feeble radiations. (3) In general, a device for measuring current or power.

BOND NOTATION SYSTEM. A system of describing cuts in *piezoelectric* crystals.

BONDED-BARRIER TRANSISTOR. A *transistor* in which material on the tip of a wire is alloyed to the base material.

BONDING ELECTRON. An *electron* which holds together two atoms in a molecule.

BONE CONDUCTION. The conduction of sound to the inner ear through the cranial bones.

BOOK CAPACITOR. A small *capacitor* with two plates hinged like the cover of a book, whose capacitance may be varied by altering the angle between the plates.

BOOLEAN ALGEBRA. A branch of symbolic logic in which logical propositions and operations are indicated by operators such as 'and', 'or', 'not', 'either', analogous to mathematical signs. It is employed in the design of *automatic computers*.

BOOM. An adjustable support for a microphone or a television camera in a broadcasting studio.

BOOST CHARGE. Of a storage battery, a partial charge, usually at a high rate for a short period.

BOOSTER. In broadcasting, a *repeater* station which amplifies a signal from a main station and re-transmits it, sometimes with a change of frequency.

BOOSTER AMPLIFIER. In audio engineering, an amplifier which increases the strength of signals from the mixer controls of an audio console so as to maintain an adequate *signal-to-noise ratio*.

BOOSTER DIODE. See *Efficiency diode*.

BOOTSTRAP AMPLIFIER. A single stage amplifier in which the output load is connected between the negative end of the H.T. supply and the cathode, and the signal voltage is applied between grid and cathode. The name is meant to indicate that a change in grid voltage 'pulls' the potential of the input source by an amount equal to the output signal.

BORO-CARBON RESISTOR. A *resistor* formed by firing a microcrystalline film of carbon and boron hydride on a ceramic-former.

BORON (B). Non-metallic element, Atomic No. 5.

BORON COUNTER. A radiation counter used to record the passage of slow *neutrons*.

BOTTOMING. The action of suppressing one alteration in the output of a non-linear device, such as a pentode valve, when an alternating signal is applied at its input, by operating the device below the *knee* of its voltage-current characteristic.

BOUND CHARGE. The portion of the charge on a conductor which does not escape to earth when the conductor is grounded.

BOX-CAR LENGTHENER. A pulse-stretching circuit in which the increase in duration is not accompanied by an increase in amplitude.

BRAIN VOLTAGE. The voltage generated by the human brain, normally between 10 and 60 microvolts in amplitude and between 1 and 60 cycles per second in frequency.

BRAIN WAVE. The characteristic waveform of the *brain voltage* which varies with the individual and with his condition.

BRAUN TUBE. An early form of *cathode-ray tube*.

BREADBOARD. In electronics, an arrangement in which components are attached temporarily to a board or chassis for experimental work.

BREAK. In a device for opening a circuit, e.g. a switch, the minimum distance between the stationary and movable contacts when the contacts are in the open position.

BREAK SHOCK. The shock in an organism in series with an electric current when the circuit is broken.

BREAKDOWN. (1) Disruptive electric discharge through an *insulator* or between electrodes in vacuum or gas-filled valves. (2) In a *semiconductor diode*, the sudden transition from a region of high dynamic resistance to one of substantially lower dynamic resistance, for increasing magnitude of bias.

BREAKDOWN VOLTAGE. The value of voltage at which *breakdown* occurs.

BREAK-IN. In communications, an automatic device which allows incoming signals to be received during breaks in transmission.

BREAK-POINT. In a *digital computer*, a point in the programme at which a special instruction enables the computer to be stopped for a visual check of progress.

BREIT AND TUVE METHOD. A method of measuring the height of certain layers of the *ionosphere* by measuring the echoes from transmitted pulses.

BRIDGE. Any one of a variety of electrical networks in which one branch – the 'bridge' proper – connects two points of equal potential and so carries no current when the circuit is suitably adjusted or 'balanced'. Bridge networks form the basis of a great variety of measuring instruments and are capable of very great precision. The following bridges are defined in this dictionary: *Anderson, Campbell, capacitance, Carey-Foster, frequency, Hay, Heaviside-Campbell, Kelvin, Maxwell, Nernst, Owen, resonance, Schering, slide-wire, Wagner earth, Wheatstone, Wien.*

BRIDGE RECTIFIER. A *full-wave rectifier* circuit in which there are four arms, each containing a rectifier.

BRIDGED T-NETWORK. A *T-network* with a fourth branch connected across the two series of the T, between input and output.

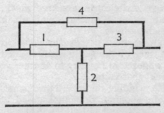

Fig. 14. Typical bridged T-network using resistors

BRIDGING. Shunting one electrical circuit by another.

BRIDGING CONNECTION. A parallel connection which allows the signal energy in the circuit to be drawn off as required with negligible effect on the normal operation of the circuit.

BRIDGING GAIN. Of an *amplifier* which bridges an *impedance* at its input, the power which the amplifier delivers to the load divided by that dissipated in the *impedance.*

BRIDGING LOSS. At a given frequency, the loss which results from bridging an impedance across a transmission system.

BRIGHTNESS. An attribute of visual perception which corresponds to the emission of more or less light from an illuminated area. Contrast *luminance,* which is a photometric quantity independent of subjective sensation.

BRIGHTNESS (BRILLIANCE) CONTROL. The manual bias control of a cathode-ray or television picture tube.

BRITISH STANDARD WIRE GAUGE (SWG). See *Wire gauge.*

BRITISH THERMAL UNIT (BTU). The energy required to raise the temperature of one pound of water by one degree Fahrenheit. 1 B.T.U. = 1055 *joules.*

BROAD BAND. In electronics, a band covering a wide range of frequencies.

BROAD-BAND TUBE (VALVE). A gas-filled tube with fixed tuning which incorporates a *band-pass filter*, used for radio-frequency switching.

BROADCAST TRANSMISSION. Radio transmission to an unlimited number of receiving stations.

BROMINE (Br). Liquid, non-metallic element, Atomic No. 35.

BRONSON RESISTANCE. The resistance formed by two electrodes in a gas which is exposed to a constant source of *ionization*.

BROWN AND SHARPE WIRE GAUGE (B & S.W.G.). See *Wire gauge*.

BROWN CONVERTER. A relay designed to operate in a direct current *chopper* amplifier, which opens and closes at the supply frequency.

BRUSH. A block of carbon or graphite, with or without copper, used to make continuous electrical contact between rotating and stationary members of electrical machines.

BRUSH DISCHARGE. An electric *discharge* which may take place in long gaps at atmospheric pressure and which often develops a tree-like or branching form.

BUBBLE CHAMBER. A kind of *cloud chamber* in which an ionized particle leaves a track of bubbles in a superheated transparent liquid.

BUCK. A voltage opposed in polarity to another voltage in series with it is said to 'buck' that voltage.

BUCKING COIL. A winding on an electromagnet which opposes the field of the main winding. Sometimes used with *electromagnetic loudspeakers* to smooth out pulsations in the supply voltage.

BUCKLEY GAUGE. A very sensitive pressure gauge based on measuring the amount of ionization produced in a gas by a specified current.

BUFFER. (1) An isolating circuit used to avoid reaction between a driving and a driven circuit. (2) A circuit with a single output and several inputs in which energizing any one of the inputs energizes the output. (3) A *memory* device used to compensate for a difference in the rate of flow of information in the different parts of a computing system.

BUFFER AMPLIFIER. An amplifier in which the reaction of variation of output load impedance on the input circuit is minimized.

BUFFER CAPACITOR. A *capacitor* connected across circuit elements to suppress voltage surges which might disturb or damage the circuit.

BUFFERED COMPUTER. A computer with a storage device which can store data temporarily so that the relatively slow speeds of input and output devices can be matched with the high speed of the computer.

BUG (ELECTRONIC). An electronically controlled keying system which converts Morse signals from a hand key into dot and dash signal elements with correct spacing and proportion.

BUILD-UP TIME. The time required for a current to rise substantially to its maximum value.

BULK LIFETIME. The average time interval between the generation and recombination of *minority carriers* in the bulk of a homogeneous *semiconductor*.

BUNCHER. The electrode of a *velocity-modulated valve* which concentrates the electrons in the constant-current electron beam into bunches.

BUNCHING. The process in a *velocity-modulated* electron stream which produces an alternating convection current as a result of differences in the velocities of electrons in the stream. See *Applegate diagram, Velocity-modulated valve.*

BUNCHING, IDEAL. In a *velocity-modulated* electron beam, the ideal condition in which all the electrons of a given bunch have a common velocity and phase.

BUNCHING ANGLE. In an electron stream, the average transit angle between the processes of velocity modulation and energy extraction at the same gap or at different gaps.

BUNCHING PARAMETER. In a *klystron*, a *parameter* which determines the degree of *bunching* and the waveform.

BURST SIGNAL. In *colour television*, the transmitted synchronizing signal which controls the phase and frequency of the colour oscillator in the receiver.

BUS. In a *computer*, one or more conductors used as a path for transmitting information from a source to a destination.

BUS-BAR. A metallic rod which carries a large current or makes a common connection between several circuits.

BUTTERFLY RESONATOR. A *resonator* or *resonant circuit* designed to work in the frequency range 100 to 3,000 megacycles per second.

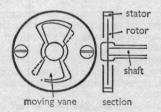

Fig. 15. Butterfly resonator

BUTTON-MICA CAPACITOR. A small capacitor with a mica *dielectric* resembling a button.

BWO. Abbreviation for Backward Wave Oscillator.

BY-PASS CAPACITOR. A *capacitor* which provides a path of comparatively low *impedance* around some element in a circuit for alternating current.

BYTE. In *data processing*: a sequence of adjacent binary digits operated on as a word and usually shorter than a word.

C

C BATTERY (U.S.A.). A battery used to supply *bias*.

C CORE. A magnetic core wound spirally from magnetic alloy strip, usually formed into a roughly rectangular shape before being accurately cut into two equal C-shaped pieces. As the magnetic grain of the material lies along the direction of the wound strip, its performance is better than an equivalent stack of interleaved laminations and it is easier to assemble with *transformer* or *magnetic amplifier* coils.

CADMIUM (Cd). White metallic element chemically similar to zinc, Atomic No. 48. Widely used as an anti-corrosion plating.

CADMIUM CELL. See *Weston normal cell*.

CAESIUM (Cs). Silver-white metallic element, Atomic No. 55. The most *electropositive* element. Used in *photoelectric cells* and as a *getter* for valves.

CAESIUM CLOCK. An extremely accurate clock in use at the National Physical Laboratory to provide a time scale uniform to 1 part in 10^{10}, i.e. 1 sec. in 300 years. It is based on the frequency of a selected spectral line from the hyper-fine spectrum of caesium and is used to calibrate standard clocks.

CALCIUM (Ca). White metallic element, Atomic No. 21. Used in the manufacture of caesium *photo cathodes* and as a *getter* in low noise valves.

CALCULATOR. (1) A small, usually mechanical, *computer*. (2) In *data processing*: a device capable of performing arithmetic. (3) Generally, a device for carrying out logical and arithmetical operations of any kind.

CALL IN. In *computer programming*: transfer control during a subsidiary operation from the main routine to a sub-routine.

CALL NUMBER. In *computer programming*, a set of symbols used to identify a sub-routine.

CALL WORD. In *computer programming*, a *call number* exactly the length of one word.

CALOTRON. A device based on the *superconductivity* of thin films of tin and lead, whose resistance can be varied by a control pulse of current, providing a multistable element.

CAMERA, ELECTRON. See *Camera, television*.

CAMERA, ELECTROPLANE. A system of optical lens elements of a motion-picture camera in which electronically produced oscillations are applied to one of the lens elements to improve the depth of field.

CAMERA, EMITRON. See *Emitron*.

CAMERA, TELEVISION. The device used in a television system to

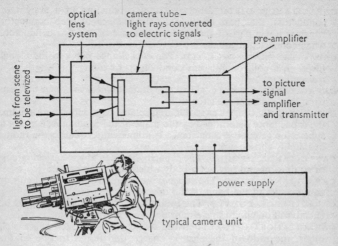

Fig. 16. Television camera – diagrammatic representation

convert the optical images from a lens into electrical signals. It consists of optical lens, *camera tube*, and *preamplifier*, housed in one container. See *Emitron*.

CAMERA SIGNAL. The *video* signal output of a *television camera*.

CAMERA SPECTRAL CHARACTERISTIC. In *colour television*, the sensitivity of each of the camera colour-separation channels with respect to wavelength.

CAMERA TUBE. An *electron beam* tube in which an optical image is converted into an electron current or pattern of electrostatic charges, and scanned in a fixed sequence to provide electric signals. See *Emitron, Iconoscope, Image dissector tube, Image iconoscope, Image orthicon, Orthicon, Plumbicon, Vidicon.*

CAMPBELL BRIDGE. A *bridge* specifically designed for comparison of *mutual inductances*.

CANAL RAYS. Rays generated behind the cathode of a discharge tube when positive ions pass through holes or tunnels bored in the cathode.

CANDELA. The unit of luminous intensity in the *SI unit* system.

CAPACITANCE. Symbol: C. The property of a system of *conductors* and *insulators* which allows them to store electric *charge* when a *potential difference* exists between the conductors. The capacitance of a *capacitor* is defined as the ratio of the electric charge transferred from one electrode to the other, to the potential difference between the electrodes which results. See also *Farad, Microfarad, Picofarad, Units.*

CAPACITANCE, DIRECT. Between two conductors in a system of conductors, that part of the *total capacitance* of each conductor which is not a part of the total capacitance of the two conductors when connected together.

CAPACITANCE, DISTRIBUTED. The capacitance between individual turns on a coil, or adjacent conductors, or along the length of a *transmission line*.

CAPACITANCE, EFFECTIVE. In a circuit, the *total capacitance* between two given points of the circuit.

CAPACITANCE, GEOMETRIC. Of an insulated conductor in a vacuum, the capacitance determined by the shape of the conductor; e.g., for a sphere the geometric capacitance is numerically equal to the radius in suitably chosen units.

CAPACITANCE, GROUND. The capacity between any electrical circuit or equipment and *earth* or a body which acts as an earth.

CAPACITANCE, HAND. The capacity between any electrical circuit or equipment and the hand of the operator.

CAPACITANCE, LUMPED. The total distributed capacity taken to be effective at one point in a circuit or on a transmission line.

CAPACITANCE, MUTUAL. Between two capacitors which affect each other, the ratio of the electric charge transferred to one to the corresponding difference of potential between the electrodes of the other.

CAPACITANCE, STRAY. Unwanted capacitance between parts of a component, circuit, or equipment.

CAPACITANCE, TOTAL. Of any conductor in a system of conductors, the capacitance between that conductor and all the others connected together.

CAPACITANCE ALTIMETER. See *Altimeter, capacitance*.

CAPACITANCE BRIDGE. A *bridge* for comparing two *capacitances*.

CAPACITANCE COEFFICIENT. Of one *conductor* in a system of insulated conductors surrounded by an insulating shell at zero potential, or on one side of an infinite conducting plane: the quantity of electricity required to raise the *potential* of the conductor to unity while all other conductors are kept at the potential of the shell or plane.

CAPACITANCE INTEGRATOR. A circuit which is basically a resistor and a capacitor in series and whose output is proportional to the current in the capacitor.

CAPACITANCE RELAY. An electronic *relay* whose operation depends on a small change of *capacitance*; e.g. that produced by bringing a hand near a metal wire or plate.

CAPACITIVE COUPLING. The association of two or more circuits by their *mutual capacitance*.

CAPACITIVE LOAD. A load which acts as a combination of resistance and capacitance.

CAPACITIVE REACTANCE. The *reactance* presented by a *capacitance* to an alternating current or direct pulsed current.

CAPACITOR. Essentially, an assembly of one or more pairs of conductors separated by insulators, used to obtain an appreciable *capacitance*, sometimes of a specified value. The two conductors are called electrodes or plates, and the insulator, which may be solid, liquid, or gaseous, is called the *dielectric*. The following types of capacitor are described: *blocking, ceramic, commutating, differential, electrolytic, fixed, guardring, guard-well, mica, non-inductive, paper, pressure-type, trimmer, variable, vibrating*.

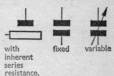

with inherent series resistance.   fixed   variable

CAPACITRON. A mercury-pool tube in which an electrode insulated from the arc starts the arc when a high voltage is applied between it and the cathode. When the electrode is outside the glass wall of the tube in the form of a band the device is called a *band-igniter tube*.

CAPACITY. (1) Synonym for *capacitance*. (2) The maximum or rated load of a machine or device. (3) Of a *computer*, the number of digits or characters the computer is able to store or process.

CAPTURE EFFECT. In *frequency modulation* reception, the suppression of a weaker signal by a stronger signal of the same frequency.

CARBON (C). Non-metallic element, Atomic No. 6. Used in the construction of some electron tubes (see *Aquadag coating*) and components such as resistors, microphones, electrodes, and batteries.

CARBON BUTTON. A small compressible container filled with granules of carbon, which is the essential part of the *carbon microphone*.

CARBON MICROPHONE. A *microphone* whose operation depends on the variation in contact resistance between carbon granules when they are subjected to varying sound pressures.

CARBON PILE REGULATOR. A variable *resistor* used for voltage and current control, consisting of a pile of discs or blocks of carbon. Compression of the pile alters the resistance over wide limits.

CARBON RESISTOR. A type of *resistor* very widely used in electronics, made of very finely ground carbon particles mixed with a ceramic binder. This material is usually moulded into small cylinders which are heat-treated and enclosed in a ceramic or plastic tube with a lead at either end.

CARBON RHEOSTAT. A variable *resistor* composed of small pieces of carbon and a means for altering the pressure on them so that the total resistance can be varied.

CARBON-FILM RESISTOR. A *resistor* with a negative temperature characteristic, made by depositing a thin film of carbon on a suitable ceramic or metallic form. See Figure 17.

CARBONIZED ANODE. The anode of a valve which has been blackened with carbon to increase its dissipation of heat.

49

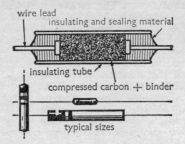

Fig. 17. Section of carbon resistor (axial leads)

CARBONIZED FILAMENT. A *thoriated-tungsten filament* coated with tungsten carbide which slows down evaporation of the thorium and so permits higher operating temperatures.

CARBORUNDUM. A crystalline *semiconductor* compound of carbon and silicon used as a *crystal detector*.

CARBURIZING. The process of producing a *carbonized filament*.

CARCINOTRON. A *backward wave* oscillator tube capable of generating frequencies of the order of 100,000 megacycles per second.

CARD FEED. In a *computer* or *data-processing system*, a device which inserts cards one at a time.

CARD PUNCH. A device for registering information on a card by punching holes in it according to a given code.

CARD READER. In a computing system, a device which transfers the information on a card to some other form of storage, processing, or display element. This is usually done by electro-mechanical feelers or 'fingers' or by photo-electric scanning.

CARDIOGRAM. See *Electrocardiograph*.

CARDIOID MICROPHONE. A *microphone* whose directional response pattern is heart-shaped.

CARDIOTACHOMETER. An electronic amplifier instrument used to time and record the heart rate.

CARDIOTRON®. A portable electronic *electrocardiograph*.

CAREY-FOSTER BRIDGE. A form of *Wheatstone bridge* designed to eliminate contact errors in measuring two nearly equal resistances.

CARMATRON. A *travelling wave tube* with a structure similar to a cylindrical *magnetron*, but with the circuit open and terminated by two coaxial feeders.

CARRIER. (1) In a *semiconductor*, mobile electrons or holes which carry charges. (2) In communications, a wave suitable for modulation, such as a simple sine wave or series of pulses.

CARRIER AMPLITUDE REGULATION (CARRIER SHIFT). A change

in the amplitude of a *carrier* wave in an *amplitude-modulated* transmitter when the modulation is symmetrically applied.

CARRIER BEAT. In *facsimile* transmission, unwanted *heterodyne* of signals from different reference oscillators which causes an interference pattern on the received copy.

CARRIER FREQUENCY. In communications: the frequency of the wave which is modulated by a signal of lower frequency.

CARRIER MOBILITY. In a *semiconductor*, the average drift velocity of *carriers* per unit electric field. See *Hall mobility*.

CARRIER MODULATION. The variation of some characteristic of a *carrier wave*.

CARRIER STORAGE. An effect which arises when the density of charge carriers in a *semiconductor* region exceeds that present under equilibrium in the zero bias condition.

CARRIER SUPPRESSION. A technique of transmission in which the *carrier wave* is not transmitted.

CARRIER SYSTEM. A system allowing simultaneous independent communications over the same circuit.

CARRIER TELEGRAPHY. *Telegraphy* in which alternating current is supplied to the line after being modulated in the transmitting apparatus.

CARRIER TELEPHONY. A form of telephony in which *carrier transmission* is employed and the carrier modulated by *voice frequency* waves. The term is usually applied to *wire telephony*.

CARRIER TRANSMISSION. Electrical transmission in which the transmitted wave results from the modulation of a single-frequency wave by a modulating wave.

CARRIER WAVE. A wave with the necessary properties to enable it to be transmitted through a selected physical system after it has been modulated.

CARRIER-TO-NOISE RATIO. The ratio of the magnitude of the carrier to that of the *noise* measured at the receiver before any non-linear process such as amplitude limiting or detection.

CARRY. In *data processing*: a character which arises in adding, when the sum of two or more digits equals or exceeds the radix of the number representation system.

CASCADE (MULTI-STAGE) AMPLIFIER. A chain of *amplifier* stages in which the output of the first supplies the input to the second, and so on.

CASCADE CONTROL. Of *computers*, a system of automatic control in which each control unit controls the succeeding unit and is itself controlled by the preceding unit.

CASCADE SHOWER. A *cosmic-ray* shower formed when a high energy electron passing through matter produces one or more *photons* whose energies are of its own order of magnitude.

CASCODE AMPLIFIER. A *cascade amplifier* made up of a grounded cathode input stage driving a grounded grid output stage. It is often used at the input of high gain receivers because of its low *noise* figure.

CATCHER. In certain types of velocity-modulated tube, e.g. a *klystron*, the electrode which receives the bunches of electrons.

CATCHING DIODE. A *diode* used to limit or clamp the voltage at some point in the circuit. See *Clamping diode*.

CATHAMPLIFIER. An amplifying circuit to allow *balanced* (push-pull) operation from a single-ended drive.

CATHODE. (1) In an electronic *valve* or *tube*, the electrode through which the primary stream of electrons enters the inter-electrode space. (2) Generally, a negative *electrode*.

symbol   symbol showing associated heater

CATHODE, INDIRECTLY HEATED. A *thermionic cathode* in the form of a thin metal tube suitably coated, with a heating filament running up inside the tube. The tube is filled with a material which is a good electrical insulator but also a good heat conductor to protect the filament and ensure good thermal conductivity to the cathode.

CATHODE, THERMIONIC. A *cathode* which provides a copious flow of emitted *electrons* when its temperature is high enough.

CATHODE, UNIPOTENTIAL. Another name for an *indirectly heated cathode* in which the whole cathode surface is effectively at the same *potential*.

CATHODE BIAS (AUTOMATIC BIAS). A way of *biasing* a valve by inserting a *resistor* in series with the *cathode*. The effect is to cause the cathode to take up a positive potential when the *grid* is connected to ground.

CATHODE DARK-SPACE (CROOKES DARK-SPACE). A dark region following the first cathode layer in a *gas discharge tube*.

CATHODE DISINTEGRATION. Destruction of the active area of the *cathode* by bombardment with positive *ions*.

CATHODE DROP (FALL). In a *glow discharge tube*, the voltage drop between the arc stream and the negative electrode.

CATHODE FOLLOWER. A circuit in which the input is applied between control grid and ground and the output is obtained from an impedance between cathode and ground. This arrangement provides a high impedance input and a low impedance output.

CATHODE GLOW. The glow observed in a discharge tube between cathode and the *cathode dark-space*. See *Gas discharge* and *Glow discharge*.

CATHODE GRID. A *suppressor grid*.

CATHODE HEATING TIME. In valves and tubes, the time from switch on required for the cathode to reach the temperature necessary for operation under prescribed conditions.

CATHODE INTERFACE IMPEDANCE. The *impedance* between the base material and the coating of a *cathode* in a valve, due to imperfect electrical contact.

CATHODE KEYING. A method of keying a radio transmitter in which the key operates in the cathode lead and opens both grid and anode circuits simultaneously.

CATHODE LUMINOUS SENSITIVITY. Of a *photo-cathode*, the photo-cathodic current divided by the luminous flux.

CATHODE MODULATION. *Modulation* by application of the modulating signal to the cathodes of valves in which the carrier is present.

CATHODE POISONING. The residual gas left in a vacuum valve may react with the free barium in the *cathode* and lead to deterioration of the emission.

CATHODE RAYS. A stream of electrons emitted by the *cathode* of a valve or tube when it is heated to cause *thermionic emission* or bombarded with positive ions.

CATHODE RESISTOR. The resistor connected in series with the cathode to produce *cathode bias*.

CATHODE SHEATH. See *Cathode dark-space*.

CATHODE SPOT. In *gas discharge tubes*, the bright spot on the cathode from which the arc appears to be generated.

CATHODE SPUTTERING. A method of depositing very thin layers of metal on the surface of a material by making the metal the *cathode* in a *vacuum tube* and bombarding it strongly with positive ions at high voltage. The object to be plated is placed inside the tube where it can receive the particles of metal projected from the cathode.

CATHODE-RAY FURNACE. A very high temperature furnace in which small objects are placed in a vacuum and subjected to high energy *electron beams*.

CATHODE-RAY LAMP. An electronic light source of great intensity, in which a small block of refractory material is heated to incandescence by high energy electrons.

CATHODE-RAY OSCILLOSCOPE (C.R.O.). A *cathode-ray tube* operated as an *oscilloscope*. Synonymous with cathode-ray oscillograph.

CATHODE-RAY TUBE (C.R.T.). A funnel-shaped vacuum tube, specially constructed to allow direct observation of the behaviour of *cathode rays*, which has numerous applications in science, engineering, and entertainment. The essential parts are: an *electron gun* in the neck which produces and projects a beam of electrons; a screen treated with a *phosphor* which receives the electron beam; and a means of producing electric or magnetic fields between the two which focus the beam on the screen and move it rapidly across the screen in any desired manner.

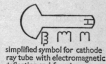

simplified symbol for cathode ray tube with electromagnetic deflection and focusing

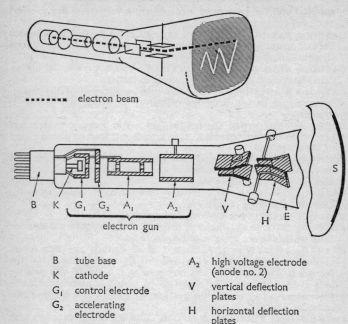

- - - - - - - electron beam

B   tube base
K   cathode
G_1 control electrode
G_2 accelerating electrode
A_1 focusing electrode (anode no. 1)

A_2 high voltage electrode (anode no. 2)
V   vertical deflection plates
H   horizontal deflection plates
E   envelope
S   fluorescent screen

Fig. 18. Cathode-ray tube (diagrammatic representation)

CATHODE-RAY TUNING INDICATOR. See *Magic eye*.

CATHODOLUMINESCENCE. Metals bombarded with *electrons* in an enclosure, as with *cathode rays*, are vaporized at their surfaces and emit luminous radiations which are characteristic of the metal.

CATHODOPHOSPHORESCENCE. *Phosphorescence* caused by cathode-ray bombardment in a similar way to *cathodoluminescence*.

CATION. A positively charged *ion* which is attracted to the *cathode*.

CATIONIC CURRENT. The part of a current which is carried by cations.

CATKIN VALVE. A vacuum valve enclosed in a metal envelope which is made the anode of the valve.

CAT'S WHISKER. A small pointed wire used to make contact with a sensitive point on the surface of a *semiconductor*.

CAVITY MAGNETRON. A *magnetron* whose anode is composed of a number of *resonant cavities* used to generate *microwave* frequencies.

CAVITY RESONATOR. A space normally bounded by conducting surfaces in which electromagnetic energy may be stored as oscillations whose frequency is determined by the shape and size of the cavity.

CELL. (1) A single source of electric *potential*. (2) In computers, an elementary storage unit. The following types of cell are described: *bias, binary, cadmium, Clark, concentration, conductivity, Daniell, dry, electrolytic, gas, half-, Kerr, lead-acid, Leclanche, mercury, nickel-iron, photoconductive, photoelectric, photoemissive, photovoltaic, primary, reversible, secondary battery, standard, storage, voltaic, Weston normal.*

CELL CONSTANT. Of an *electrolytic cell*, the resistance in ohms of the cell when it is filled with a liquid of unit resistivity.

CELL RESISTANCE, INTERNAL. The resistance offered to the passage of current through the *cell* by the materials inside the cell. Mathematically, the instantaneous value of the ratio of the difference between the open-circuit *e.m.f.* generated and the potential difference at the terminals with current flowing, to the current.

CELL-TYPE TUBE. A radio-frequency switching tube designed to fit into a *resonant cavity*, consisting of a *spark gap* enclosed in a gas at low pressure.

CELLULOID. An insulating material synthesized from guncotton and camphor. As it is highly inflammable its use in electronics is now very restricted.

CELOTEX®. A patented wall board made from dried wood which is an excellent heat and sound insulator.

CENTIMETRIC WAVES. *Microwaves* with frequencies between 3 and 30 Gigacycles per second, corresponding to wavelengths between 10 and 1 centimetres.

CENTIPEDE®. A high-power *travelling wave tube* developed at Stanford University, U.S.A.

CENTRE TAP. A connection placed at the electrical mid-point of a resistor, transformer, or other electrical device.

CERAMIC CAPACITOR. A *capacitor* whose *dielectric* is made of a ceramic material. Such a capacitor may be much smaller than other types of a similar capacitance because of the very high *permittivity* of the ceramics used. See *Barium titanate*. See Figure 19.

CERAMIC PHOTOCELL. A *photocell* which employs a light-sensitive resistor composed of titanium dioxide and various titanates and metallic oxides.

CERAMIC VALVE (TUBE). A valve, usually for operation at high frequency and high power, in which the glass envelope is replaced by ceramic. See also *Stacked-ceramic valve*.

CERAUNOGRAPH. An electronic instrument for recording lightning discharges.

CERIUM (Ce). Metallic element, Atomic No. 58.

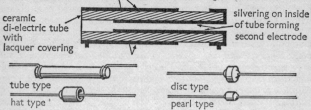

Fig. 19. Construction and representative types of ceramic capacitor

CERMET. An alloy of a metal and a ceramic which retains its useful properties at extremely high temperatures.

CERROBEND. A soldering alloy with the very low melting point of about 70°C.

CERROSEAL. A low temperature soft solder containing indium; melting point about 120°C.

CHAD. In *data processing*: the piece of material removed when forming a hole or notch in a storage medium such as paper tape or punched cards.

CHANNEL. (1) In electrical communications, a band of frequencies or a specified path for the transmission and reception of electric signals. (2) In a computer or data-processing system, a path or route along which information may flow and be stored.

CHANNEL CAPACITY. The maximum number of symbols per second which can be transmitted along a given channel.

CHANNEL EFFECT. In *junction transistors*, a current observed between the *emitter* and *collector* which does not flow through the base region and is thought to be due to leakage along surface paths.

CHAPERON RESISTOR. A resistor whose element is wound in a special way with doubled wire so that it has low distributed inductance and capacitance.

CHAPMAN REGION. An idealized region in the *ionosphere* in which the variation of the electron density with height follows an approximately parabolic law.

CHARACTERISTIC CURVE. A graph which represents the relation between two magnitudes which are characteristic of the behaviour of any device or apparatus. In electronics the most frequently used of these curves are those for *valve characteristics*. See Figure 20.

CHARACTRON. A *cathode-ray tube* in which the electron beam is directed to excite the suitably treated phosphors on the screen so that visible digits or characters appear.

CHARGE. (1) Of a conductor, the total quantity of electricity thereon.

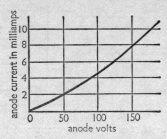

Fig. 20. Characteristic curve of a typical diode valve

(2) Of a capacitor, the quantity of electricity on either electrode.
(3) Of a storage battery, the conversion of electric energy into chemical energy in the cell.

CHARGE CARRIER. See *Carrier (1)*.

CHARGE-STORAGE TUBE. A *storage tube* which retains the information on its active surface in the form of electric charges.

CHECK DIGIT. In *computers*, one or more digits added to a word or other information unit solely in order to act as a check on the word; e.g. on the number of digits in the word.

CHECK PROBLEM (ROUTINE). The accuracy of a *computer* may be checked by supplying it with a problem whose answer is known. Faults in the computer may sometimes be diagnosed by a study of the errors in the computer solution.

CHEMICAL BATTERY. A device for converting chemical into electrical energy.

CHLORINE (Cl). Gaseous element, Atomic No. 17.

CHOKE (CHOKING COIL). An *inductor* designed to present a relatively high impedance to alternating current. See *Audio-frequency choke, Radio-frequency choke, Smoothing choke, Swinging choke.* See Fig. 21.

CHOKE (MICROWAVE). A groove approximately one quarter of a wavelength deep cut into a metal surface (of a *waveguide*) to prevent escape of *microwave* energy.

CHOKE COUPLING. (1) The use of a choke in the output of an amplifier stage to replace the primary of an interstage transformer. (2) In *waveguides*, an indirect coupling designed to prevent high-frequency leakage, using microwave chokes.

CHOKE JOINT. A method of joining sections of transmission line or waveguide for *microwaves* which prevents leakage and high *homic losses* by making the joint in the form of a *choke coupling.*

CHOPPER. A device for interrupting a current or a beam of light at regular intervals. The interrupted signals are more easily amplified,

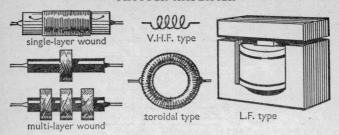

single-layer wound

V.H.F. type

multi-layer wound

toroidal type

L.F. type

Fig. 21. Typical chokes (not to scale)

especially in the presence of noise, than the continuous ones. See *Vibrator*.

CHOPPER AMPLIFIER. An amplifier in which the input signal is first converted into alternating current by a system of *relays* or by rotating *chopper* contacts so that normal ac technique may be used for amplifying small direct current signals.

CHROMA CONTROL. In *colour television*, a control for adjusting the degree of saturation of the colours in the reproduced picture.

CHROMATON. A smaller version of the *chromatron*.

CHROMATRON (CHROMOSCOPE). A *cathode-ray tube* employing four screens used in colour television.

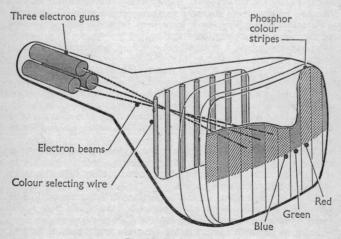

Three electron guns

Phosphor colour stripes

Electron beams

Colour selecting wire

Red

Green

Blue

Fig. 22. Chromatron

CHROMINANCE CHANNEL. In *colour television*: signals which represent hue and saturation but not luminance or brightness.

CHROMINANCE SUBCARRIER. In *colour television*, the *carrier* whose modulation *sidebands* are added to the monochrome signal to convey the colour information.

CHROMIUM (Cr). Hard, grey, metallic element, Atomic No. 24.

CHROMOPHORIC ELECTRONS. Electrons in the outer shells of certain atoms which readily absorb radiations in the visible spectrum.

CHRONOGRAPH. An instrument for recording graphically short intervals of time.

CHRONOMETER, ELECTRONIC. An accurate electronic clock usually based on a frequency standard.

CHRONOPHER. An instrument which produces accurately timed signals from a standard clock.

CHRONOSCOPE, ELECTRONIC. See *Chronometer*.

CHRONOTRON. An electronic device for determining the time between two events by measuring the positions along a transmission line of pulses initiated by the events.

CIRCLE DIAGRAM. A diagram which provides a graphical solution of the equations which represent the characteristics and behaviour of a *transmission line*.

CIRCLE-DOT MODE. A method of storing binary digits in a *storage tube* in which one kind of digit is represented by a small circle on the screen, and the other kind by a similar circle with a concentric dot.

CIRCUIT. An electrical *network* in which there is at least one path which can be closed.

CIRCUIT ANALYSER. A combination of measuring instruments or measuring circuits, usually in one container or rack, for determining the values of two or more quantities in a circuit.

CIRCUIT BREAKER. A device for making and breaking a circuit under normal or abnormal conditions, but not frequently.

CIRCUIT DIAGRAM. A symbolic representation of the functions and interconnections of the parts of an electrical or electronic device or apparatus. See *Block diagram, Graphical symbols, Wiring diagram*.

CIRCUIT EFFICIENCY. Of valves or tubes, the ratio of the power at the desired frequency delivered to the load of an oscillator or amplifier, to the power at the desired frequency delivered by the electron stream to the output circuit.

CIRCUIT ELEMENT. Any constituent part of a circuit excluding interconnections.

CIRCUIT MAGNIFICATION METER. See *Q meter*.

CIRCUIT PARAMETERS. See *Parameters*.

CIRCUIT RINGING. Low frequency oscillations set up in a radio receiver, which are heard as a resonant tone, due to reception of a pulse of radio-frequency energy.

CIRCULAR ELECTRIC WAVE

CIRCULAR ELECTRIC WAVE. A wave with circular electric lines of force.

CIRCULAR MAGNETIC WAVE. A wave with circular magnetic lines of force.

CIRCULAR MIL. A unit of area equal to the area of a circle whose diameter is one-thousandth of an inch: i.e. 0·000000785 sq. in. Used in the measurement of the cross-section of wires and tubes.

CIRCULAR SWEEP. In *cathode-ray tube* displays, a *sweep* circuit which provides a circular *time base*.

CIRCULAR TRACE. A trace on the screen of a *cathode-ray tube* which is circular.

CIRCULARLY POLARIZED WAVE. An elliptically *polarized wave* in which the ellipse is a circle in the plane perpendicular to the plane of propagation.

CIRCULATING MEMORY. An electronic *memory* which includes a means for delaying the transmission of information and a means for regenerating and re-inserting the information into the delaying system. See *Delay line, Mercury tank*.

CLAMPING CIRCUIT. (1) An electronic circuit which maintains either of the amplitude limits of a waveform at a specified potential. (2) A circuit which clamps the base of a waveform at a specified potential or current.

CLAMPING DIODE. A *diode* used to clamp a voltage at some point in a circuit.

CLARK CELL. A standard cell for precise measurements where the current drain is negligible, as in *bridges*. It is a mercury-zinc cell and produces a voltage of (1·434) at 15°C.

CLASS A AMPLIFIER. A valve amplifier in which the *grid bias* and alternating grid voltages are such that, for any particular valve, *anode current* flows all the time.

CLASS AB AMPLIFIER. A valve amplifier in which *grid bias* and alternating grid voltage are such that the *anode current* in any particular valve flows for more than half but for less than the whole of the electrical cycle.

CLASS B AMPLIFIER. A valve amplifier in which the *grid bias* is adjusted approximately to the *cut-off* value so that *anode current* is nearly zero in the absence of any signal voltage on the grid, and flows for approximately one half of the electrical cycle in any particular valve to which an alternating grid voltage is applied.

CLASS C AMPLIFIER. A valve amplifier in which the *grid bias* is adjusted to a value greater than *cut-off* so that *anode current* is zero in the absence of alternating grid voltage, and, in any particular valve, flows for less than one half of each electrical cycle when an alternating voltage is applied to the grid.

CLASS D AMPLIFIER. An amplifier which uses *pulse width modulation*.

CLEAR. Restore a *storage* or *memory* device to a defined state, usually the one which indicates zero.

CLEARING FIELD. An electrostatic field used in a *cloud chamber* to clear the chamber of *ions* which are not desired.

CLIPPER. A device which automatically limits its output to a specified maximum value at any instant.

CLIPPING CIRCUIT. (1) A circuit in which the peak amplitude of an electrical signal is limited to a specified value. (2) In *pulse* circuits, a method of reducing the amplification of frequencies below a specified frequency or for removing the tail of a *pulse* after a fixed time. (3) In a voice-operated telephone the loss of initial or final syllables due to imperfections in the voice-operated device.

CLIPPING TIME. The *time constant* of a *clipping circuit*.

CLOCK. In *data processing*: a device which generates periodic signals for synchronization.

CLOCK FREQUENCY. Of a *computer*, the master frequency of the timing pulses which govern the schedule of operations.

CLOCK PULSE. In a *computer*, a pulse applied to *logical elements* to effect *logical operations*.

CLOCKSPRING CORE. A core wound from thin magnetic tape into a tight coil.

CLOSE COUPLING. A degree of electrical coupling greater than the

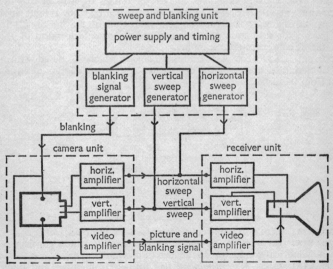

Fig. 23. Closed-circuit television system (diagrammatic)

critical coupling, e.g. as with an open circuit and a closed circuit directly connected together.

CLOSED CIRCUIT. Any path or loop through which current can pass without a break.

CLOSED SUB-ROUTINE. Of a *computer*, a *sub-routine* which is stored separately from the main routine.

CLOSED-CIRCUIT TELEVISION SYSTEM. A *television* system in which the television camera, receiver, and associated controls are directly linked by cables and there are no aerials or open circuits. See Figure 23.

CLOSED-LOOP CONTROL. A system of *servo controls* which obeys some arbitrary command and governs its own behaviour according to the results of its control operations. See *Feedback* loop.

CLOSED-LOOP TEST. Of a complex *servo* system (e.g. the guidance system for a guided missile), a test in which all the *feedback loops* are closed.

CLOUD CHAMBER. A device for studying the behaviour of *ionized particles* by photographing the vapour trails formed on them by condensation in a supersaturated vapour.

CLOUD PULSE. In a *charge-storage tube*, the pulse due to *space-charge* effects produced by turning the electron beam on or off.

CLUTTER. Unwanted signals, images, or echoes on a *radar* display.

COARSE SCANNING. (1) In *television, scanning* with a light spot whose diameter is comparable to the detail of the image. (2) In *radar*, the technique of obtaining an approximate location of the target before accurate scanning.

COATED CATHODE. A thermionic *cathode* which is coated with a material, such as barium oxide, to increase the emission of electrons from its surface.

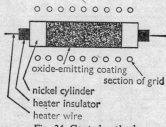

Fig. 24. Coated cathode

COAXIAL CABLE. A cable consisting of one or more *coaxial lines*.

COAXIAL CAVITY. A *resonator* for *microwaves* which consists of a cylindrical conductor on the axis of a cylindrical cavity.

COAXIAL LINE. A *transmission line* formed from inner and outer cylindrical conductors with a common axis.

COAXIAL TERMINAL. A terminal to which a *coaxial line* may be connected.

COBALT (Co). Hard, grey, metallic element, Atomic No. 27. Strongly magnetic and an important constituent of magnetic alloys, stainless steels, and high temperature alloys.

COBOL. COmmon Business Orientated Language: developed in the U.S.A. as a possible international common language *programme* for commercial applications of *computers*.

COCKROFT-WALTON CIRCUIT. See *Voltage-multiplier rectifier*.

COCOMAT®. A computer-assisted program developed by Rolls-Royce Ltd for their own use.

CODAN. An electronic circuit used to make a receiver inactive in the absence of a signal.

CODE. In *computer* technique, a system of symbols which represent information in a form convenient for the computer.

CODE ALPHABET. A list of short, easily recognized words which identify letters in radio or telephone messages.

CODE CONVERTER. A device which transforms a digital input in one code to a digital input in another code.

CODE DELAY. An arbitrary interval inserted between signal pulses from master and slave transmitters.

CODE DECIMAL. See *Binary coded decimal*.

CODED PROGRAMME. Instructions, usually in the form of a list coded in the machine's language, for solving a problem with a *digital computer*.

CODER. An electronic device for producing *pulse code modulation* which can convert electrical speech signals into code characters and subsequently decode and reproduce them to provide a good approximation of the originals.

CODING. The design and application of a *coded programme*.

CODING TUBE. A special *cathode-ray tube* used as a *coder*.

COED. Computer Operated Electronic Display.

COERCIMETER. A meter which measures the *coercive force* associated with magnetic materials.

COERCIVE FORCE. The reverse field in the *magnetization cycle* which is necessary to reduce the intensity of magnetization to zero.

COERCIVITY. The value of the *coercive force* when the initial magnetization has the saturation value for the material.

COHERENT OSCILLATOR. One capable of producing very pure and well-defined oscillations, as in, e.g. a *laser*.

COHO. Abbreviation for COHerent Oscillator.

COIL. A compact arrangement of convolutions of one or more conductors to provide an *inductance*. See *Inductor*.

COIL CONSTANT. Of an *inductance* coil at a given frequency, the ratio of its reactance to its effective resistance. See *Q*.

COIL LOADING. The insertion of *coils* in a line at intervals to improve its transmitting performance.

COINCIDENCE COUNTER. A circuit which records the radiation exciting two or more *counters* simultaneously, used to determine the direction of radiation and for the detection of cosmic ray showers.

COINCIDENCE GATE. An electronic circuit which produces an output pulse only when each of two or more input circuits receives pulses simultaneously or within a specified time interval.

COINCIDENCE TUNING. Tuning all the stages of an amplifier to the same frequency in contrast to *band-pass tuning*.

COLD CATHODE. (1) A *cathode* whose operation is not dependent on raising its temperature above the ambient. (2) An electrode used to provide electrons by secondary emission.

COLD-CATHODE COUNTER TUBE. See *Dekatron*.

COLD-CATHODE RECTIFIER. A *cold-cathode* gas tube in which the electrodes differ greatly in size so that electron flow is much greater in one direction than in the other; e.g. a *mercury-pool rectifier*. Also called a gas-filled rectifier.

symbol

COLD-CATHODE TUBE (VALVE). An electron tube or valve containing a *cold cathode*.

COLIDAR. COherent LIght Detection And Ranging.

COLLATOR. In computing and data-handling, a punch card machine at which a card may be compared or collated with other cards so as to determine its proper pocket.

COLLECTOR. (1) In a *transistor*, an electrode through which a primary flow of *carriers* leaves the inter-electrode region. (2) An electrode which collects electrons or ions which have fulfilled their function in an electron tube or valve.

COLLECTOR CAPACITANCE. The *depletion-layer capacitance* associated with the *collector* of a transistor.

COLLECTOR CURRENT. In a *transistor*, the current which flows at the *collector* when suitable bias is applied.

COLLECTOR CURRENT RUNAWAY. Continuing increase in *collector current* due to a rise in temperature in the *collector junction* as the current increases.

COLLECTOR EFFICIENCY. Of a *transistor*: the ratio of the useful power output to the dc power input.

COLLECTOR JUNCTION. Of a *semiconductor*, usually a transistor: a junction through which the current carriers flow from a region in which they are *majority carriers* to one in which they are *minority carriers*.

COLLECTOR RING. See *Slip ring*.

COLLISION FREQUENCY. The number of collisions per unit time between an electron and a molecule of gas.

COLLISION IONIZATION. *Ionization* produced when a particle with sufficient energy collides with an atom and removes an electron.

COLOUR BREAK-UP. In *colour television*, any transitory separation of the picture into its colour primaries due to rapid changes in viewing; e.g. blinking.

COLOUR BURST. In *colour television*, the part of the composite colour signal which is used to establish a reference for demodulating the chrominance signal.

COLOUR CELL. In a *colour picture tube*, the smallest area which includes a complete set of all the primary colours in the repeating pattern.

COLOUR CENTRE. A vacancy in a crystal lattice which attracts electrons and absorbs light of a particular wavelength.

COLOUR CODE. In electronic engineering the values, tolerances, and ratings of components such as resistors, capacitors, and small transformers are indicated by bands or spots of colour painted

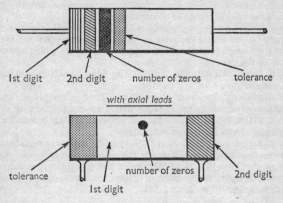

with axial leads

with radial leads

Colour	First Digit	Second Digit	Zeros Following	Tolerance
BLACK	—	0	—	—
BROWN	1	1	0	1%
RED	2	2	00	2%
ORANGE	2	3	000	3%
YELLOW	4	4	0,000	4%
GREEN	5	5	00,000	—
BLUE	6	6	000,000	—
PURPLE	7	7	0,000,000	—
GREY	8	8	00,000,000	—
WHITE	9	9	000,000,000	—
GOLD	—	—	—	5%
SILVER	—	—	—	10%
NO COLOUR	—	—	—	20%

Fig. 25. Colour code for tubular resistors and capacitors

on the most readily visible surfaces. The colour code in current use for tubular fixed resistors and capacitors is shown in Fig. 25.

COLOUR CODER. In *colour television*, an apparatus for producing the colour picture signal from camera signals and the *chrominance subcarrier*.

COLOUR CONTRAST. The ratio of the intensities of the sensations caused by two colours.

COLOUR DECODER. In *colour television*, an apparatus which obtains signals for the colour display device from the colour picture signal and the *colour burst*.

COLOUR FLICKER. In *colour television*, the flicker due to fluctuations in both chromaticity and luminance.

COLOUR FRINGING. In *colour television*, spurious colour at the boundaries of objects in the colour image.

COLOUR GATE. In *colour television*, a circuit which allows a single primary colour signal to arrive at the *electron gun* when the electron beam is oriented to strike the corresponding colour phosphor on the screen of the tube.

COLOUR GRID. An arrangement of wire screens at the viewing end of a *chromatron*.

COLOUR KILLER. In *colour television*, a biasing circuit which suppresses monochrome signals which would appear tinged with colour.

COLOUR PHASE. The phase relation between the chrominance television signal and the *colour burst signal*.

COLOUR PICTURE TUBE. A special type of *cathode-ray tube* which provides an image in colour by the *scanning* of a *raster* and by varying the intensity of excitation of the phosphor on the screen to produce light of the chosen primary colours. See *Chromatron, Colourtron*, and Figs. 26 and 141.

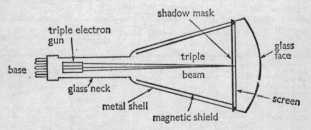

Fig. 26. Principle of colour picture tube (see also Fig. 141)

COLOUR PURITY MAGNET. A magnet placed near the neck of a *colour picture tube* which modifies the path of the electron beam so as to improve the purity of the colour display.

COLOUR RESPONSE. The relative sensitivity of photoelectric devices and of the human eye to different colours.

COLOUR SIGNAL. Any signal in a *colour television* system used to control the chromaticity values of the picture.

COLOUR TELEVISION. The transmission of pictures in colour. All the systems in use depend on the fact that any desired colour can be obtained by an appropriate mixture of the primary colours: red, green, and blue. Thus, a colour television system picks up and transmits the red, green, and blue components of the televised scene and at the receiver the three colour images are reproduced and combined to give a colour picture. See *Television*.

COLOUR TELEVISION MASK. A perforated metal disc which directs a particular colour electron beam to the corresponding phosphor dot in each cluster on the screen of a *colour picture tube*.

COLOUR TEMPERATURE. The temperature of the *black body* radiator required to produce the same chromaticity as the light under consideration.

COLOURTRON. A special type of *cathode-ray tube* for display of colour images which has several improvements over earlier types. It has three electron guns, one for each primary colour, a phosphor dot screen mounted directly on a glass plate, and an accurately aligned aperture mask. See Figs. 26 and 141.

COLPITTS OSCILLATOR. An oscillator in which a tuned *tank circuit* is connected between the grid and anode of a valve, or between base and collector of a transistor, and in which the tank capacitance is made up of two capacitors in series with their common connection at cathode or emitter potential.

COLUMN. In *data processing*: a vertical arrangement of characters or other expressions.

COMA. In a *cathode-ray tube*: a defect of the image which causes a spot on the screen to appear comet-shaped.

COMMAND GUIDANCE. Electronic guidance of guided missiles or aircraft in which signals from an outside agency cause the guided object to follow a directed path in space.

COMMAND SIGNAL. In *computers*, one of a set of several signals which occur as a result of an *instruction*.

COMMON-BASE CONNECTION. A method of operating a *transistor* in which the signal is fed between the emitter and base, the output is between collector and base, and the base is grounded.

COMMON-COLLECTOR CONNECTION. A method of operating a *transistor* in which the signal is fed between base and collector, the output is between emitter and collector, and the collector is grounded. Usually known as *emitter follower*.

COMMON-EMITTER CONNECTION. A method of operating a *transistor* in which the signal is fed between the base and the emitter, the output is between the collector and emitter, and the emitter is grounded.

COMMON MODE REJECTION (RATIO). A figure of merit for *differential amplifiers*, indicating their performance in suppressing voltages or circuits which are alike in the two inputs.

COMMUNICATION. The transmission of information by means of *electro-magnetic* waves or by signals along wires. See *Telecommunication*.

COMMUNICATION RECEIVER. A *radio receiver* which can receive continuous waves, interrupted continuous waves, and amplitude-modulated waves.

COMMUNICATIONS NETWORK (-SYSTEM). A number of stations interconnected by communications links which may be open or closed circuit.

COMMUTATING CAPACITOR. A *capacitor* in the circuit of a *rectifier* valve which prevents the application of a large negative voltage to the anode after the valve has been cut off.

COMMUTATING MACHINE. An electrical machine for converting alternating into direct current or vice versa, provided with a *commutator*.

COMMUTATING REACTANCE. An inductance in series with the load of a *mercury-arc rectifier* to prevent the arc extinguishing on one anode before it is established on the other.

COMMUTATION. The conversion of alternating to direct current or vice versa with a *commutator*.

COMMUTATION SWITCH. In *pulse communication* systems, a mechanical or electrical device which performs repetitive sequential switching.

COMMUTATOR. A cylindrical assembly of conductors which are individually insulated and connected to the sections of a winding. Brushes sliding on the conductor surfaces make connection in turn with external electrical circuits.

COMMUTATOR RIPPLE. Small pulsations in the voltage and current of dc generators due to the fact that the output is composed of a series of mechanically rectified alternating currents.

COMPACTRON®. Two or more conventional tubes in a single unit with a 12-pin base.

COMPANDER. A combination of a *compressor* in a communication path to reduce the signal volume range, followed by an *expander* to restore the original volume range, its purpose being to improve the ratio of signal to interference in the path.

COMPANDING. The process of *compression* followed by *expansion*.

COMPARATOR. A circuit which compares two signals and indicates the result of the comparison.

COMPASS. See *Gyro flux-gate compass, Magnetic compass, Radio compass*.

COMPATIBILITY. Of a *colour television* system, the property which permits normal monochrome reception of the transmitted signal by unaltered monochrome receivers.

COMPENSATING COILS. A set of coils arranged to alter the distribution of a *magnetic flux* in a desired manner.

COMPENSATING VOLTMETER. A *valve voltmeter* in which a resistor bridge supplies the current to cancel out the quiescent current and so bring the needle to zero for the original condition.

COMPENSATOR. In television, the inductance coils used in the grid and anode circuits of the *video* amplifier to compensate for the attenuation of the higher frequencies.

COMPILE. To convert a program in *autocode* to a program written in the *machine code* of a particular computer.

COMPILER. A programme which compiles.

COMPLEMENTARY TRANSISTORS. A *p-n-p* and *n-p-n transistor* pair used to obtain a *push-pull* output from a common signal input.

COMPLEMENTING CIRCUIT. In *digital computers*: a circuit used to perform subtraction without the use of negative numbers.

COMPLEX TONE. A sound wave produced by the combination of simple *sinusoidal* components of different frequencies.

COMPOSITE COLOUR SIGNAL. In *colour television*, the colour picture signal plus *blanking* and *synchronizing* signals.

COMPOSITE CONDUCTOR. A conductor in which two or more strands of different metals, such as copper and steel, are assembled and operated in parallel.

COMPOSITE CONTROLLING VOLTAGE. Of a multi-electrode valve, the voltage of the anode of an *equivalent diode* combining the effects of all individual electrode voltages in establishing the *space-charge current*.

COMPOSITION RESISTOR. A *resistor* composed of a carbon compound. See *Carbon resistor*.

COMPOUND CIRCUIT. A series circuit which includes more than one source of voltage and more than one branch consuming current.

COMPOUND GENERATOR. A shunt-wound generator whose regulation is improved by an additional winding of a few turns in series with the armature or the load.

COMPOUND MODULATION. A series of *modulation* processes in which the modulated wave of one becomes the modulating wave of the next.

COMPOUND MOTOR. A motor which has both series and shunt windings.

COMPRESSION. A process in which effective gain applied to a signal depends on the magnitude of the signal, being greater for small than for large signals.

COMPRESSION RATIO. The ratio of the magnitude of the gain at a reference signal level to its magnitude at a higher stated signal level.

COMPRESSIONAL WAVE. In an elastic medium, a wave which causes an element of the medium to change its volume without rotation.

COMPRESSOR. A *transducer* whose range of output voltages is smaller in amplitude than the range of its input voltages.

COMPTON EFFECT. The elastic scattering of *photons* by *electrons*.

COMPTON ELECTROMETER. A sensitive type of *quadrant electrometer*.

COMPTON (RECOIL) ELECTRON. An electron set in motion through interaction with a photon in the *Compton effect*.

COMPTON SCATTERER. A hypothetical medium composed of free-electron gas.

COMPUTER. A device which can accept and supply information and in which the information supplied is derived from the information accepted by logical processes. The two main types are *digital computers* and *analogue computers*, and they make extensive use of electronic devices and circuits. See *Data-handling system*.

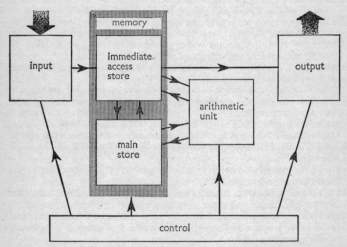

Fig. 27. Simplified diagram of the elements of a digital computer.

COMPUTER CODE. A system of characters and rules to represent information. See *Language*.

COMPUTER CONTROL. Those parts of a *digital computer* which effect the carrying out of instructions in proper sequence, the interpretation of each instruction, and the application of the proper signals to the arithmetic unit and other parts.

COMPUTER-ASSISTED PROGRAM. A program which allows detailed calculations of machine operations to be arrived at from simple *instructions*.

CONCENTRATION CELL. A cell which contains two solutions of the same salt in different concentrations and in which a piece of the same metal forms the electrode in each solution.

CONCURRENT WORKING. Any method of working in a *data processing* system in which more than one operation or sequence of operations is executed at the same time.

CONDENSER. Obsolescent term for *capacitor*.

CONDITIONAL BREAKPOINT INSTRUCTION. A *conditional jump* which causes a computer to stop if a specified switch is set.

CONDITIONAL JUMP. In *digital computers*: an instruction which ensures that the correct one of several *addresses* is used to obtain the next instruction.

CONDUCTANCE. With direct current, the reciprocal of *resistance*. With alternating current the resistance divided by the square of the *impedance*; or the real part of the *admittance*. Symbol: G, practical unit: *mho*.

CONDUCTIMETER. An instrument for determining electrical conductivity.

CONDUCTING MATERIAL. A material such that when a potential is applied between any two points on it a relatively large flow of *conduction current* takes place. *Metals* and strong *electrolytes* are conductors in all applications. Cf. *Semiconductor, Dielectric*.

CONDUCTION. The process by which energy is transmitted in a substance while the substance as a whole does not move.

CONDUCTION BAND. In the energy spectrum of a solid, a range of energies in which electrons can move freely under the influence of an electric field. See *Energy band*.

CONDUCTION CURRENT. An electric current composed of the movement of free charges. Mathematically, over any surface it is the integral of the conduction *current density* over that surface. Symbol: I.

CONDUCTION ELECTRONS. The electrons in the *conduction band* of a substance which are free to move under the influence of an electric field.

CONDUCTIVITY. The reciprocal of *volume resistivity*. Symbol: κ.

CONDUCTIVITY CELL. A cell used in measuring the *conductivity* of a liquid, comprising a liquid container and two electrodes.

CONDUCTIVITY METER. A bridge circuit, basically of the *Wheatstone bridge* type, in which the conductivity to be measured forms one arm of the bridge and constitutes a *conductivity cell*.

CONDUCTIVITY MODULATION. Of a *semiconductor*, the variation of the conductivity of the semiconductor by variation of the charge *carrier* density.

CONDUCTOR. A body or substance which offers a low *resistance* to the passage of *electric current*; *conducting material*.

CONE LOUDSPEAKER. A *loudspeaker* in which a paper or fibre cone is mechanically coupled to a magnetic driving unit.

CONFUSION REFLECTOR. Any kind of reflector of electromagnetic waves, such as strips of metal foil or paper, which can cause false signals on enemy radar when dropped from aircraft. See *Rope*, *Window*.

CONICAL SCANNING. *Radio* or *radar scanning* in which the direction of maximum response generates a cone.

CONJUGATE BRANCHES. Any two branches of a *network* such that an electromotive force in one branch produces no current in the other branch.

CONJUGATE IMPEDANCE. Of a given *impedance* an impedance having an equal *resistance* component, and a *reactance* component equal in magnitude but opposite in sign.

CONSONANCE. *Resonance* in a primary circuit employing a capacitor.

CONSTANTAN. An alloy of copper (60 per cent) and nickel (40 per cent) with a very low temperature coefficient of resistance, used in the manufacture of precision wire-wound resistors and thermo-couples.

CONSTANT-CURRENT CHARACTERISTIC. The graph of the relation between the voltages of two valve electrodes when the current of one of them, and all other voltages, are kept constant.

CONSTANT-CURRENT GENERATOR. A valve circuit, usually employing a *pentode*, whose *anode resistance* is so high that anode current is substantially independent of load variations.

CONSTANT-CURRENT MODULATION. A method of *amplitude modulation* in which a source of constant current supplies a radio-frequency generator and a modulation amplifier in parallel, so that current variations in one cause equal and opposite variations in the other, and so modulation of the carrier output is effected.

CONSTANT-CURRENT REGULATION. Of an electric generator, regulation in which the regulator maintains the current output constant regardless of the load.

CONSTANT-CURRENT TRANSFORMER. A transformer which maintains a constant current in the secondary circuit whatever the load impedance, when supplied from a constant source of potential.

CONSTANT-K NETWORK. A *ladder network* whose product of series and shunt impedances is independent of frequency over the working range.

CONSTANT-LUMINANCE TRANSMISSION. The system of transmission of *colour television*, currently in use in the U.S.A., in which the luminance or brightness is controlled solely by the black and white signals and is not affected by the signals which carry the colour information.

CONSTANT-POTENTIAL TRANSFORMER. A transformer which maintains a constant ratio of secondary to primary voltages, regardless of the load.

CONSTANT-VOLTAGE CHARACTERISTIC. A graph of voltage against current, for which the voltage is practically constant for all values of the current.

CONSTANT-VOLTAGE TRANSFORMER. See *Constant-potential transformer.*

CONTACT. Momentary, temporary, or permanent bringing together of the surfaces of two *conductors* so that current may flow.

CONTACT BLOCK. A block of carbon or other conducting material used in a *relay.*

CONTACT BREAKER. A device which automatically interrupts a *contact.*

CONTACT CURRENT. The current through the contacts of a *relay* when actuating a protective device.

CONTACT E.M.F. An *electromotive force* which may arise when two conductors of different materials are placed in contact.

CONTACT FOLLOW. In a *relay,* the distance two contacts travel together just after touching.

CONTACT GETTERING. Absorption of residual gas in a vacuum tube by direct contact of the gas with the *getter* after it has been dispersed in the tube.

CONTACT MICROPHONE. A *microphone* designed to pick up mechanical vibrations directly and convert them into corresponding electrical currents or voltages.

CONTACT POTENTIAL. See *Contact e.m.f.*

CONTACT RESISTANCE. The resistance at the surface of contact between two conductors.

CONTACT SEPARATION. The minimum distance between closing relay contacts when the contacts are in the open position.

CONTACT WEAR ALLOWANCE. The thickness of material which may be worn away before the electrical contact surfaces become inadequate to carry the rated current.

CONTACTOR. A device, not operated by hand, for repeated making and breaking of a circuit.

CONTENTS. Of a *computer,* the information stored in the *memory.*

CONTINUITY. The condition existing in a circuit when the electrical path is closed and current can flow.

CONTINUOUS CONTROL. A system of control in which the controller is supplied continuously with an actuating signal and the quantity controlled is monitored continuously.

CONTINUOUS CURRENT. See *Direct current.*

CONTINUOUS DUTY. A condition of service which demands operation at constant load for an indefinite time.

CONTINUOUS RATING. The rating which defines the load which can be supported by a device for an indefinitely long time.

CONTINUOUS WAVES (CW). *Radio* or *radar waves* which maintain a constant amplitude and constant frequency.

CONTRAST. In *television,* the ratio between maximum and minimum brightness values in the picture.

CONTRAST CONTROL. See *Automatic contrast control.*

CONTROL CIRCUIT. See *Computer control.*

CONTROL ELECTRODE. An *electrode* used to start or vary the current between other electrodes.

CONTROL GRID. (1) A *grid*, usually between cathode and anode in a valve, used as a *control electrode*. (2) In a *klystron*, the electrode in the gun which controls the beam current.

CONTROL RATIO. In a *gas tube*, the ratio of the change in anode voltage to the corresponding change in grid voltage, all other conditions being maintained constant.

CONTROL REGISTER. Of a *computer*, the *register* which stores the current instructions governing the computer for a given cycle.

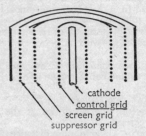

cathode
control grid
screen grid
suppressor grid

Fig. 28. Cross-section of a typical pentode valve showing position of control grid

CONTROL SEQUENCE. Of a *computer*, the order in which instructions are executed.

CONTROL SIGNAL. In a control system, the signal applied to the controlled equipment which effects the corrective changes.

CONTROL SYSTEM RESET RATE. The number of corrections per minute made by the control system.

CONTROL UNIT. See *Computer control*.

CONTROL VALVE (TUBE). A valve in an *automatic gain control* circuit.

CONTROL WINDING. Of a *saturable reactor* in a magnetic amplifier, the winding which applies a controlling *magnetomotive force* to the core.

CONTROL-CIRCUIT VOLTAGE. In *magnetic amplifiers*, an alternating voltage provided by the control winding which resets the core flux in response to an input signal or a resetting signal.

CONTROLLED-CARRIER MODULATION. A system of *modulation* in which the carrier is amplitude-modulated both by the signal frequencies and by the envelope of the signal so that the modulation factor remains constant whatever the signal amplitude.

CONVECTION CURRENT. In an electron stream, the time rate at which charge is transported through a given surface.

CONVECTIVE DISCHARGE. A discharge in air from a high-voltage source which consists of visible or invisible charged particles.

CONVERGENCE. In multibeam *cathode-ray tubes*, the condition for the intersection of the electron beams at a specified point. See *Beam convergence, Dynamic convergence*.

CONVERGENCE COILS. The focusing electromagnets in a *colourtron* which ensure that the beam converges at all points on the mask.

CONVERGENCE CONTROL. In *colour television*, a manual control on the receiver which adjusts the potential on the *convergence electrodes* and so determines the degree of convergence of the electron beams in the colour picture tube.

CONVERGENCE ELECTRODE. An electrode whose electric field converges two or more electron beams.

CONVERGENCE MAGNET. A magnet whose magnetic field converges two or more electron beams.

CONVERGENCE SURFACE. In multibeam *cathode-ray tubes*, the surface generated by the point of intersection of two or more electron beams during the scanning process.

CONVERSION CONDUCTANCE. In an electronic *frequency changer*, the ratio of the output current at the intermediate frequency to the input voltage at the signal frequency.

CONVERSION ELECTRON. An electron emitted in the de-excitation of a nucleus by direct coupling between an excited nucleus and an extra-nuclear electron.

CONVERSION GAIN. In a *superheterodyne receiver*, the ratio of the intermediate frequency output voltage to the input signal voltage of the first detector.

CONVERSION GAIN RATIO. In a *frequency converter*, the ratio of the signal power available at the output to that available at the input.

CONVERSION TRANSCONDUCTANCE. Of a *heterodyne conversion transducer*, the ratio of the magnitude of the desired output-frequency component to that of the signal component of voltage when the impedance of the output termination is negligible.

CONVERSION TRANSDUCER. A *transducer* in which the input and output frequencies are different.

CONVERSION VOLTAGE GAIN. Of a *conversion transducer*, the ratio of output frequency voltage across the output termination, with the transducer between the input frequency generator and the output termination, to input frequency voltage across the input termination of the transducer.

CONVERTER (CONVERTOR). (1) A machine which changes current of one kind into current of another kind. (2) A device which changes one frequency to another. (3) A device which changes information coded in one form into the same information coded in another form. In this sense, since the codes may be 'languages', a converter may be a translating device.

The following kinds of converter are described in this book: *analogue-to-digital, arc, Brown, direct-current, facsimile, field-control, frequency, inverter, mercury-arc, motor, pentagrid, Poulsen arc, quenched-spark, radio-frequency, rotary, short-wave, superheterodyne, synchronous, waveguide.*

CONVERTER VALVE (TUBE). The valve which combines the functions of *mixer* and *local oscillator* in a heterodyne conversion transducer.

COOLIDGE TUBE. A type of X-ray tube invented by W. D. Coolidge in 1913 which employs a thermionic cathode and an air- or water-cooled anode.

COOPER-HEWITT LAMP. A mercury vapour ultra-violet lamp, usually in the form of a long glass tube.

CO-ORDINATE INSPECTION MACHINE. A machine used to check the dimensions of a workpiece by means of a probe, where co-ordinates are measured by an electronic system and displayed on a *digital read-out*.

CO-ORDINATED TRANSPOSITION. In transmission systems, periodic transpositions of lines in electric supply or communication circuits, or in both, to reduce *inductive coupling* between them.

COPLANAR ELECTRODES. *Electrodes* mounted in the same plane in a valve or tube assembly.

COPPER (Cu). Bright, reddish, metallic element, Atomic No. 29. Second only to silver as an electrical conductor and so very extensively used in electronics and electrical engineering.

COPPER-OXIDE METER. An instrument which measures alternating currents and voltages by rectifying them with a copper-oxide rectifier and indicating with a direct-current meter.

COPPER-OXIDE RECTIFIER. See *Rectifier, dry-disc.*

COPY. With reference to *computers*, transfer information contained in one register into another register, leaving the information in the original register unchanged.

CORE. The material that acts as a path for the *magnetic flux* in *transformers* and *coils.*

CORE LOSS. The energy loss in the ferromagnetic core of an inductor or transformer due mainly to *eddy currents* and *hysteresis.*

CORE STORE. See *Magnetic memory.*

CORKSCREW RULE. If a corkscrew is pointed along a conductor in the direction of the current, the direction of the magnetic field is the direction in which the handle turns if the corkscrew is then driven into an imaginary cork.

CORNER ADMITTANCE. In *coaxial line resonators*, the admittance presented to the line by a *reflex klystron.*

CORNER REFLECTOR. A reflector consisting of three triangular metal surfaces, arranged like the inside corner of a box, used in radar for testing and calibration.

CORONA. A luminous discharge which appears round the surface of a conductor when the voltage gradient exceeds a critical value and causes ionization of the air. A corona discharge is used in certain types of *gas discharge tube*.

COROSIL®. An alloy of iron and silicon used as a magnetic core material for transformers, etc.

CORRECTIVE NETWORK. An electric *network* designed to improve the transmission or impedance properties (or both) of an electric circuit.

COSECANT-SQUARED BEAM. A beam of electrons shaped so that it has nearly the same intensity for near and far objects.

COSINE POTENTIOMETER. See *Potentiometer, cosine*.

COSINE WINDING. A specially shaped winding in some *magnetic deflection coils* for cathode-ray tubes which improves the focusing of the electron beam.

COSMIC NOISE. Radio *noise* caused by some source outside the earth, e.g. sunspots.

COSMIC RAYS. Extremely penetrating radiation originating outside the earth's atmosphere which produces ionizing events when it passes through air or other matter. The primary cosmic rays probably consist mainly of *protons*, but these are nearly all absorbed in the atmosphere and the secondary rays which reach the earth's surface consist of all kinds of elemental particles with lower energy than those of the primary rays.

COSMIC RAY SHOWER. The simultaneous appearance of a number of light ionizing particles as indicated by a *cloud chamber* or separated *counters*.

COTTON BALANCE. An apparatus for determining the intensity of a magnetic field by measuring the vertical force on a current-carrying wire placed perpendicular to the field.

COULOMB. The unit of *electric charge* in the *SI system*.

COULOMB, ABSOLUTE. A measure of the quantity of electricity which passes a given point in a given time. See *Units*.

COULOMB ENERGY. That part of the binding energy of the atoms of a solid which is due to electrostatic attraction between electrons and ions.

COULOMB FIELD. An *electrostatic* field which behaves as if it were due to a charge entirely concentrated at one point.

COULOMB FRICTION. A retarding force developed in sliding electrical parts which is opposite to the direction or rotation of movement.

COULOMB'S LAW. Of *electrostatic* attraction, the force of attraction or repulsion between two charged bodies (whose charges behave as though concentrated at a point) is proportional to the magnitude of the charges and inversely proportional to the square of the distance between them.

COULOMETER. An electrolyte *cell* used to measure a quantity of electricity by the amount of a substance which is liberated electrochemically. Synonymous with voltameter.

COUNT. The external indication of any device which counts ionizing events caused by radiation.

COUNTER. (1) A device for counting ionizing events, or the detector part of such a device. (2) An electro-mechanical, photoelectric, or electronic device for counting electric pulses. A wide variety of counters exist, their design depending on the number and repetition rate of the pulses to be counted and how long the information has to be stored. The counters described in this dictionary include the following: *alpha, boron, coincidence, computer, crystal, electro-mechanical, frequency and time measurement, gas-filled, gas radiation, Geiger, ionization, long, proportional, radiation, ring, scale-of-two, scintillation, thin-wall.*

COUNTER, COMPUTER. In mechanical *analogue computers* a means for measuring the angular displacement of a shaft.

COUNTER FIELD EMISSION. In a *radiation counter*, the release of an *electron* from the counter *cathode* surface by an approaching ionic charge.

COUNTER/FREQUENCY METER. An instrument in which a reference standard, usually a quartz crystal oscillator, is used to measure frequency and period. For direct frequency measurement it counts the number of cycles of the unknown in a standard time interval. For period measurement it counts the number of standard pulses in one or more periods of the unknown frequency.

COUNTER LAG TIME. Of a *radiation counter*, the delay between the primary ionizing event and the occurrence of the count.

COUNTER OPERATING VOLTAGE. Of a *radiation counter*, the voltage between anode and cathode when the counter is operative.

COUNTER STARTING POTENTIAL. Of a *Geiger counter*, the lowest counter voltage at which ionizing radiation produces current pulses in the circuit.

COUNTER TUBE. A detector which can provide electrical pulses in response to bursts of radiation or charged particles incident on it, usually because of the *ionization* in the tube.

COUNTER TUBE HYSTERESIS. Of a radiation-counter tube, a temporary change in the counting rate versus voltage characteristic caused by previous operation.

COUNTING RATE. The average rate of occurrence of *ionizing events*, observed by a counting system.

COUNTING-RATE METER. A device which provides a continuous indication of the average rate of ionizing events.

COUPLED CIRCUITS. Two or more circuits connected in a network in such a way as to react on each other.

COUPLED CIRCUITS, FORCED. In electronic equipment coupled circuits are usually driven by potentials which originate in the system and are sustained by it, so that the behaviour of the circuit is forced.

COUPLED CIRCUIT, FREE. A coupled circuit set into *oscillation* by a transient electrical disturbance oscillates freely until the oscillations are gradually dissipated in the resistance of the circuit.

COUPLING. In electronic circuits, the association of two or more circuits or systems in such a way that power may be transferred from one to another.

COUPLING CAPACITOR. A *blocking capacitor* used to join two stages in a circuit.

COUPLING COEFFICIENT. In *resistive, capacitive*, and *inductive coupling*, the ratio of the coupling *impedance* to the geometric mean of the impedance of corresponding elements in the two meshes.

COUPLING COIL. A coil used to transfer electrical energy between two systems.

COUPLING TRANSFORMER. A *transformer* used to couple two circuits by means of its *mutual inductance*.

COVALENT BOND. A chemical bond between two atoms in which two electrons are shared.

C.P.S. EMITRON®. See *Emitron, cathode potential stabilized*.

CRACK DETECTOR, ELECTROMAGNETIC. A device which detects flaws in iron or steel by producing a strong magnetic field in the material so that fine magnetic powder conglomerates along the outline of the flaw.

CREST FACTOR. See *Peak factor*.

CREST VALUE. See *Peak value*.

CREST VOLTMETER. A *voltmeter* which reads the *peak value* of the voltage at its terminals.

CRIPPLED LEAPFROG TEST. In computer programming, a variation of the *leapfrog test* in which the test from a single group of storage positions is repeated and does not 'leap'.

CRITICAL ANGLE, IONOSPHERE. The angle of radiation of a transmitted wave which will not be reflected from the *ionosphere*.

CRITICAL ANODE VOLTAGE. See *Anode breakdown voltage*.

CRITICAL COUPLING. The greatest coupling between two *coupled circuits* for which there is only one condition for a maximum in the secondary circuit.

CRITICAL DAMPING. The least value of *damping* to prevent *oscillation*.

CRITICAL GRID CURRENT. Of a multi-electrode gas tube, the grid current corresponding to the *critical grid voltage*, before anode breakdown.

CRITICAL GRID VOLTAGE. Of a gas or mercury tube, the value of rising grid voltage at which the anode current is established.

CRITICAL HIGHPOWER LEVEL. In *attenuator valves*, the radio-frequency power level at which ionization is produced without a control electrode discharge.

CRITICAL POTENTIAL. A measure of the energy required to raise the energy level of an orbital electron to a higher energy band in the atom.

CRITICAL RESISTANCE. In an oscillator or other type of electric circuit the value of resistance at which *critical damping* occurs.

CRO. Abbreviation for Cathode Ray Oscilloscope.

CROOKES DARK-SPACE. See *Cathode dark-space.*

CROOKES TUBE. An early *discharge tube* invented by Sir William Crookes for the study of cathode rays.

CROSS COUPLING. Unwanted *coupling* between two communication channels or their component parts.

CROSS MAGNETIZING. The effect of armature reaction on the magnetic field of a generator.

CROSS MODULATION. *Intermodulation* due to modulation of the desired signal carrier by an undesired signal.

CROSS NEUTRALIZATION. A method of *neutralization* in push-pull amplifiers in which a part of the anode-cathode ac voltage of each tube is fed to the grid-cathode circuit of the other tube through a neutralizing capacitor.

CROSS TALK. The unwanted sound in an electro-acoustic receiver – e.g. a telephone receiver – connected with a given transmission channel caused by *crosscoupling* to another transmission channel carrying sound-controlled electric waves. Also, the electric waves in the disturbed channel which produce the unwanted sound.

CROSSFIRE. Interfering currents in a telegraph or signalling channel due to telegraph or signalling currents in another channel.

CROSSHATCH PATTERN. A test pattern on a *television* picture tube, consisting of vertical and horizontal lines, produced by a special generator.

CROSS-OVER AREA. In a *cathode-ray tube*, the point at which the electron beam comes to a focus inside the accelerator anode.

CROSS-OVER FREQUENCY. In electric *dividing networks*, the frequency at which the powers delivered to each of the adjacent frequency channels, terminated with the correct loads, are equal.

CRT. Abbreviation for Cathode Ray Tube.

CRYOGENICS. The study of materials at temperatures near absolute zero. See *Cryotron, Meissner effect, Superconductivity.*

CRYOMETER. A thermometer for very low temperatures.

CRYOMITE®. A miniature refrigerator for maintaining temperatures in the range 25°K to 150°K.

CRYOSAR. In *computers*: a component whose operation depends on the impact *ionization* of impurities in germanium.

CRYOSISTOR. A low-temperature three-terminal switch.

CRYOSTAT. (1) A refrigerating unit for the production of temperatures approaching absolute zero. (2) A thermostat for low temperatures.

CRYOTRON. A normally resistive element which can be maintained at the threshold of *superconductivity* by surrounding it with a strong magnetic field. Used in memory and switching devices in computer networks.

CRYSTAL, IDEAL. A crystal which is perfectly regular in structure and contains no foreign atoms.

CRYSTAL ANALYSIS. See *X-ray crystallography*.

CRYSTAL BURN-OUT. Permanent damage to the structure of a crystal by exposure to excessive radio-frequency power.

CRYSTAL CONTROLLED OSCILLATOR. An oscillator whose frequency of oscillation is governed by a *piezoelectric crystal* unit.

CRYSTAL CONTROLLED TRANSMITTER. A radio transmitter whose carrier frequency is directly controlled by a *crystal oscillator*.

CRYSTAL COUNTER. In nucleonics, a counter employing a type of crystal which can be made momentarily conducting by an ionizing event.

CRYSTAL CUTTER. A cutter for gramophone recording in which the mechanical displacements of the recording stylus are derived from a *piezoelectric crystal*.

CRYSTAL DETECTOR. A *detector* composed of a *cat's whisker* in contact with a crystal or a *semiconductor junction* whose resistance to a current is much higher in one direction than in the other.

CRYSTAL DIAMAGNETISM. The property of negative *magnetic susceptibility* exhibited by some crystals, e.g. silver and bismuth.

CRYSTAL DIODE. See *Crystal rectifier*.

CRYSTAL ELECTROSTRICTION. The changes which occur in the dimensions of a *dielectric* under the influence of an external electric field.

CRYSTAL FILTER. A *filter network* which includes one or more *piezoelectric* crystals to provide resonant or anti-resonant circuits.

CRYSTAL GROWING. See *crystal pulling*.

CRYSTAL IMPURITY. An atom inside a crystal which is foreign to it.

CRYSTAL LATTICE. The arrangement of atoms within the crystal structure.

CRYSTAL LOUDSPEAKER. A loudspeaker in which the mechanical displacements are produced by *piezoelectric* action.

CRYSTAL MICROPHONE. A microphone which operates by the generation of electric charge on *piezoelectric crystal* subjected to the sound waves.

symbol

CRYSTAL MIXER. A device in a receiver in which a *crystal detector* provides the non-linear characteristic required for mixing the received signal and the local oscillator signal. It can be used at high frequencies and high power levels.

CRYSTAL OSCILLATOR. An *oscillator* whose frequency is determined by a *piezoelectric crystal*, usually of quartz. Such an oscillator attains a high order of stability and provides a laboratory frequency standard.

CRYSTAL OVEN. In very accurate frequency standards the crystal oscillator is usually enclosed in a small oven maintained at constant temperature to reduce frequency drift.

CRYSTAL PICK-UP. A gramophone *pick-up* whose operation depends on the generation of electric charges by the deformation of a *piezoelectric crystal*.

CRYSTAL PULLING. The technique of producing large single crystals, usually for *semiconductor* devices, by withdrawing the developing crystal slowly from the melt.

CRYSTAL RECTIFIER. A rectifying element using a *semiconductor diode* with two terminals, designed to be used in a manner similar to that of diode valves.

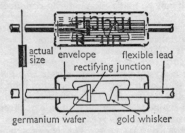

Fig. 29. Crystal rectifier (gold-bonded germanium type)

CRYSTAL SET. A radio receiver using a *crystal detector* but no vacuum valves.

CRYSTAL SPECTROMETER. An instrument for measuring the wavelengths of X-rays or gamma rays by diffraction through a crystal.

CUBEX STEEL®. A silicon iron alloy in which the crystals are so oriented that magnetization can be produced in four directions.

CUPROUS OXIDE. One of the oxides of copper formed by partial oxidation of the metal, used in *copper-oxide rectifiers*.

CURIE BALANCE. A torsion balance invented by P. Curie to measure the force exerted on a non-ferromagnetic material by a non-uniform field.

CURIE POINT. (1) In *ferromagnetic* materials, the temperature above which the magnetization is lost. (2) In *ferroelectric* materials the temperature above which the polarization is lost.

CURIE'S LAW. The *susceptibility* of a *paramagnetic* substance is inversely proportional to the absolute temperature.

CURIUM (Cm). Radioactive element, Atomic No. 96.

CURRENT (ELECTRIC CURRENT). The rate of transfer of electricity. The movement of charged particles is always involved in a current and these may be *electrons, protons, ions,* etc. If the charges are not all of the same kind the net current is the algebraic sum of the charges. Symbol: *I*, unit: *ampere*. See *Units*.

CURRENT AMPLIFICATION. (1) In a *transducer*, the ratio of the current in a specified load to the current in the input circuit of the transducer. (2) In a *photomultiplier*, the ratio of the signal output current to the photoelectric signal current from the cathode at constant electrode voltages.

CURRENT ANTINODE. A point of maximum current along a *transmission line* or aerial, with *standing waves*.

CURRENT AVERAGE (HALF PERIOD). Of a symmetrical *alternating current*, the algebraic average of the current values during a half period, starting with zero value of current.

CURRENT DENSITY. (1) At a point, a vector with the same direction as the current, and magnitude equal to the ratio of the current flowing through a very small area around and perpendicular to the point, to that area. (2) On an electrode in an electrolytic cell, the current per unit area of that electrode. See also *Conduction current* and *Displacement current*.

CURRENT EFFICIENCY. For a specified process, the proportion of the current which effectively carries out the process in accordance with *Faraday's law*.

CURRENT FEED. An aerial feed in which the feeder is connected at a *current antinode*.

CURRENT GENERATOR. A two-terminal circuit element whose terminal current is independent of the voltage at its terminals.

CURRENT LIMITER. A device which restricts the flow of current to a specified amount whatever the applied voltage. See also *Current regulator*.

CURRENT NODE. A point of zero electric current along a *transmission line* or aerial with *standing waves*. Cf. *Current antinode*.

CURRENT REGULATOR. An electronic device designed to maintain a constant current in a circuit independent of supply voltage variations and sometimes of load and temperature variations.

CURRENT SATURATION. The condition in the operation of a *thermionic valve* where an increase in anode voltage does not result in an increase in anode current. See *Temperature-limited emission*.

CURRENT TRANSFORMER. An *instrument transformer* for the transformation of current.

CURTIS WINDING. A technique of winding low-capacitance, low-inductance *resistors*, in which the wire is periodically reversed by passing alternate turns through a slot in the resistor.

CUT-OFF. Of a *valve*, the minimum negative dc voltage (or bias) required on the control grid to stop the flow of anode current.

CUT-OFF FREQUENCY. Of an electrical *network*, the frequency at which the insertion loss between specified impedances exceeds by a specified amount the loss at a reference point in the transmission band; or the frequency where the attenuation constant changes from zero to a positive value or vice versa. The latter definition is usually termed the theoretical cut-off frequency, and the former, the effective cut-off frequency.

CUT-OFF VOLTAGE. Of a *valve*, the *electrode* voltage which reduces the dependent variable of one characteristic to some specified low value.

CUT-OUT. An electrical device which interrupts the flow of current either automatically or by hand.

CYBERNETICS. 'The field of control and communication theory whether in the machine or the animal' (Norbert Wiener).

CYCLE. (1) An orderly series of changes regularly repeated. (2) The complete series of changes taking place in the value of a recurring variable quantity during a period, e.g. an alternating current passes through its cycle of values once in every period. (3) In computers, a set of operations which may be treated as a unit and repeated as a unit. (4) In computer arithmetic, the process in which characters are taken from the end of a word and inserted in correct sequence at the beginning.

CYCLE CRITERION. In *computer* programming, the number of times a particular cycle is to be performed; or the storage element which contains that number.

CYCLE INDEX. In *computer* programming, the number of times a cycle has been performed, or the difference between that number and total number of repetitions desired.

CYCLE RESET. In *computer* programming, setting a cycle index back to its initial value.

CYCLE TIME. Of a *magnetic memory*: the minimum time interval between the starts of successive read–write cycles.

CYCLE TIMER. A control mechanism for opening and closing electrical contacts to a preset time schedule.

CYCLIC MAGNETIZATION. See *Magnetic hysteresis*.

CYCLOGRAM. A graphical representation of the performance of *negative resistance oscillators*.

CYCLOGRAPH. An electron-optical system in which an electron beam moves in two directions at right-angles.

CYCLOIDOTRON. A proposed cross-field fast-wave millimetre tube.

CYCLOPHON. A *cathode-ray tube* in which the *electron beam* moves round a series of target anodes or collectors arranged in a circle. The tube can be used as a multi-position switch, each collector being switched on when the beam is focused on it.

CYCLOTRON. An *accelerator* invented by E. O. Lawrence, designed to provide a beam of high energy protons or deuterons. The particles are accelerated between two large semi-circular electrodes or 'dees' by the electric field from a radio-frequency oscillator, and are maintained in a nearly circular orbit by the magnetic field from large electromagnets.

D

DAG®. See *Aquadag coating*.

DAMPED. A term applied to any system which is capable of oscillation but in which the amplitude of the free oscillation steadily decreases through the dissipation of energy. The cause of the dissipation of energy may be internal or external.

DAMPED CURRENT. An alternating current whose amplitude steadily diminishes.

DAMPED IMPEDANCE. See *Blocked impedance*.

DAMPED NEEDLE. The needle or pointer of an indicating instrument which quickly comes to rest.

DAMPED OSCILLATION. An oscillation in which the amplitude of the oscillating quantity progressively decreases.

DAMPED VIBRATION. See *Damped oscillation*.

DAMPED WAVE. A wave in which the amplitude of each *sinusoidal* component decreases with time.

DAMPER. Of an instrument, a device for diminishing the oscillation of the moving part.

DAMPER VALVE. An electronic circuit for switching on or off a damping resistance for an oscillatory circuit.

DAMPER WINDING. In electric machines, a permanently short-circuited winding arranged so that it opposes rotation or pulsation of the magnetic field.

DAMPING. The effect on the motion of a body or system which is capable of free oscillations or vibrations of internal or external agencies which hinder or prevent those oscillations.

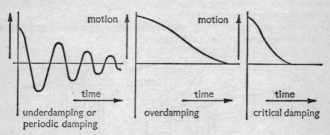

Fig. 30. Damping

DAMPING, APERIODIC. The condition of a system in which the *damping* is so large that, if the system is subjected to a single disturbance of any kind, it comes to a position of rest without continuing through the rest position.

DAMPING, CRITICAL. (1) The amount of *damping* which just prevents oscillation. (2) In an instrument or control system: the amount of damping below which *overshoot* occurs.

DAMPING, PERIODIC (UNDERDAMPING). The condition of damping in which the amount of damping is less than that required for *critical damping*.

DAMPING, RELATIVE. The ratio of the actual *damping factor* to the one which would produce critical damping.

DAMPING FACTOR. Of any underdamped motion, the ratio of the *logarithmic decrement* to the time required for a complete oscillation.

DAMPING FACTOR, INSTRUMENT. The ratio of the deviations of the pointer or indicator in two consecutive swings from the equilibrium position, the larger deviation being divided by the smaller.

DAMPING MAGNET. A permanent magnet used to damp the motion of a system by positioning it in relation to a moving conductor so as to produce a force opposing motion between them.

DANIELL CELL. A two-fluid zinc-copper primary cell; one form consists of an outer vessel containing a zinc plate in dilute sulphuric acid in which is immersed a porous pot containing a copper plate in a saturated solution of copper sulphate.

DARAF. *Farad* spelt backwards: the practical unit of *elastance*, the reciprocal of *capacitance*.

DARK CONDUCTION. Electrical conduction in a photosensitive material which is not exposed to radiation.

DARK CURRENT. See *Electrode dark current*.

DARK DISCHARGE. An invisible electric discharge in a gas.

DARK RESISTANCE. The resistance of a *photocell* when it is not illuminated, which is greater than the resistance under illumination.

DARK-SPACE. See *Cathode dark-space*.

DARK-TRACE TUBE. See *Skiatron*.

D'ARSONVAL CURRENT. In electrobiology, a high-frequency current of low voltage and high amperage.

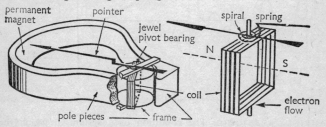

Fig. 31. D'Arsonval movement in a permanent-magnet moving-coil meter

D'ARSONVAL GALVANOMETER. A direct-current *galvanometer* in which a rectangular coil pivoted between the poles of a horseshoe magnet reacts with the magnetic field, producing a torque. The angle through which the coil rotates as a result of the torque is a measure of the current flowing through the coil, and the deflection is observed either by a pointer attached to the coil or by an optical system using a light beam and a small mirror on the coil. The D'Arsonval movement is used on nearly all galvanometers and commercial *voltmeters* and ammeters.

DASHPOT. An appliance for preventing sudden or oscillatory motion of a moving part of any electrical or mechanical device, by the friction of a gas (usually air) or a liquid.

DATA LINK. Equipment, e.g. a wide-band cable, which permits the transmission of information in data form.

DATA LOGGING. The recording of data in a form in which it can be used readily by a computer. In modern systems the data is fed into solid-state transducers and converted to signals which are recorded on punched or magnetic tape, strip-printers, or typewriters.

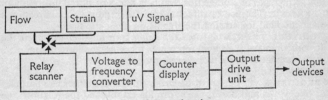

Fig. 32. Data logging

DATA PROCESSING. The operations performed on data in any data-handling system.

DATA REDUCTION. The process of reducing the quantity of, or of rearranging, experimentally obtained data so that conclusions may be drawn more readily from it.

DATA RESOLUTION. The resolution of a control system.

DATA STABILIZATION. In a data-handling system, establishment of references by which the data may be calibrated or assessed.

DATA-HANDLING CAPACITY. (1) Of a computing system, the number of *bits* of information which may be stored at one time and the rate at which these bits may be fed, manually or automatically, to the input. (2) Of a *telemetering* system, the maximum amount of information which can be transmitted and received over a specified link.

DATA-HANDLING SYSTEM. Automatic or semi-automatic equipment which can collect, receive, transmit, and store numerical data and, if required, perform mathematical operations on the stored data and tabulate or indicate the results. The equipment may also be designed to transmit results to other systems and to control their operation. Data may be handled in discrete steps, as digital or binary signals, or continuously, as *analogue* or position signals.

DATAMATIC 1000. A large U.S. commercial electronic digital computer and data-handling system, with very large storage capacity and fast operation.

DAYLIGHT. The total radiation from the sky and sun, equivalent to that of a *black body* at 6,500°K.

DB. Abbreviation for *decibel*.

DBM. Abbreviation for *decibels referred to one milliwatt*.

DC. Abbreviation for *direct current*.

DC AMPLIFIER. See *Direct-coupled amplifier*.

DCC. Abbreviation for *double-cotton-covered wire*.

DC COMPONENT. The average value of a signal.

DC COUPLING. See *Direct coupling*.

DC DUMP. In a *computer*: the removal – intentional or accidental – of all dc power from a computer system or an essential part of it. For some types of storage this results in a loss of stored information.

DC ERASING. The use of direct current to produce a unidirectional magnetic field for erasing the signals on magnetic tape.

DC PICTURE TRANSMISSION. *Television* transmission in which the dc component of the signal represents the average illumination.

DCR. Abbreviation for *direct-current restorer*.

DDA. Digital Differential Analyser.

DE BROGLIE WAVELENGTH. *Electrons* can behave either like particles or like waves and de Broglie's formula established the relation: the product of the wavelength and the momentum of an electron is equal to *Planck's constant*, from which the wavelength can be determined if the momentum is known.

DE-ACCENTUATOR. A device used in *frequency modulation* receivers to de-emphasize the higher frequencies in the received signal and so restore their original relative amplitude. See *De-emphasis*.

DEAD END. The end of a wire to which no electrical connection is made; the part of a coil not in use.

DEAD LOAD. See *Artificial load*.

DEAD ROOM. A room which has an unusually large degree of sound absorption.

DEAD SHORT. A short circuit of very low resistance.

DEAD TIME. In an electrical circuit, valve, or instrument, the time immediately after receiving a stimulus during which it is insensitive to another impulse or stimulus.

DEAD-BEAT INSTRUMENT. An instrument whose pointer goes immediately to its true position without wavering.

DEAD-TIME CORRECTION. In a radiation counter, the correction to the observed counting rate to allow for probable events during the *dead time*.

DEATHNIUM (CENTRE). A zone of irregularity in a *semiconductor* crystal lattice, caused by disorders in the arrangement of the atoms such as grain boundaries and interstitial atoms, which is believed to play an important part in the process of generation and recombination of electron-hole pairs.

DEBUG. To isolate and remove the mistakes from a computer *routine*.

DEBUNCHING. In *velocity-modulated valves* (tubes), e.g. *klystron*, the bunching of electrons tends to be counteracted by the electrostatic repulsion between electrons in a bunch.

DEBYE UNIT. A unit of *dipole moment* equal to 10^{-18} statcoulomb-cm.

DECADE. In electric circuit theory, the frequency interval between two frequencies having a ratio of 10 to 1.

DECADE BOX (UNIT). A self-contained device for introducing *resistance, capacitance*, or *inductance* in definite steps into test and measuring circuits. A box consists of two or more sections, each section having ten switched positions and ten times the value of the preceding section. The switches can thus be set to any integral value within the range of the box.

DECADE SCALER. A *scaler* with a scaling factor of ten.

DECAY. In a *charge-storage* tube, the decrease in the amount of stored charge not due to erasing. See *Decay time*.

DECAY CHARACTERISTIC. The relation, on a graph, of the electrical or magnetic energy dissipated in a system to the time elapsed after the source of power, which built up the energy, has been removed.

DECAY TIME. In a *charge-storage* tube, the time required for the stored charge to decay to a specified fraction of its original value. The fraction is normally the reciprocal of e, the base of natural logarithms.

DECCA (NAVIGATOR). A radio aid to navigation which employs multiple receivers to indicate and measure the relative phase of continuous wave signals from several synchronized transmitters.

DECELERATING ELECTRODE. An electrode used to slow down electrons or other charged particles, usually consisting of a metal plate or wire in or near the path of the particles.

DECIBEL (DB). The ratio of two values of power. The number of decibels is expressed as ten times the logarithm to the base ten of the power ratio. Commonly used for expressing transmission gains, losses, and levels.

DECIBELS REFERRED TO ONE MILLIWATT (DBM). The *decibel* is a number with no dimensions so that if actual power levels are of interest a reference level is required. This is usually taken to be one *milliwatt*. Thus 30 dbm represents a power level of 1 *watt*.

DECIMAL-TO-BINARY CONVERSION. The process of converting a decimal number to a *binary* number.

DECINEPER. One tenth of a *neper*.

DECISION ELEMENT. In computing systems, a circuit whose output is the result of a *logical operation* on one or more binary digits.

DECLINATION. The angular difference between the directions of magnetic north and true north (magnetic declination) or between grid north and true north (grid declination).

DECLINOMETER. An instrument for measuring the direction of a magnetic field in relation to astronomical or survey coordinates.

DECODER. (1) An electronic device for decoding pulsed signals in a radar *beacon* or an *interrogator*. (2) In *computers*, a network in which several inputs are excited together to produce a single output.

DECOMPOSITION VOLTAGE. In electrolysis, the voltage at which electrode current begins to increase abruptly, i.e. the minimum voltage which will produce continuous electrolysis.

DECOUPLING. The reduction of *coupling*, especially unwanted coupling.

DECOUPLING NETWORK. A network designed to prevent the inter-action of two electric circuits. Decoupling networks usually consist of *resistance-capacity* or *inductance-capacity* filters and their most frequent application is to prevent undesirable coupling between valves connected to a common power supply.

DECREMENT. Of a damped harmonic system, the ratio of the amplitude of oscillation at the beginning of any period to the amplitude at the end of that period. See *Damping factor, Logarithmic decrement*.

DECREMETER. An instrument for measuring the *logarithmic decrement* of a train of waves.

DEE. A hollow D-shaped electrode used as an *accelerating electrode* in a *cyclotron*.

DE-EMPHASIS (POST-EMPHASIS). The restoration of a pre-empha-sized signal to its original form by introducing a frequency response complementary to that produced by *pre-emphasis*.

DE-ENERGIZE. Disconnect a circuit from its power source.

DEFECT CONDUCTION. In a *semiconductor*, a hole conduction in the *valence band*. See *Acceptor impurity*.

DEFINITION. In television, the fidelity of reproduction of the images.

DEFLAGRATOR. An electric cell with very low internal resistance which can therefore provide a very high output current.

DEFLECTING COILS. See *Deflecting yoke*.

DEFLECTING ELECTRODE. An electrode used to deflect *electrons* or other charged particles from their normal paths, especially in *cathode-ray tubes*.

DEFLECTING PLATE. One of the *deflecting electrodes* in a cathode-ray tube.

DEFLECTING YOKE. An assembly of one or more coils which carry the current to provide the field for magnetic deflection of electron beams. The most common application is around the neck of a *cathode-ray tube*. See *Electromagnetic deflection*.

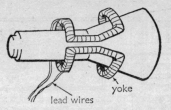

Fig. 33. Deflecting yoke in position on a cathode-ray tube

DEFLECTION. In a *cathode-ray tube*, the bending of an electron beam after the focused beam leaves the electron gun.

DEFLECTION ANGLE. In a *television* tube, the angle through which the beam is bent vertically, horizontally, or diagonally.

DEFLECTION DEFOCUSING. A progressive broadening of the spot due to the electron beam in a television tube as it is deflected farther from the middle of the picture.

DEFLECTION SENSITIVITY. In a *cathode-ray rube,* the ratio of the displacement of the electron beam at its point of impact to the change in the deflecting field. Usually expressed in millimetres/volt applied between deflecting electrodes, or millimetres/gauss of the deflecting magnetic field.

DEFLECTOR. See *Deflecting electrode.*

DEFORMATION POTENTIAL. An electric *potential* due to deformation of the crystal lattice of semiconductors and conductors which affects the behaviour of *free electrons* in the lattice.

DEGASSING. Of a vacuum valve or tube, the removal of occluded and residual gases from the materials inside the enclosure and from the glass or metal envelope. See *Getter.*

DEGAUSSING. Of a ship, the creation of a sustained magnetic field to counteract the ship's own field and so avoid detonation of magnetic mines.

DEGENERACY. In a resonant device, the condition where two or more modes have the same frequency.

DEGENERATE SEMICONDUCTOR. A *semiconductor* in which the number of electrons in the *conduction band* approaches that of a metal.

DE-IONIZATION. The return of an *ionized gas* to its neutral or un-ionized condition after the removal of all sources of ionization.

DE-IONIZATION POTENTIAL. The potential at which the ionization of a gas – in a gas-filled tube – ceases, and conduction stops.

DE-IONIZATION TIME. Of a gas tube, the time required for the grid to regain control after interruption of the anode current. See *Thyratron.*

DEKATRON®. A cold-cathode *scaling tube* with, normally, a radially

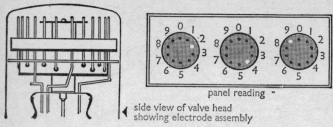

side view of valve head
showing electrode assembly

panel reading

Fig. 34. Dekatron, showing valve head and panel of three counters

symmetrical array of ten *cathodes* and associated transfer electrodes surrounding the anode. The tube operates by transfer of a *glow discharge* from one cathode to the next in response to an impulse and can be used for switching or counting with a visible display of fast counts in the decimal system.

DELAY. (1) The retardation of the time of arrival of a signal after transmission through a physical system. (2) A circuit designed to introduce a known delay between its input and output terminals.

DELAY CABLE. A length of concentric cable, usually connected between a *surge generator* and an object under test to delay the arrival of the surge.

DELAY DISTORTION. Departure from flatness in the *phase delay* or *envelope delay* of a system over the frequency range required for transmission or the effect of such departure on a transmitted signal.

DELAY-EQUALIZER. A corrective network designed to make the *phase delay* or *envelope delay* of a system substantially constant over a desired frequency range.

DELAY GENERATOR. A phase-shifting device – usually a variable delay line – designed to delay a signal without introducing phase distortion.

DELAY LINE. A real or artificial transmission line or equivalent device designed to introduce *delay*.

DELAY LINE, ARTIFICIAL. (1) In general, any device for producing time delay of a signal. (2) A circuit with *lumped* elements which duplicates the characteristics of a transmission line including the time-delay characteristic.

DELAY LINE, SONIC (ACOUSTIC). A device which can transmit delayed sound pulses by recirculation of wave patterns containing information, usually in binary form.

DELAY LINE, ULTRASONIC. A device in which the time required for propagation of elastic waves in a medium (usually mercury) is used to obtain a *delay*. See *Mercury tank.*

DELAY LINE MEMORY. In computers, a *circulating memory* which employs a *delay line*, as the major element in the circulating path.

DELAY TIME. See *Delay*.

DELAY TIME REGISTER. An electric or acoustic *delay line* together with its input, output, and circulation circuits.

DELAYED AUTOMATIC GAIN CONTROL. In radio receivers, a technique to achieve full amplification of very weak signals while maintaining a constant output at the detector for all signals above a certain field strength.

DELAYED BIAS. See *Delayed automatic gain control*.

DELAYED PULSE GENERATOR. A *pulse generator* in which the main pulse can be delayed, in relation to an internal trigger, or a secondary pulse can be delayed in relation to the main pulse.

DELLINGER EFFECT. Radio *fading*.

DELTA CONNECTION. Connection of a three-phase electrical system so that the corresponding windings of the transformers form a triangle.

DELTA NETWORK. A set of three branches connected in series to form a mesh.

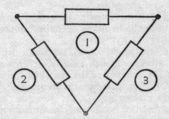

Fig. 35. Delta network using resistors

DELTA RAY. An *electron* ejected by recoil when a rapidly moving charged particle passes through matter.

DELTA WAVE. The lowest frequencies – under 10 cycles per second – in the frequency spectrum of a *brain wave*.

DEMAGNETIZATION CURVE. A portion of the *magnetic hysteresis loop* which shows the peak value of *residual magnetism* and how the magnetization reduces to zero as the demagnetizing force is applied.

DEMAGNETIZATION ENERGY. The energy which would be released when a magnetized body is completely demagnetized.

DEMAGNETIZED FIELD. The field created by the magnetic poles which develop on a magnetic substance when a magnetic field is applied to it. The demagnetizing field acts in opposition to the magnetizing field.

DEMAND METER. A device which indicates or records the demand or maximum demand for electrical power.

DEMODULATION. In *communication* or *radar* receivers, the process of separating the original information from a *modulated* signal wave. The circuits or equipments used for this purpose are called de-modulators or *detectors*. See also *Frequency discriminator*.

DEMODULATOR. See *Demodulation*.

DEMOUNTABLE VALVE (TUBE). An electronic valve or tube which may be disassembled and reassembled for the renewal of electrodes or the insertion of components and materials required for processing or experiment.

DENDRITE. In metallurgy, a crystal usually produced by solidification of a liquid, characterized by a many-branched structure like a tree.

DENSITOMETER. An instrument which employs a light-sensitive element to measure the optical density of a material by the change in the quantity of light transmitted or reflected by it.

DENSITY MODULATION. In a *microwave resonator* tube, modulation of the beam current by the *radio-frequency* voltage across the *interaction space*.

DEPLETION LAYER. In a *semiconductor*, a region in which the density of mobile charge carriers is insufficient to neutralize the fixed charge density of donors and acceptors.

DEPLETION-LAYER CAPACITANCE. In a *semiconductor diode*, the rate of charge with applied voltage of ionized donors and acceptors in the *depletion layer*.

DEPLETION-LAYER TRANSISTOR. A transistor which depends on the movement of carriers through a *depletion layer*, e.g. a *spacistor*.

DEPOLARIZATION. (1) Removal of the gas which collects on the plates of an electric *cell* during charging or discharging. (2) Decrease in the *polarization* of a *dielectric*.

DEPOLARIZER. A substance producing *depolarization*.

DEPTH FINDER. See *Echo sounder*, *Sonar*.

DERATING. When a component or equipment is operated under unusual or extreme conditions it is the usual practice to reduce its maximum performance ratings so as to leave an adequate safety margin. Derating curves, e.g. for the permissible dissipation of a resistor at high ambient temperatures, are sometimes provided to indicate the proposed ratings for the particular conditions.

DERIVATIVE ACTION. In an *automatic control system*, operation in which the speed of correction depends on how fast a condition is departing from normal.

DERIVED CIRCUIT. A *parallel* electric circuit whose current is obtained from another circuit.

DERIVED UNITS. Units which are derived from *fundamental units* by the application of physical laws.

DETECTION. The process of extracting information from the fluctuations with time of an electromagnetic wave.

DETECTOR. (1) A device whose main function is *detection*. (2) A device which performs the function of identification and/or location. See the following types of detector (*discriminator*): *Carborundum, Crack, Crystal, Diode, Electrolytic, Foster-Seeley, Frequency, Gate, Grid-leak, Heat, Infra-red, Linear, Phase, Power, Pulse, Square-law, Standing-wave, Ultrasonic.*

DETECTOR, FIRST. In a *superheterodyne* receiver, the stage in which a local oscillator is used to obtain an *intermediate frequency* (i.f.) from the received radio-signal frequency.

DETECTOR, SECOND. In a *superheterodyne receiver*, the stage in which the intelligence frequency is separated from the intermediate frequency.

DETECTOR VALVE (TUBE). An electronic *valve* or *tube* used as a *detector*.

DETUNE. Tune out or change the frequency of a *tuned circuit* so that it differs from that of the desired signal.

DEUCE. A high speed digital computer built at the National Physical Laboratory, Teddington.

DEVIATION. (1) In *frequency modulation*, the amount by which the carrier frequency alters when modulated. (2) In automatic control systems, the difference between the actual value of the controlled variable and the value to which the controlling mechanism is set. (3) The difference between the setting of a compass needle and magnetic north, due to a local magnetic disturbance.

DEVIATION ABSORPTION. Absorption of radio waves at frequencies near the critical frequency of the *ionosphere*.

DEVIATION DISTORTION. Distortion in an *FM receiver* due to inadequate bandwidth, inadequate amplitude-modulation rejection, or inadequate linearity in the *discriminator*.

DEVIATION LOSS, ANGULAR. Of a *transducer* for sound emission or reception, the ratio, in decibels, of the reference response at the principal axis to the response of the transducer at a specified angle from the principal axis.

DEVIATION RATIO, F-M. In a frequency-modulation system, the ratio of the maximum frequency *deviation* to the maximum modulating frequency of the system under specified conditions.

DEW LINE. A line of radar stations along the seventieth parallel of the North American continent, designed for Distant Early Warning.

DF. Abbreviation for *direct finder*.

DIACRITICAL CURRENT. The current in a coil which produces a flux in the core equal to half the flux for saturation.

DIAGNOSTIC ROUTINE. In *computers*, a routine designed to reveal a fault in the computer or a mistake in coding.

DIALOGUE EQUALIZER. In high-fidelity *audio amplifiers*, a filter attenuator designed to reduce the amplification of speech at low frequencies.

DIAMAGNETIC. Having a *permeability* less than that of a vacuum.

Diamagnetic materials tend to move away from a strong magnetic field in which they are placed.

DIAPHRAGM. (1) In a *microphone*, a thin metal disc which can be vibrated by sound waves or which can produce sound waves by vibration as in a telephone ear-piece. (2) In an electrolytic cell, a porous partition which permits the passage of ions, while preventing the mixing of the conducting solutions.

DIATHERMANOUS. A material which transmits, with little attenuation, thermal or infra-red radiation.

DIATHERMIC SURGERY. A technique used in surgery in which a high-frequency electric arc is used to cut living tissue with the minimum amount of bleeding.

DIATHERMY. The treatment of diseases by the use of sustained and undamped current oscillation at frequencies ranging from several hundred thousand to several millions of cycles per second. The application of these currents to living tissues is accompanied by a marked heating effect.

DIATHERMY MACHINE. A machine, consisting chiefly of an oscillator with its power supply and control equipment, for generating the radio-frequency power used in *diathermy*.

DICKE'S RADIOMETER. An instrument for the precise measurement of *microwave noise* power, by comparing an unknown noise source with the noise from a specially shaped modulation wheel rotating in a slot in a waveguide system.

DIELECTRIC. A material which offers relatively high resistance to the passage of an electric current but through which *magnetic* or *electrostatic lines of force* may pass. Most insulating materials e.g. air, porcelain, mica, glass, are dielectrics and a perfect vacuum would constitute a perfect dielectric.

DIELECTRIC ABSORPTION. A phenomenon occurring in imperfect *dielectrics* where positive and negative charges separate and then accumulate in certain regions in the body of the dielectric. This effect is observed as a gradually decreasing current after the application of a fixed dc voltage.

DIELECTRIC AMPLIFIER. An amplifier using a *ferroelectric* capacitor whose variation of capacitance with applied voltage is used to provide amplification.

DIELECTRIC BREAKDOWN TESTS. Tests in which a higher voltage than the rated voltage is applied for a specified time to determine the margin against breakdown of insulating materials and assemblies containing them.

DIELECTRIC CONSTANT. The ratio of the electric *flux density* produced in the *dielectric* to that produced in free space by the same electric force. The dielectric constant of the dielectric of a *capacitor* is proportional to its *capacitance*. Synonymous with relative *permittivity*.

DIELECTRIC CONSTANT, ABSOLUTE. Synonymous with *permittivity.*

DIELECTRIC CURRENT. The current flowing at any instant through a *dielectric* placed in a changing electric field may consist of four components: *displacement current*; *absorption current*; *conduction current*; and decaying conduction current. The displacement current is the only one always present.

DIELECTRIC DIODE. A capacitor whose negative electrode can emit electrons into a special dielectric – such as cadmium sulphide crystals – so as to provide current flow in one direction.

DIELECTRIC DISPERSION. The variation in the magnitude of the *dielectric constant* with increasing frequency.

DIELECTRIC FATIGUE. In some *dielectrics* resistance to breakdown decreases after prolonged application of a voltage.

DIELECTRIC HEATING. The heating of an insulating material in an alternating electric field, which results from the periodic displacement of electrons within the atoms and molecules of the material. In industrial dielectric heating equipment the work is usually placed between the plates of a *capacitor* connected to a high-frequency source of power.

DIELECTRIC HYSTERESIS. The phenomenon by which the *electric flux* in a *dielectric* depends not only on the present value of the electric force but also on the previous state of the material. See *Ferroelectrics.*

DIELECTRIC LENS. A lens made of a suitable *dielectric* to refract radio waves before they reach a microwave aerial.

DIELECTRIC LOSS. In a *dielectric* subjected to a varying electric field, the time rate at which electrical energy is converted into heat.

DIELECTRIC LOSS ANGLE. The difference between ninety degrees and the *dielectric phase angle.*

DIELECTRIC PHASE ANGLE. The angular difference in phase between the *sinusoidal* potential difference applied to a *dielectric* and the component of the resulting alternating current having the same period as the potential difference.

DIELECTRIC PHASE CONSTANT. Per unit length in a *dielectric*, 2π divided by the wavelength of the electromagnetic wave in the dielectric.

DIELECTRIC POLARIZATION. The *dipole moment* per unit volume of a *dielectric.*

DIELECTRIC RELAXATION. The inability of the molecules of a dielectric to become or remain aligned with the direction of a field applied to it, when the period of the applied field is sufficiently short.

DIELECTRIC STRAIN (STRESS). The strain set up in a dielectric by the potential difference due to opposing charges in a *capacitor.*

DIELECTRIC STRENGTH. Of an insulating material, the *potential gradient* at which electrical failure or breakdown occurs. See *Breakdown voltage.*

DIELECTRIC SUSCEPTIBILITY. A dimensionless unit, the ratio of the magnitude of the *dielectric polarization* to the product of the *permittivity* of free space and the *field strength*.

DIELECTRIC WAVEGUIDE. A *waveguide* in which the waves travel through solid dielectric material.

DIELECTRIC WEDGE. A wedge-shaped piece of *dielectric* material used to match the impedance of an air-filled waveguide to a *dielectric waveguide*.

DIELECTRIC WIRE. A filament or thin rod of *dielectric* material.

DIFFERENCE DETECTOR. A *detector* whose output is related to the difference between the amplitudes of two input signals.

DIFFERENCE TRANSFER FUNCTION. In a *feedback control loop*, the transfer function relating a loop difference signal to the corresponding loop input signal.

DIFFERENTIAL. In electronics, referring to a circuit, device, or machine whose principle of operation depends on the difference between two opposing effects.

DIFFERENTIAL AMPLIFIER. An amplifier having two similar input circuits connected so as to respond to the difference between two voltages or currents and effectively suppress voltages or currents which are alike in the two input circuits.

DIFFERENTIAL ANALYSER. An *analogue computer* used for solving differential equations.

DIFFERENTIAL CAPACITOR. A *variable capacitor* in which the *stator* is split into two sections insulated from each other. As the rotor is turned the capacity of one section increases as the capacity of the other decreases.

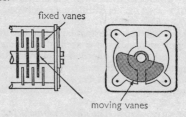

Fig. 36. Construction of differential capacitor

DIFFERENTIAL DISCRIMINATOR. A *discriminating* circuit, used, e.g. in pulse height analysers, which passes pulses above a specified minimum value but not greater than a somewhat higher value.

DIFFERENTIAL GAIN. In a *television video* system, the difference in the gain in decibels for a small high-frequency signal at two specified levels of a low-frequency signal on which it is superimposed.

DIFFERENTIAL GAIN CONTROL. A method of adjusting the gain of a radio receiver to correspond with an anticipated change of signal level, so that the amplitude differences between signals at the output of the receiver are reduced.

DIFFERENTIAL GALVANOMETER. A *galvanometer* whose deflection is proportional to the differential field due to two currents of opposite polarity flowing through two identical coils.

DIFFERENTIAL INSTRUMENT. An instrument designed to measure the difference between two similar quantities. For example, in a current differential meter the currents flow in two circuits arranged to counterbalance when the currents are equal, but any difference in the current causes movement of an indicator.

DIFFERENTIAL RESISTANCE. Of a *diode*: the resistance measured between the terminals of the diode under small-signal conditions.

DIFFERENTIAL SELSYN (SYNCHRO). In control circuits, a *selsyn* or *synchro* in which both stator and rotor are wound to produce rotating magnetic fields, so that a change in the position of the rotor has the effect of introducing a differential angle into the control system.

DIFFERENTIAL TRANSFORMER. A position *transducer* consisting of a transformer with two secondary windings, in which alternating voltages are produced depending on the position of the primary winding or of the magnetic core.

DIFFERENTIAL WINDING. A winding, e.g. in a compound motor, which opposes the action of another winding.

DIFFERENTIATOR (DIFFERENTIATING CIRCUIT, NETWORK). A *transducer*, device, or circuit whose output waveform is designed to be equivalent to the time rate of change of its input waveform. In computers, especially the analogue type, such devices are frequently used to provide an output proportional to the derivative, or time rate of change, of an input signal.

DIFFRACTION. The process of interference which affects all wave motions encountering an opaque obstacle or a discontinuity between two different mediums. In particular, electromagnetic waves and electron beams are bent and produce *diffraction patterns*. See *Electron diffraction*.

DIFFRACTION ANGLE. In electromagnetic diffraction, the angle between the diffracted ray and the normal to the surface at the point of incidence.

DIFFRACTION GRATING. A device for separating incident radiation into its component frequencies by causing diffraction. An optical diffraction grating consists of a number of closely spaced fine lines accurately engraved on a mirror or highly polished metal. A *microwave* diffraction grating can take the form of a fine wire mesh.

DIFFRACTION PATTERN. A pattern of concentric rings, similar to the pattern obtained when light passes through a small aperture produced by beams of electrons and other atomic particles when they are directed against the faces of crystals or thin metal foils.

DIFFUSED-BASE TRANSISTOR. A *transistor* in which a non-uniform base region is produced by gaseous diffusion.

DIFFUSED JUNCTION. See *Semiconductor junction, diffused*.

DIFFUSION CAPACITANCE. Of a *semiconductor diode*, the rate of change of injected charge with applied voltage.

DIFFUSION CONSTANT. In a *semiconductor*, the ratio of diffusion current density to the gradient of charge carrier concentration. Equal to the product of *drift mobility* and the average thermal energy per unit charge of the *carriers*.

DIFFUSION LENGTH. In a *semiconductor*, the average distance to which the *majority carriers* diffuse between generation and recombination.

DIFFUSION POTENTIAL. The potential difference across the boundary of an electrical double-layer in a liquid.

DIFFUSION PUMP. The type of pump used for producing high-vacuum systems such as those of vacuum valves and tubes. The principle of operation is the formation of a molecular jet of mercury or of certain oils into which the gas molecules diffuse and are carried out through the exhaust.

DIGILOCK®. A pulse code digital telemetry system, developed for missiles and rockets.

DIGITAL COMPUTER. A *computer* which operates with information, numerical or otherwise, which is represented in the form of digits. Thus the simplest form of digital computer is the human hand, and the use of fingers to represent numbers and numerical operations the simplest form of the explicit use of a *language* which is characteristic of digital computers. In modern use the term generally means electronic digital computer; the following types are described in this dictionary: *Ace, Apex, Atlas, Datamatic 1000, Deuce, Eniac, George, Leo, Maniac, Mercury, Pegasus, Ramac, Rascal, Stretch, Univac, Whirlwind*.

DIGITAL COUNTER. A device, usually solid-state, for counting the discrete electrical pulses given out by digital transducers.

DIGITAL DIFFERENTIAL ANALYSER. A *digital computer* which uses built-in integrators to carry out the mathematical process of integration.

DIGITAL INPUT. Input information in the form of digits which are fed in automatically by punched tape or punched card.

DIGITAL VOLTMETER. A *voltmeter* which displays the measured values as numbers composed of digits.

DIGITIZER. A device which converts analogue measurements into digital form.

DIGITRON®. A *cold-cathode* gas-filled tube in which the cathodes are shaped to form characters, usually the digits 0 to 9. A selected cathode is made to glow by a switching connection to one side of a power supply.

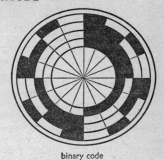

Gray code   binary code

Fig. 37. Two ways of coding optical discs which convert angular movements of a shaft into digital measurements

DIHEPTAL BASE. A base with fourteen pins used for *cathode-ray tubes.*

DIODE. A *valve* containing only two *electrodes*: an *anode* and a *cathode.*

DIODE CHARACTERISTIC. (1) A graph of the anode current of a *diode* as a function of the anode voltage. (2) Of a multi-electrode valve, the composite electrode characteristic measured with all the electrodes except the cathode connected together.

symbol

DIODE CLIPPER. A *clipping circuit* which uses a *diode.*

DIODE DETECTOR. A *detector* circuit employing a *diode* valve or crystal.

DIODE DETECTOR, CURRENT SENSITIVITY. In a low-level crystal diode detector, the ratio of the rectified current to the radio-frequency power dissipated.

DIODE MIXER. A *mixer* in the form of a cathode and anode in a glass envelope which is fitted into a radio-frequency line and performs the same function as a *crystal mixer.*

DIODE RECTIFIER. A *rectifier* which employs a *diode* as the uni-directional device.

DIODE-CAPACITY MEMORY. In high-speed *computers,* a rapid-access memory in which *diodes* and *capacitors* are the basic storage units.

DIODE-PENTODE. A valve consisting of a *diode* and a *pentode* in the same envelope.

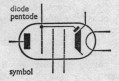

diode pentode

symbol

diode triode

symbol

DIODE-TRIODE. A valve consisting of a *diode* and a *triode* in the same envelope.

DIP (INCLINATION). The angle, measured in a vertical plane, between the direction of the earth's magnetic field and the horizontal.

DIPLEX OPERATION. In communications, the simultaneous or alternate transmission or reception of two messages using a common feature such as a single carrier or a single aerial.

DIPLEX TELEGRAPHY. The transmission by *telegraphy* of two messages in the same direction and at the same time without interference.

DIPLEXER. A system of coupling which enables a communications *transmitter* and a *radar* to be used with the same aerial.

DIPOLE. See *Electric dipole, Electric doublet, Magnetic moment.*

DIPOLE AERIAL. A straight radiator, usually half a wavelength long, and fed at the centre to produce the maximum radiation in the plane normal to its axis.

DIPOLE MOLECULE. A molecule which has a *dipole moment* because of the permanent separation of the effective centre of positive charge from that of negative charge.

DIPOLE MOMENT. Of two equal charges, the product of one of the charges and the vector separating them.

DIPOLE MOMENT, INDUCED. A *dipole moment* in an atom or molecule which exists by virtue of its presence in an electric or magnetic field.

DIRECT CAPACITANCE. See *Capacitance, direct.*

DIRECT CONNECTION. Of two or more electric machines, connection so that they rotate at the same speed, usually with their shafts connected end to end.

DIRECT COUPLING. In electrical *networks*, association of two or more meshes which have only one branch in common.

DIRECT-COUPLED AMPLIFIER. An amplifier in which the anode of one stage is coupled directly to the grid of the following stage, or is coupled directly through a chain of resistors. It is capable of amplifying direct current.

DIRECT CURRENT (DC). Symbol: *I.* Undirectional current which does not change in value, or in which the changes are negligibly small.

DIRECT-CURRENT AMPLIFIER. Misnomer for *direct-coupled amplifier.*

DIRECT-CURRENT BALANCER. A method of coupling and connecting two or more similar direct-current machines so that the conductors connected to the junction points of the machines are kept at constant potential.

DIRECT-CURRENT CONDUCTANCE. See *Conductance.*

DIRECT-CURRENT CONVERTER. A converter which changes direct current from one voltage to another.

DIRECT-CURRENT ELECTRON-STREAM RESISTANCE. The ratio of *electron-stream potential* to the dc component of the stream current.

DIRECT-CURRENT GENERATOR. A rotary machine which converts mechanical into direct-current power.

DIRECT-CURRENT MAGNET. An *electromagnet* whose windings carry direct current.

DIRECT-CURRENT MOTOR. A *motor* for operation by direct current.

DIRECT-CURRENT RECEIVER. A receiver designed to operate from direct-current domestic mains supply.

DIRECT-CURRENT RESISTANCE. *Resistance* to an unchanging current. It is the ratio of the direct voltage across the *resistor* to current flowing through it: $R = E/I$ where R is the resistance, E the voltage, I the current.

DIRECT-CURRENT RESTORER. In a circuit which cannot transmit slow variations but which readily transmits fast ones: a means by which a dc or low-frequency component may be restored after transmission.

DIRECT-CURRENT TELEGRAPHY. The form of telegraphy in which the signals are formed by direct current supplied to the line.

DIRECT-CURRENT TRANSFORMER. A device for measuring large values of direct current, based on the measurement of current in an armature rotated inside a magnet which is energized by the large direct current.

DIRECT GRID BIAS. The direct component of grid voltage, usually called '*grid bias*'.

DIRECTION FINDER, AUTOMATIC (ADF). A system for finding the direction of a radio-frequency source automatically, employing a rotating aerial and an indicator, such as a cathode-ray tube, showing strength of received signal against direction. If the aerial is too large to be rotated conveniently, two stationary aerials at right angles may be used.

DIRECTION FINDER, RADIO. A system for determining the origin of a transmitted signal, employing a radio receiver, a directional aerial, and associated equipment.

DIRECTION OF CURRENT. Through a surface, the direction of motion of the *positive electricity* when it is the predominating part of the motion; and the direction opposite to the direction of motion of the *negative electricity* when electrons are the predominating part of the motion.

DIRECTIONAL COUPLER. A transmission coupling device for sampling or exciting an incident or reflected wave in a *coaxial transmission line* or waveguide.

DIRECTIONAL GYROSCOPE. An instrument which indicates direction; it contains a free gyroscope holding its position in azimuth and which therefore indicates any angular deviation from the reference position.

DIRECTIONAL HOMING. Following a path so that the object is always at the same relative bearing.

DIRECTIONAL PATTERN. Of an aerial, a graph of the radiation or reception as a function — of direction.

DIRECTIONAL REFERENCE FLIGHT. Guided flight, as with a *guided missile,* which is controlled by external signals varied as required.

DIRECTIONAL SELECTIVITY. Of an aerial or microphone, the preferential sensitivity of the device to radiation from a particular direction.

DIRECTLY GROUNDED. Connected to *ground* without the intentional insertion of *impedance* in the connection.

DIRECTLY-HEATED CATHODE. A filamentary *cathode* which carries its own heating current as distinct from an *indirectly heated* cathode.

DIRECTOR. Of an aerial, an auxiliary structure mounted in the direction of the major lobe of radiation of the aerial which increases the radiation in that direction.

DIRECT RAY. See *Direct wave.*

DIRECT VOLTAGE. A unidirectional voltage in which the changes in value are negligible. The popular expression 'dc voltage' is not very logical.

DIRECT WAVE. A wave which is propagated directly through space, i.e. the one which follows the shortest path between source and receiver.

DIRECT-READING INSTRUMENT. An electrical instrument which indicates the true reading of the quantity being measured without the necessity for any calculation or conversion of units.

DIRECT-WIRE CIRCUIT (SINGLE-WIRE CIRCUIT). A simple signalling circuit with a high degree of reliability, consisting of one metallic conductor with a ground return and receiving equipment which responds to an increase or decrease in current.

DISC ARMATURE. The *armature* of a generator or motor in which the coils are wound on a flat disc instead of a drum.

DISCHARGE. (1) In general, to remove or reduce an *electric charge.* (2) In a *gas tube,* the passage of electric current through the gas. (3) Of a *capacitor,* the release of the stored energy of the capacitor in an external circuit. (4) Of a battery, the conversion of the chemical energy of the battery into electrical energy.

DISCHARGE LAMP. A lamp in which light is produced by the passage of electricity through a metallic vapour or a gas enclosed in a tube or bulb.

DISCRIMINATION. The selection of a signal of a particular frequency, amplitude, phase, or other characteristic by eliminating all other signals which symbol for symmetrical tube enter the *discriminator.*

DISCRIMINATOR. A device in which the properties of a signal, such as frequency or phase, are converted into amplitude variations. See the following types of discriminator: *Differential, Frequency, Gated-beam tube, Phase, Pulse, Time.*

DISC-SEAL(ED) VALVE (TUBE). A valve constructed so that its electrodes are annular discs or thick cylinders in place of wires, to reduce lead inductances and inter-electrode coupling.

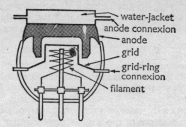

Fig. 38. Cross-section of a disc-seal triode

DISH. A *microwave* reflector with a concave surface, e.g. part of a hollow sphere.

DISINTEGRATION VOLTAGE. The lowest anode voltage at which destructive positive ion bombardment of the cathode occurs in a hot-cathode gas tube.

DISPATCHER. Of a *digital computer*, the part which performs the switching operations which determine the sources and destinations for the transfer of words.

DISPENSER CATHODE. See *Impregnated cathode*.

DISPERSION. The separation of any kind of radiation in relation to some variable of the radiation such as energy or frequency. See *Dielectric dispersion*.

DISPLACEMENT CELL. A battery cell in which the ions of one element go into solution by displacing ions of a second element in the solution, the latter being then deposited. See *Daniell cell*.

DISPLACEMENT CURRENT. A concept invented by Maxwell to account for the flow of current in non-steady-state conditions, e.g. through the *dielectric* of a *capacitor* in a series alternating circuit. Mathematically, it is proportional to the time rate of change of electric *displacement flux* through the surface of a dielectric. See *Electric displacement, Maxwell's equations*.

DISPLACEMENT FLUX. Through a surface in a dielectric, the integral over the surface of the normal component of the displacement.

DISPLAY. Information presented on a *radar* screen or *cathode-ray tube* with dials representing the information.

DISRUPTIVE DISCHARGE. The sudden and large increases of current passing through an insulating material which accompanies the breaking down of the material under electric stress.

DISSECTOR TUBE. A *camera tube* with a continuous *photocathode* on which a photoelectric emission pattern is formed by moving its electron-optical image over an aperture. See *Image dissector tube*.

DISSIPATION. The loss of energy through resistive forces, in particular electrical *resistance*. The loss usually appears as heat and is not available for doing useful work. In a resistive circuit the loss is equal to I^2R, where I is the current and R the resistance.

DISSIPATION FACTOR. Of a *dielectric*, the relation between *conductivity* and *permittivity* at a given frequency. Dissipation factor $D = \dfrac{\kappa}{\omega\epsilon}$ where κ is the conductivity, ω is 2π times the frequency, and ϵ is the permittivity. It is also the reciprocal of the Q factor.

DISSIPATIONLESS LINE. An ideal *transmission line* in which there is no energy loss.

DISTANCE MARK. In *radar*, a mark on a cathode-ray screen which represents the distance of the target from the radar set.

DISTANT-READING COMPASS. A compass in which the indicator is remote from the sensing device, e.g. *gyro flux-gate compass*.

DISTORTION. An unwanted change in waveform. The principal sources of distortion are: a non-linear relation between input and output of a component; non-uniform transmission at different frequencies; and phase shift which is not proportional to frequency. The following types of distortion are described in this dictionary: *amplitude, attenuation, audio-frequency harmonic, barrel, delay, deviation, envelope delay, geometric, harmonic, hysteresis, inter-modulation, keystone, non-linear, peak, phase, pincushion, quantization*.

DISTORTION AND NOISE METER. An instrument which measures *audio-frequency distortion, noise,* and *hum* in an amplitude-modulated or frequency-modulated signal.

DISTORTION FACTOR, AMPLIFIER. Expressed as a percentage, equal to 100 times the square root of the ratio of the sum of the squares of the amplitudes of *harmonics*, to the square of the *fundamental* of the output signal.

DISTORTION FACTOR METER. An instrument used to measure *distortion factor* in amplifiers.

DISTRIBUTED AMPLIFIER. A *wideband amplifier* in which the valves or transistors are distributed uniformly along artificial delay lines and the gain can be increased indefinitely by adding more valves etc.

DISTRIBUTED CAPACITANCE. The inherent capacitance of electrical systems considered as distributed along their length or other dimensions. See *Capacitance, distributed*.

DISTRIBUTED CONSTANTS. Inherent constants, such as *inductance, capacitance,* or *resistance*, distributed throughout a unit in an electrical system.

DISTRIBUTED INDUCTANCE. See *Inductance, distributed*.

DISTRIBUTION CABLE. In communications, a cable extending from a feeder cable into a defined area for which it provides service.

DISTRIBUTION CONTROL. A control which varies the distribution of scanning speeds during the *trace interval*.

DIVERSITY GAIN. Gain in reception as a result of the use of two or more receiving aerials. See *Diversity reception*.

DIVERSITY RECEPTION. A method of radio reception which minimizes the effects of *fading* by combining or selecting two or more received signals from different sources which carry the same intelligence but may differ in strength or *signal-to-noise ratio* at any instant. The sources may differ in frequency, or aerials which are differently located or polarized may be used.

DIVIDING NETWORK. For loudspeakers, a frequency selective network which divides the frequency range to be radiated into two or more parts to the appropriate loudspeakers.

D-LAYER. The lowest layer of the *ionosphere*, at between twenty-five and fifty miles above the earth, which absorbs some of the radio energy reflected by the E and F layers.

DOBA'S NETWORK. A network designed to improve the fidelity in linear *pulse amplifiers* when *pulse rise times* of less than a tenth of a microsecond are involved.

DOMAIN WALL. See *Bloch wall*.

DOMINANT MODE. In *waveguide* transmission, the *mode* with the lowest *cut-off frequency*.

DONOR. In a *semiconductor*, an imperfection which permits electronic conduction by the donation of *electrons*. It gives rise to *n-type conductivity*.

DOORKNOB TRANSFORMER. A device for converting from *coaxial line* transmission to rectilinear *waveguide* transmission.

DOORKNOB VALVE (TUBE). A vacuum valve with very small, closely spaced electrodes, which is shaped like a doorknob. It is designed for ultra-high-frequency transmitters.

DOPED SEMICONDUCTOR. See *Doping*.

DOPING. The addition of small quantities of impurities to a *semiconductor* to achieve a desired characteristic.

DOPING COMPENSATION. The addition of *acceptor impurities* to an *N-type semiconductor* or of *donor impurities* to a *P-type semiconductor* to compensate for the effect of the impurity first present.

DOPPLER EFFECT. The apparent change in the frequency of vibrations, electromagnetic or sound, of a source when that source is moving relative to the observer. The effect is used in radio and radar to obtain information. See *Doppler radar, Radio-Doppler*.

DOPPLER RADAR. A *radar* system which employs the *Doppler effect* to distinguish between fixed and moving targets, and to provide information on the velocity of moving targets by measuring the frequency shift between direct and reflected waves.

DOSIMETER. Any instrument which measures radiation dose.

DOT GENERATOR. In *television*, a test generator which produces evenly spaced bright dots or small squares against a dark ground, or vice-versa, used to adjust the *convergence* of the picture tube.

DOT-SEQUENTIAL COLOUR TELEVISION. A system of *colour television* in which each of the three cameras, red, green, and blue transmits for the duration of one scanning dot or picture element in turn. The primary colour elements are then transmitted in sequence over a single channel and presented on the viewer's screen in sequence. Owing to persistence of vision the viewer sees a complete image in colour.

DOUBLE BRIDGE. See *Kelvin bridge*.

DOUBLE DIODE. A valve consisting of two *diodes* in the same envelope.

DOUBLE IMAGE (GHOST). In a *television* receiver, the effect caused by reception of picture signals via direct and reflected paths, the latter arriving slightly later on the screen.

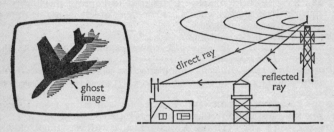

Fig. 39. Double image

DOUBLE LIMITING. In radar, the process of *limiting* both the positive and the negative amplitudes of a wave.

DOUBLE MODING. Of a *magnetron*, abrupt and irregular changes from one frequency to another.

DOUBLE MODULATION. A system of *modulation* in which a signal wave first modulates a *carrier* on one frequency, and the resultant wave then modulates a second carrier of another frequency.

DOUBLE SUPERHETERODYNE RECEPTION. A method of reception in which two *frequency converters* are employed before final detection.

DOUBLE TRIODE. A valve which incorporates two *triodes* in the same envelope.

DOUBLE TUNING. In a *superheterodyne receiver*, undesired reception of the same station at two different settings on the tuning scale.

DOUBLE-BASE JUNCTION TRANSISTOR. A *tetrode transistor* which has two base connections on opposite sides of a central region.

DOUBLE-BEAM C.R.T. A *cathode-ray tube* having two independent electron beams which can produce separate or overlapping traces on the screen.

DOUBLE-COTTON-COVERED WIRE. A conductor, usually copper wire, covered with two layers of cotton insulation.

DOUBLE-LENGTH NUMBER. In computers, a number with twice as many digits as are normally used in a particular computer.

DOUBLE-POLE SWITCH. A switch capable of operating in two separate electric circuits simultaneously, or in both lines of a single circuit.

DOUBLER. See *Frequency doubler, Voltage doubler (rectifier)*.

DOUBLE-SIDEBAND TRANSMISSION. Transmission of both *sidebands* which result from *modulation* of the *carrier* by the modulating signal.

DOUBLE-STREAM AMPLIFIER. A *travelling-wave* amplifier used for *microwaves* in which the interaction of two electron beams at different velocities is used to produce amplification.

DOUBLE-SURFACE TRANSISTOR. A transistor in which the emitter and collector connections to the semiconductor are on opposite sides of the base region.

DOUBLE-THROW SWITCH. One with which connections can be made by closing the switch blade into either of two sets of contacts.

DOUBLE-TUNED CIRCUIT. One in which two circuit elements may be tuned separately and which usually consists of *coupled* circuits of two meshes.

DOUBLE-WOUND COIL. A winding which consists of two parts wound on the same core.

DOUGHNUT. A hollow, evacuated toroid used to accelerate electrons or other particles, e.g. in a *betatron*.

DOWNWARD MODULATION. *Modulation* in which the instantaneous carrier amplitude is always less than the amplitude of the unmodulated carrier.

DP. Abbreviation for Data Processing.

DRIFT. In electronics, the change with time of any function or characteristic of a circuit. Drift is most commonly encountered during the warming up of an instrument or equipment but it also occurs independently of temperature.

DRIFT MOBILITY. In a *semiconductor*, the velocity per unit electric field of the free-charge carriers. See also *Hall mobility*.

DRIFT SPACE. (1) In an electron tube, a region free of external alternating fields in which electrons become re-positioned. (2) The space between the *buncher* and the *catcher* in a *velocity-modulated tube*, e.g. a *klystron*.

DRIFT TRANSISTOR. A *transistor* in which the resistivity of the material of the base increases smoothly between emitter and collector junctions.

DRIFT VELOCITY. The average velocity in the direction of an applied electric field of electrons in a *conductor*, or of ions in a gas. Applied to charge carriers in a *semiconductor*, it is termed *drift mobility*. See *Ion mobility*.

DRIFT-TUBE TUNNEL. In linear *accelerators* and certain *klystrons*, a tunnel or conducting tube through which the charged particles drift while the accelerating rf voltage changes.

DRIVE PATTERN. In *facsimile* transmission, a pattern due to variation of density caused by periodic errors in the position of the recording spot.

DRIVER. An electronic circuit which provides the input for another electronic circuit, especially the amplifier stage preceding the output stage of a transmitter or receiver.

DRIVING-POINT IMPEDANCE. At any pair of terminals of a *network*, the ratio of an applied potential difference to the resultant current at these terminals, with specified terminations at all terminals.

DROP-IN. In *digital computers*, a spurious *bit* due to dust particles or raised areas on the surface of the magnetic medium moving past the reading head.

DROP-OUT. Failure to read a *bit*, for the same reason as with a *drop-in*.

DROPPING RESISTOR (DROPPER). A *resistor* whose function in the circuit is to reduce a given *voltage* by a voltage drop along the resistor.

DRUM CONTROLLER. A controller in which the moving contact parts are arranged on a cylindrical surface.

DRUM MEMORY. See *Magnetic drum*.

DRUM SPEED. In *facsimile* transmission, the speed in revolutions per minute of the transmitter or recorder drum.

DRY CELL. A portable cell in which the electrolyte is in the form of a jelly or in a porous medium so that it is non-spillable. The best-known form is the portable dry-battery used for torches, portable radios, etc.

DRY DISC RECTIFIER. See *Rectifier, dry-disc*.

DRY ELECTROLYTIC CAPACITOR. See *Electrolytic capacitor*.

DRY FLASHOVER VOLTAGE. Of an insulator, the voltage at which the air surrounding a clean dry insulator breaks down completely between electrodes.

DRY JOINT (CONTACT). An imperfect soldered joint through which direct current does not flow easily.

DRY REED. See *Reed relay*.

DTL. Abbreviation for Diode Transistor Logic, descriptive of a type of *integrated circuit logic element*.

DUAL MODULATION. The process of modulating a common *carrier wave* or *subcarrier* by two different types of *modulation* (e.g. a-m and f-m) each conveying separate information.

DUAL-CHANNEL SOUND. In *television* receivers, a technique employing a separate *intermediate-frequency* stage for sound and video signals after the common first detector stage.

DUANE AND HUNT'S LAW. The maximum *photon* energy in an X-ray spectrum is equal to the kinetic energy of the electrons producing the X-rays.

DUBBING. The process of combining two or more sources of sound, at least one of which is a recording, into a composite recording.

DUDDLE ARC (SINGING ARC). A direct current *arc* which produces an audio-frequency current and corresponding sound waves if a suitable tuned circuit is connected across it.

DUMMY LOAD. An element designed to be substituted for the normal load or output circuit of a generator and used to absorb the output power of the generator.

DUMP. In *data processing*: to copy the contents of all or part of a *storage*.

DUODIODE. A *double diode*.

DUPLET. An electron pair which constitutes a bond between two atoms.

DUPLEX OPERATION. In communications, simultaneous operation of correlated transmitting and receiving apparatus at either end of a communication link.

DUPLEX RADIO TRANSMISSION. Simultaneous transmission of two radio signals using a common *carrier* wave.

DUPLEX TELEGRAPHY. *Telegraphy* designed to operate in both directions.

DUPLEXER. In *radar* practice, a device in which a *transmit-receive switch* functions during the finite delay between a transmitted pulse and its return echo to connect the transmitter and receiver to a common aerial system.

DUPLICATION CHECK. In *computers*, verification that the results of two independent performances of the same operation are identical.

DURCHGRIFF. See *Penetration factor*.

DUST CORE. In magnetic devices, a core made of finely crushed magnetic powder, cemented into a compact form.

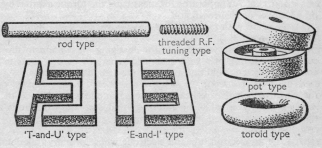

rod type

threaded R.F. tuning type

'pot' type

'T-and-U' type   'E-and-I' type   toroid type

Fig. 40. Types of dust core

DYNAMIC BEHAVIOUR. The behaviour of a control system or an individual component in relation to operational time.

DYNAMIC CHARACTERISTICS. See *Valve characteristics, dynamic*.

DYNAMIC CONVERGENCE. In a *colour television* picture tube, the convergence of the three electron beams at openings away from the centre of the *shadow mask*.

DYNAMIC FOCUSING. In a *colour television* picture tube: the process of varying the voltage on the focusing electrode automatically so that the beam spots remain in focus as they sweep across the flat screen.

DYNAMIC IMPEDANCE. The *impedance* of a parallel tuned circuit at its frequency of resonance.

DYNAMIC RANGE. Of a transmission system, the difference in decibels between the noise level of the system and its overload level.

DYNAMIC REGULATOR. A *transmission regulator* whose adjusting mechanism is in self-equilibrium at only one or a few settings and requires control power to maintain it at other settings.

DYNAMIC RESISTANCE. (1) The resistance of a device, generally a valve, under operating conditions. (2) The *dynamic impedance* when this is purely resistive.

DYNAMIC SENSITIVITY. Of a *phototube*, the ratio of the alternating component of anode current to the alternating component of incident radiant flux.

DYNAMIC TRANSFER CHARACTERISTICS. See *Valve characteristics, dynamic*.

DYNAMO. A *generator* for the production and delivery of direct current.

DYNAMO-ELECTRIC AMPLIFIER. A generator used as a power amplifier at low frequencies. The input signal is used to control the excitation of the stationary field and the output is taken from the rotating armature.

DYNAMOMETER. An apparatus for measuring the torque exerted by a prime mover or electric motor.

DYNAMOMETER WATTMETER. See *Electrodynamometer, Wattmeter, Dynamometer*.

DYNAMOTOR. A rotary converter for changing alternating to direct current, consisting of a motor and dc generator to one shaft in a common magnetic field but with separate armature windings.

DYNATRON OSCILLATOR. A *negative-resistance oscillator*, based on the effective negative resistance between anode and cathode of a *screen-grid tetrode* valve, operating so that secondary electrons at the anode are attracted to the screen-grid.

DYNE. See *Units*.

DYNODE. Of an electron tube, an electrode whose primary function is to provide *secondary emission* of electrons. See *Photomultiplier*.

DYNODE SPOTS. In *image orthicons*, spurious signals may be caused by variations in the secondary-emission ratio over the surface of a *dynode* which is scanned by the electron beam.

E

EAR, ARTIFICIAL. See *Artificial ear*.

EAR MICROPHONE. A contact *microphone* shaped to fit into the ear where it picks up the voice of the wearer. N.B. Its function is to transmit not to receive sounds.

EARLY BIRD (HS 303). The first synchronous communications satellite (see *Syncom*) to carry regular commercial telecommunications traffic. It is about the size of a side-drum and was launched by NASA in June 1965 to hover about 22,000 miles above mid-Atlantic. It is an active 2-way satellite and contains two frequency-changing *transponders*.

EARLY-WARNING RADAR. A *radar* set or system designed to detect the approach of aircraft or missiles at the greatest possible distance.

EARPHONE. A *transducer* for converting electric signals to acoustic vibrations, designed to be closely coupled to the ear. See *Loudspeaker, Telephony*.

EARPHONE COUPLER. A specially shaped cavity, provided with a microphone, for the testing of earphones.

EARTH (GROUND). (1) The conducting mass of the earth or of any conductor in direct electrical connection with it. (2) A connection, whether intentional or accidental, between a conductor and the earth. (3) As a verb, connect any conductor with the earth. (4) At zero potential with respect to the earth.

EARTH CAPACITANCE. See *Ground capacitance*.

EARTH CURRENT. (1) Current flowing in any part of a circuit or equipment connected to earth. (2) Any current flowing in the earth.

EARTH ELECTRODE. A conductor embedded in the earth, used to maintain conductors connected to it at earth potential and for dissipating current conducted to it into the earth.

EARTH PLATE. A plate of conducting material buried in the earth to serve as an *earth electrode*.

EARTH TERMINAL. A terminal provided on the frame of a machine, or on a piece of apparatus or component, for the purpose of making an *earth* connection.

EARTHED (GROUNDED). Connected to *earth* or to some conducting body which serves as earth.

EARTHED CIRCUIT. A circuit of which one or more points is intentionally connected to *earth*.

EARTH-RETURN CIRCUIT. A circuit which has an insulated conductor, or more than one in parallel, between two points and which is completed through the *earth*.

EARTH-WIRE. A *conductor* leading to *earth*, usually situated near the associated line conductors.

E-BEND. In a *waveguide*, a smooth bend in the direction of the axis of the waveguide, throughout which the axis remains in a plane parallel to the direction of polarization.

EBONITE. An insulating material of hard rubber used for panels and coil formers.

ECCLES-JORDAN CIRCUIT. See *Trigger circuit*.

ECG. Abbreviation for *Electrocardiogram*.

ECHO. (1) In radio, a wave returned to the transmitter with sufficient magnitude and delay to be distinguishable from the directly transmitted wave. (2) In radar, the portion of the energy of the transmitted pulse reflected back to the receiver.

ECHO BOX. A device for testing the overall performance of a radar system, comprising an adjustable *cavity resonator* of high *Q* which receives some of the pulse energy from the transmitting system and retransmits it to the receiving system as a slowly decaying transient.

ECHO CHAMBER. A room or enclosure designed to produce artificial echo effects.

ECHO CURRENT. A signal reflected back along a *transmission line* by the encounter of the original signal with a discontinuity in the line.

ECHO FLUTTER. A rapid succession of reflected *radar pulses* resulting from a single initial pulse.

ECHO RANGING. A method of measuring range and direction of submerged, or partly submerged, objects by the time of arrival and direction of the echoes from sonic and supersonic pulses reflected from the objects. See *Sonar*.

ECHO SOUNDER. A device for measuring the depth of water beneath a ship by observing the time between the emission of electrically or mechanically propagated sound waves downwards and the return of their echoes from the sea bottom.

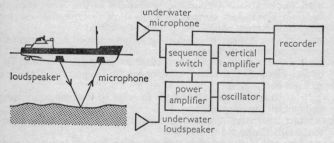

Fig. 41. Principle of echo sounder

ECHO SUPPRESSOR. (1) A voice-operated device for telephone two-way circuits which attenuates echo currents in one direction caused by telephone currents in the other direction. (2) In navigation equipment, a circuit which desensitizes the equipment for a short period after reception of a pulse so that delayed pulses arriving by indirect reflection paths are rejected.

ECM. Abbreviation for Electronic Counter-Measures.

EDDY CURRENTS. Local currents induced in the material of a conducting body by a varying or relatively-moving magnetic field.

EDDY-CURRENT LOSS. The loss of energy which results from the flow of *eddy currents* and usually appears as heat. Where the variation of the *magnetic flux* becomes rapid, as in alternating currents, the eddy current loss becomes more important and at radio-frequencies it becomes the major factor in the design of magnetic cores. See *Dust core, Powdered-iron core*.

EDGE EFFECT. In an electric field created by a potential difference between parallel plates, the non-uniformity at the edges of the plates caused by the bulging out of the lines of *electric flux*.

EDISON BATTERY. A storage battery using nickel peroxide and iron plates in an electrolyte of potassium hydroxide.

EDISON EFFECT. The phenomenon, first discovered by Edison, of electrical conduction between an incandescent filament and an independent cold electrode in the same envelope, when the cold electrode is made positive with respect to the filament. See *Thermionic emission*.

EDIT. Arrange or rearrange (usually for printing) the information from the output unit of a *digital computer*.

EDP. Abbreviation for *Electronic data processing*.

EEG. Abbreviation for *Electroencephalograph*.

EFFECTIVE CAPACITANCE. See *Capacitance, effective*.

EFFECTIVE MASS. A loosely defined concept in the band theory of solids which takes account of the fact that electrons near the bottom of a *band*, and *holes* near the top, often behave like particles with masses different from the mass of a *free electron*.

EFFECTIVE RESISTANCE. The total *resistance* offered to the flow of alternating current: including the dc or ohmic resistance, and resistance due to *eddy current, hysteresis, dielectric*, and *corona* losses. Also called ac resistance, high-frequency resistance, and rf resistance.

EFFECTIVE VALUE. Of recurring variable quantities such as amperes, volts, resistance, etc., in periodic currents. See *Root-mean-square value*.

EFFICIENCY DIODE (BOOSTER DIODE). In a television receiver, a *diode rectifier* connected so as to increase the beam deflection voltage of the cathode-ray tube during the forward period of operation.

EFFLUVE. The *corona* discharge from a high-frequency apparatus, or an electrostatic generator, used to stimulate the human skin.

EHF. Abbreviation for *Extremely High Frequency*.

E.H.T. Extra High Tension: usually referring to the high voltage supply for a cathode-ray or television tube.

E.I.A. STANDARD CODE. A standard code for machine-tool *numerical control* systems proposed by the U.S. Electronic Industries Association, using 8-track paper tape.

EIDOPHOR SYSTEM. A projection *television* system which employs diffraction effects to produce the modulation of light for the picture.

EINTHOVEN GALVANOMETER. See *String electrometer*.

ELASTANCE. In *electrostatics*, the reciprocal of *capacitance*. Symbol: *S*, Unit: *daraf*.

ELASTIVITY. Of a *dielectric*, the reciprocal of *permittivity*.

ELECTRET. The electrical analogue of a permanent magnet: a *dielectric* body with separate electric *poles* of opposite sign and a stable existence.

ELECTRIC(AL). Containing, producing, carrying, arising from or actuated by *electricity*. Related to, pertaining to, or associated with electricity. N.B. 'Electric' and 'electrical' are used interchangeably in this dictionary.

ELECTRIC ATTRACTION. See *Attraction, electrical*.

ELECTRIC AXIS. The X-axis of a *piezoelectric crystal*.

ELECTRIC BRAZING. Brazing processes in which the heat is obtained from an electric current.

ELECTRIC CELL. A device in which chemical energy and electrical energy are interchanged. See *Battery, Cell*.

ELECTRICAL CHARGE. The quantity of electricity contained in or on a body. A positive charge indicates that *positive electricity* is in excess, and a negative charge that *negative electricity* is in excess. Symbol: *Q*. See *Atomic charge, Bound charge, Electron charge, Free charge, Space charge, Units*.

ELECTRIC CONDUCTION. The process of transmission of electricity in matter. In metals it consists substantially of the migration of *electrons*; in gases and liquids, of the migration of ionized atoms or molecules.

ELECTRIC CONDUCTOR. See *Conducting material, Conductor*.

ELECTRIC CONTROLLER. In *automatic control* systems, a control element actuated by electric power.

ELECTRIC CURRENT. See *Current*.

ELECTRIC DEGREE. One 360th part of an alternating-current cycle.

ELECTRIC DIPOLE. Two equal and opposite charges very close together.

ELECTRIC DISCHARGE. See *Discharge*.

ELECTRIC DISPLACEMENT. The *electric flux* per unit area normal to the direction of the flux.

ELECTRIC DOUBLET. A system with a finite *electric moment*, mathematically equivalent to two infinite charges of opposite sign, an infinitesimal distance apart.

ELECTRIC EYE. Colloquial term for *photoelectric* or *photovoltaic cell*.

ELECTRIC FIELD. The space around a charged body, or a varying magnetic field, in which an electric charge would experience a mechanical force.

ELECTRIC FIELD STRENGTH (FORCE, INTENSITY). At a point, a vector whose magnitude and direction are equal to the mechanical force per unit charge on a very small charged body at the point.

ELECTRIC FLUX. The quantity of electricity displaced across a given area in a *dielectric*. The total flux displaced across a surface enclosing a charge is equal to the charge.

ELECTRIC FLUX DENSITY. See *Electric displacement*.

ELECTRIC FORMING, SEMICONDUCTOR. Application of electrical energy to a *semiconductor device* to modify its electrical characteristics permanently.

ELECTRIC HEATING. The transformation of electrical into heat energy by the flow of current through a *resistance*; through an *arc*; by *induction*; or by *dielectric heating*.

ELECTRIC HORN. A horn with a diaphragm which is electrically vibrated.

ELECTRIC INTERFERENCE. *Interference* caused by the operation of electrical apparatus other than radio stations.

ELECTRIC LOAD. See *Load*.

ELECTRIC MOMENT. See *Dipole moment*.

ELECTRIC POLARIZATION. See *Dielectric polarization*.

ELECTRIC POTENTIAL. Of a point, the *potential difference* between the point and some equipotential surface, usually that of the earth which is chosen to have zero potential.

ELECTRIC SCREEN. A conducting screen which reduces the penetration of an *electric field* into an assigned region.

ELECTRIC SPECTRUM. The colour spectrum of an electric *arc*.

ELECTRIC STRENGTH. See *Dielectric strength*.

ELECTRIC TRANSDUCER. A *transducer* in which all the waves concerned are due to electric fields.

ELECTRICITY. A general term for the phenomena associated with *electrons* and *protons* at rest or in motion. At rest these particles are characterized by an *electric field* and in motion by an electric and a *magnetic field*.

ELECTROBALLISTICS. The measurement of the velocity of projectiles by electronic methods.

ELECTROBIOLOGY. The electrical phenomena associated with living organisms.

ELECTROCAPILLARITY. Capillary effects in liquids resulting from the passage of electric currents or from electric charges.

ELECTROCARDIOGRAM. The record produced by an *electrocardiograph*.

ELECTROCARDIOGRAPH. An instrument for measuring and recording the current and voltage waveforms associated with the action of the heart. See Figure 42.

ELECTROCATAPHORESIS (ELECTROPHORESIS)

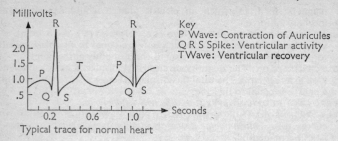

Fig. 42. Electrocardiogram

ELECTROCATAPHORESIS (ELECTROPHORESIS). The migration of microscopic particles under the influence of a suitable electric field; the migration of *ions* through a membrane under the influence of an electric current.

ELECTROCHEMICAL EQUIVALENT. The mass of a substance undergoing electrochemical change due to the passage of one *coulomb*.

ELECTROCHEMICAL SERIES. A classification of the elements in order of the electrode potential developed when an element is immersed in a solution of normal ionic concentration.

ELECTROCHEMICAL VALVE. See *Electrolytic rectifier*.

ELECTROCHEMISTRY. The branch of science and technology which deals with the interchange between chemical and electric energy.

ELECTROCHRONOGRAPH. An electrically driven clock with an electromagnetic recorder for recording short intervals of time.

ELECTRODE. (1) A conductor, which may be non-metallic, by means of which a current passes into or out of a liquid, a gas, or an insulating material. (2) Of a *semiconductor device*, an element which performs one or more of the functions of emitting or collecting electrons or holes, or controlling their movements by an electric field. (3) In an *electrolytic cell*, a metallic conductor at which there is a change from conduction by electrons to conduction by ions or other charged particles.

ELECTRODE CHARACTERISTIC. In a valve, a relation between the electrode voltage and current to the electrode, the voltages on all other electrodes being maintained constant.

ELECTRODE CURRENT. Of valves or tubes, the net current from an electrode into the inter-electrode space.

ELECTRODE DARK CURRENT. Of a phototube or camera tube, the electrode current which flows when there is no radiation incident on the photocathode, under specified conditions of radiation shielding and temperature.

ELECTRODE DISSIPATION. The power dissipated in the form of heat

118

by an electrode as a result of electron or ion bombardment, and radiation from other electrodes.

ELECTRODE DROP. The voltage drop in the electrode due to its resistance.

ELECTRODE INVERSE CURRENT. Current flowing through an electrode in the direction opposite to that for which the valve was designed.

ELECTRODE VOLTAGE. The voltage between an electrode and the cathode or a specified point of the cathode if it is not at uniform potential. The terms *anode voltage, grid voltage*, etc., designate the voltages between these specific electrodes and the cathode.

ELECTRODELESS DISCHARGE. An internal luminous discharge which can appear in a gas-filled tube when it is placed in a high-frequency electric field.

ELECTRODEPOSITION. The depositing of a substance upon an electrode by *electrolysis*. This includes *electroforming* and *electroplating*.

ELECTRODERMAL REACTION. The change in the electrical resistance of the skin of a live animal which may accompany emotional stress.

ELECTRODYNAMIC. Referring to *electricity* in motion.

ELECTRODYNAMIC INSTRUMENT. An instrument whose operation is based on the reaction between current in one or more movable coils and current in one or more fixed coils.

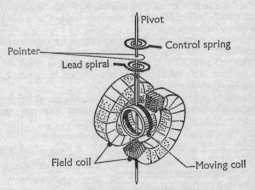

Fig. 43. Electrodynamic movement

ELECTRODYNAMIC MACHINE. An electric *generator* or *motor* having a rotating armature provided with windings and a *commutator*, whose output voltage is produced primarily by the effects of currents in the armature winding.

ELECTRODYNAMICS. The phenomena, science, and applications of moving electric charges, i.e. of *electricity* in motion. Cf. *Electrostatics.*

ELECTRODYNAMOMETER. An instrument which uses the interaction of the magnetic fields of fixed and movable sets of coils to produce a square-law deflection. It is frequently used as a laboratory standard for current, voltage, or power measurements for both dc and ac. When used as a *wattmeter* one coil is connected in series with the load and the other across the load.

ELECTROENCEPHALOGRAPH. An instrument used in the recording of *brain voltages,* consisting of a sensitive voltage or current detector, a very stable *direct-current amplifier,* and an electronic recording system.

ELECTROFLUOR. Transparent material which has the ability to store electrical energy and later release it as *fluorescent* (visible) *light.*

ELECTROFLUORESCENCE. The production of *fluorescence* in a solid by conversion of electrical power independently of irradiation or other excitation. Cf. *Electroluminescence.*

ELECTROFORMING. The production or reproduction of articles by *electrodeposition.*

ELECTROGEN. A molecule which emits *electrons* when illuminated.

ELECTROGRAPH. Any graph or tracing produced electrically. In particular, equipment for this purpose used in *facsimile* transmission.

ELECTROHYDRAULIC. Relating to devices which use both electrical and fluid power, usually for control purposes.

ELECTROKINETICS. A branch of *electrodynamics* concerned with the heating effects of conduction currents and the distribution of currents in a network.

ELECTROKYMOGRAPH. An electronic recording instrument used in *electrobiology.*

ELECTROLUMINESCENCE. *Luminescence* resulting from the application of an electric potential to certain solids, e.g. a dielectric *phosphor,* as in the *electroluminescent lamp.*

ELECTROLUMINESCENT LAMP. A luminescent panel lamp consisting of a phosphorescent powder sandwiched between transparent

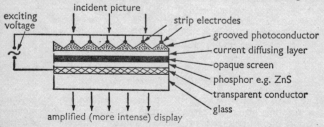

Fig. 44. Section of an electroluminescent display panel used to increase the intensity of a picture

conducting sheets which produces light when excited by an alternating voltage.

ELECTROLYSIS. The production of chemical changes by the passage of current through an *electrolyte*.

ELECTROLYTE. A substance which produces a conducting medium when dissolved in a suitable solvent (usually water).

ELECTROLYTIC CAPACITOR. A *fixed capacitor* in which the insulating material between the electrodes consists of a thin film of metal oxide formed on one or both plates by electrochemical action in an electrolyte. The electrolyte is contained between the electrodes and also serves to make contact with the outer surface of the oxide film. This describes the 'wet' type which is generally used. In the 'dry' electrolytic, the electrolyte is contained in saturated paper and a thin film of gas from a moist paste forms the dielectric. In recently developed types, tantalum sheets are used as electrodes, immersed in an electrolyte of sulphuric acid. Electrolytic capacitors provide a high capacity in a limited space.

symbol

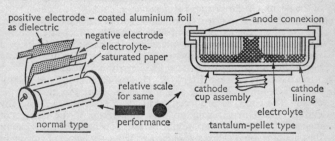

Fig. 45. Construction of two types of electrolytic capacitor

ELECTROLYTIC CATHODE. An electrode through which current leaves a conductor of the non-metallic class, and at which positive ions are discharged and negative ions formed.

ELECTROLYTIC CELL. A system of electrodes and one or more electrolytes designed to carry out an electrochemical reaction.

ELECTROLYTIC DETECTOR. See *Electrolytic rectifier*.

ELECTROLYTIC DISSOCIATION. The reversible resolution of certain substances into oppositely charged *ions*.

ELECTROLYTIC POLISHING. The production of a smooth, lustrous surface on a metal by making it the anode in an electrolytic solution and preferentially dissolving the protuberances.

ELECTROLYTIC RECTIFIER. A *rectifier* which employs electrolytic action to convert alternating to direct current.

ELECTROLYTIC TANK. A tank filled with a poorly conducting fluid in which scale models of a system of electrodes – e.g. the electrodes of a thermionic valve – may be immersed. By applying appropriate voltages to the electrodes and tracing out equipotential lines in the fluid with a probe the behaviour of the model can be used to predict the performance of the original system.

ELECTROMAGNET. An apparatus embodying a *ferromagnetic* core which is strongly magnetized only when an electric current passes through a winding surrounding the core.

ELECTROMAGNETIC BRAKING. A system in which a braking action is applied to or removed from a motor or moving system by means of an *electromagnet*.

ELECTROMAGNETIC CRACK DETECTOR. See *Crack detector, electromagnetic*.

ELECTROMAGNETIC DEFLECTION. One of the two methods of controlling the beam in a *cathode-ray tube*, in which currents with a saw-tooth waveform are applied to pairs of electromagnets on the outside of the glass envelope of the tube. One pair of electromagnets provides vertical displacement and the other horizontal displacement of the beam. Cf. *Electrostatic deflection*.

ELECTROMAGNETIC FOCUSING. Of a cathode-ray electron beam, the use of magnetic coils to provide a magnetic focusing field for the beam and counteract the spreading of the beam due to its own space charge. The technique is used extensively in the *electron microscope* and *cathode-ray tube*. Cf. *Electrostatic focusing*.

ELECTROMAGNETIC INDUCTION. The production of an *electromotive force* in a circuit by a change of *magnetic flux* through the circuit. The e.m.f. so produced is called an *induced e.m.f.* and any current which may result, an *induced current*. If the change of magnetic flux is due to variation in current flowing in the same circuit the phenomenon is called *self-induction*; if due to change of current in a different circuit it is called *mutual induction*.

ELECTROMAGNETIC LENS. An *electron lens* which employs *electromagnetic focusing technique*.

ELECTROMAGNETIC LOUDSPEAKER. See *Magnetic armature loudspeaker*.

ELECTROMAGNETIC MIRROR. A surface from which electromagnetic radiation is reflected, e.g. a radar *dish*.

ELECTROMAGNETIC SCREEN. A screen of conducting and possibly magnetically permeable material which reduces the penetration of a magnetic and/or electric field in an assigned region.

ELECTROMAGNETIC SPECTRUM. See chart for *electromagnetic waves* See Figure 46.

ELECTROMAGNETIC SYSTEM OF UNITS (E.M.U.). A system of absolute electrical units, based on the centimetre-gram-second (c.g.s.) system, whose primary electrical unit is the unit *magnetic pole*. It involves the choice of the *permeability* of free space as the fourth fundamental unit. See also MKSA, *Units*.

ELECTROMAGNETIC WAVES. Waves of associated *electric* and *magnetic fields,* in which each field is always at right angles to the other and to the direction of propagation. See Fig. 46 for the classification of e.w. See also *Maxwell's equations.*

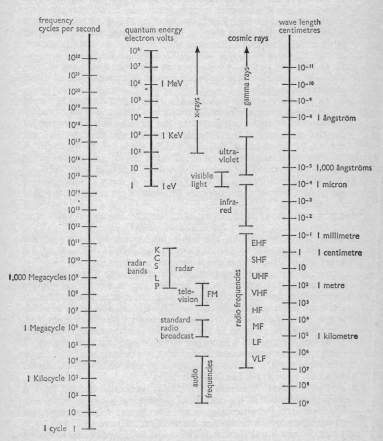

Fig. 46. Frequency spectrum of electromagnetic waves

ELECTROMECHANICAL COUNTER. A device for counting the pulses from an electronic circuit, consisting of a set (usually four) of geared and numbered discs (registers) actuated by a *solenoid.*

ELECTROMECHANICAL RECORDER. Any device which transforms electric signals into mechanical motion of similar form and cuts or embosses such motion in a suitable medium.

ELECTROMETER. A type of *voltmeter* in which *potential difference* or *electric charge* is measured by the mechanical forces between electrically charged bodies. The instrument is usually provided with a calibrated scale for direct reading. Cf. *Electroscope.* See the following types of electrometer: *Compton, Hoffman, Lindemann, Quadrant, String, Vibrating reed, Wulf-string, Electrostatic voltmeter.*

ELECTROMETER VALVE. A high-vacuum valve with a very low *control-electrode conductance* for the measurement of extremely small direct currents or voltages.

ELECTROMOTIVE FORCE (E.M.F.). The property of a physical source which causes a movement of electricity in a circuit. It is measured by the amount of energy generated by the transfer of unit quantity of positive electricity in the direction of the current flow. In a circuit containing a number of different e.m.f.s the e.m.f. of the circuit is their algebraic sum, and this is equal to the line integral of the *electric field strength* round the circuit. Symbol: E, Practical unit: Volt.

ELECTROMOTIVE FORCE, COUNTER (BACK E.M.F.). The effective *e.m.f.* in a circuit which opposes the flow of current in a specified direction.

ELECTROMOTIVE FORCE, INDUCED. See *Electromagnetic induction.*

ELECTROMOTIVE SERIES. See *Electrochemical series.*

ELECTROMYOGRAPH. An electronic instrument for recording the currents generated by contractions in muscles. In the case of the heart muscles the instrument is an *electrocardiograph.*

ELECTRON. An elementary particle, assumed to be a constituent of every atom, which contains the smallest known electric *charge* (see *Physical constants*). The charge may be positive or negative, although the term 'electron' is commonly used for the negative particle, which is also called the *negatron.* The positive electron is called the *positron.* The mass of an electron at low velocities is approximately equal to 1/1837 of the mass of a hydrogen atom. In addition to its mass and charge, the electron has associated with it a definite amount of *spin.* The word is often used as an adjective synonymous with *electronic.*

ELECTRON AFFINITY. See *Work function.*

ELECTRON BEAM. A stream of *electrons,* generally emitted from a single source such as the *cathode* of a *cathode-ray tube,* moving with approximately the same velocity and direction in neighbouring paths so that they form a beam. See *Electron gun, Klystron.*

ELECTRON BEAM RECORDING. The use of an *electron beam* to record data from a computer directly on to microfilm.

ELECTRON(IC) CHARGE. The charge of a single electron; the fundamental unit of electrical charge. See *Physical constants*.

ELECTRON CHARGE/MASS RATIO. The ratio of the *electronic charge* to the *rest mass* of the electron, which forms a fundamental *physical constant*.

ELECTRON COUPLING. The technique, as in an *electron-coupled oscillator*, of coupling two circuits with a single multi-grid tube in which electron currents between electrodes form the coupling link.

ELECTRON CURRENT. *Current* due to a flow of electrons.

ELECTRON DENSITY, EQUIVALENT. In an *ionized gas,* the product of ion density and the ratio of the mass of an electron to an ion of the gas.

ELECTRON DEVICE. One in which *conduction* is principally by *electrons* moving through a vacuum, gas, or *semiconductor*.

ELECTRON DIFFRACTION. The behaviour of electrons, analogous to the *diffraction* of visible light, when they are directed on to the faces of crystals or thin metal foils. The resulting *diffraction patterns* are used to investigate crystal structure.

ELECTRON DRIFT. In a *conductor,* the actual transfer of *electrons* as opposed to the transfer of energy collisions between succeeding electrons.

ELECTRON EMISSION. The liberation of *electrons* from a surface into the surrounding space. Quantitatively, the rate at which electrons are emitted from an electrode. See *Emission*.

ELECTRON GAS. The aggregate of *free electrons* moving in a vacuum, gas, or semiconductor.

ELECTRON GUN. An electrode structure which produces and may deflect, focus, and control the position and intensity of an *electron beam*. An essential part of many electron devices, e.g. *cathode-ray tube, electron microscope, iconoscope*.

ELECTRON LENS. An electromagnetic device consisting of magnetic coils and electrodes used to reflect and focus *electron beams* in a

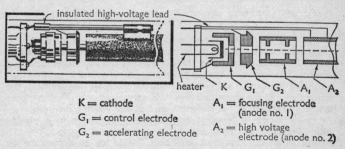

K = cathode
G_1 = control electrode
G_2 = accelerating electrode

A_1 = focusing electrode (anode no. 1)
A_2 = high voltage electrode (anode no. 2)

Fig. 47. Electron gun of a typical T.V. picture tube (pictorial and diagrammatic)

125

manner closely resembling the action of an optical lens on beams of light. See *Electron gun, Electron microscope.*

ELECTRON MAGNETIC MOMENT. See *Physical constants.*

ELECTRON MASS. See *Physical constants.*

ELECTRON MICROSCOPE. An instrument designed for the detailed examination of electron-emitting surfaces in order to study the emission of electrons, and for the investigation of extremely small structures for which the magnification and resolving power of a light microscope is inadequate. The essential parts are an electron

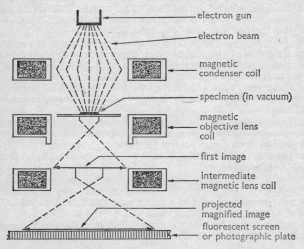

electron gun

electron beam

magnetic condenser coil

specimen (in vacuum)

magnetic objective lens coil

first image

intermediate magnetic lens coil

projected magnified image

fluorescent screen or photographic plate

Fig. 48. Principle of magnetic electron microscope (in section)

source (see *Electron gun*), an electron lens system, which may be electromagnetic or electrostatic, analogous to an optical lens system, and a fluorescent screen or stage on which the magnified image is projected. A television type of scanning technique is also used so that amplification can be applied to a modulated signal derived from the image; and, in the emission types of electron microscopes, the object is also the source of the electrons used as the viewing medium.

ELECTRON MIRROR. One of the reflecting electrodes in a photo-multiplier tube. See *Dynode.*

ELECTRON MULTIPLIER. A structure, inside an electron tube or valve, which employs *secondary emission* from solids to produce current amplification.

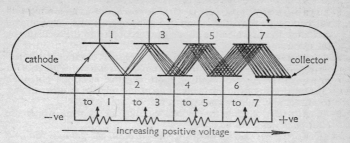

Fig. 49. Principle of secondary-emission electron multiplier. 1 to 7 are secondary electron emitters, also called dynodes

ELECTRON MULTIPLIER PHOTOTUBE. See *Photomultiplier*.

ELECTRON OPTICS. The science and application of the physical and optical properties of *electron beams* under the influence of magnetic or electric fields but not materially influenced by the presence of atoms.

ELECTRON PARAMAGNETIC RESONANCE (EPR). *Paramagnetic resonance* in which the resonance effect is due to *conduction electrons* in metals or *semiconductors*.

ELECTRON SPIN. The rotation of an *electron* about its axis, which is supposed to have a constant angular momentum.

ELECTRON SPIN RESONANCE (ESR). See *Electron paramagnetic resonance*.

ELECTRON TELESCOPE. A telescope in which the infra-red image of a distant object is focused on the photo-sensitive cathode of an *image converter tube* and the resultant electron image enlarged by *electron lenses* and displayed on a fluorescent screen.

ELECTRON (IC) TUBE (VALVE). A device in which the movement of *electrons* through a gas or a vacuum, between two electrodes, takes place in a sealed (or continuously exhausted) envelope. See *Valve*.

ELECTRON-BEAM DC RESISTANCE. The quotient of the *electron-beam voltage* and the direct current output of the electron beam.

ELECTRON-BEAM INSTRUMENT. An instrument which is based on the deflection of a beam of electrons by an electric and/or magnetic field, causing it to strike a fluorescent screen. See *Cathode-ray tube, Electron microscope*.

ELECTRON-BEAM TUBE (VALVE). An electronic tube or valve in which several electrodes form, control, and direct one or more *electron beams*. See *Cathode-ray tube, Cyclophon, Klystron*.

ELECTRON-BEAM VOLTAGE. At any point in an electron beam: the time average of the voltage between that point and the electron-emitting surface.

ELECTRON-COUPLED OSCILLATOR. An *oscillator* using a multi-grid valve in which the cathode and two grids operate as a conventional oscillator and in which the anode load is coupled to the oscillator by the electron stream.

ELECTRONEGATIVE. (1) Charged with *negative* electricity. (2) Referring to an element above hydrogen in the *electrochemical series,* or in a relative sense to signify a more negative electrode potential. (3) Descriptive of the tendency of an atom, relative to other atoms, to attract and hold valence electrons.

ELECTRONIC. Relating to devices, circuits, or systems which use *electron devices.*

ELECTRONIC CHARGE, SPECIFIC. See *Electron charge/mass ratio.*

ELECTRONIC COMMUTATOR. (1) An electronic switching circuit. (2) A radial-beam electronic tube. See *Cyclophon.*

ELECTRONIC COMPUTER. See *Computer.*

ELECTRONIC CONTROL. Application of electronic techniques to the control of machines, power, data and material processing, projectiles and satellites, moving vehicles, etc. See also *Automation, Computer, Cybernetics.*

ELECTRONIC DATA PROCESSING. *Data processing* performed largely by electronic means.

ELECTRONIC EFFICIENCY. In an *oscillator* or *amplifier,* the ratio of the power at the desired frequency delivered by the electron stream to the circuit, to the average power supplied to the stream.

ELECTRONIC SWITCH. An electronic device used as a switch, usually a high-speed switch.

ELECTRONIC TUNING. The process of changing the operating frequency of a system by changing the characteristics, e.g. the velocity, density, or geometry, of a coupled electron beam.

ELECTRONICS. That branch of science and technology which is concerned with the study of the phenomena of conduction of electricity in a vacuum, in a gas, and in semiconductors; and with the application of devices using those phenomena.

ELECTRON-INDICATOR TUBE. See *Electron-ray tube.*

ELECTRON-POSITRON PAIR. See *Pair production.*

ELECTRON-RAY TUBE. A device which gives a visual indication of correct tuning in radio receivers and is also used as an indicator for voltmeters and galvanometers. It consists of a triode and a simplified *cathode-ray tube,* in which a metal fin used as a control electrode causes a fluorescent pattern to open or close.

ELECTRON-STREAM POTENTIAL. At any point in an *electron beam*: the time average of the potential difference between that point and the electron-emitting surface.

ELECTRON-VOLT. The energy gained by an *electron* in passing through a *potential difference* of one *volt.* Symbol: *ev.* 1 *mev*=one million electron volts. See *Physical constants.*

ELECTROSTATIC LENS

ELECTRON-WAVE TUBE (VALVE). See *Travelling wave tube.*

ELECTRO-OPTICAL EFFECT. The interaction between the refractive properties of some transparent dielectrics and a strong electric field in which they are placed. See *Kerr effect, Pockels effect.*

ELECTROPATHOLOGY. The study of human diseases by electrical means.

ELECTROPATHY. See *Electrotherapy.*

ELECTROPHONIC EFFECT. The aural sensation produced when alternating current of certain frequencies and magnitudes is passed through the human body.

ELECTROPNEUMATIC. Relating to devices which are both electrical and air power, usually for control purposes.

ELECTROPOLAR. Having magnet poles or positive and negative electric charges.

ELECTROPOLISHING. See *Electrolytic polishing.*

ELECTROPOSITIVE. In general, referring to an element below (more positive than) hydrogen in the *electrochemical series.* In a relative sense, a more positive electrode potential.

ELECTROSCOPE. An electrostatic device for indicating a potential difference or electric charge by the repulsion of two similarly charged bodies. The essential part of an *electrometer.* See *Gold-leaf electroscope, Lauritsen electroscope.*

ELECTROSTATIC. Referring to *electric charges* at rest.

ELECTROSTATIC ACCELERATOR. An *accelerator* which depends on an *electrostatic field* due to a high direct voltage.

ELECTROSTATIC CHARGE. A quantity of electricity at rest on the surface of an *insulator* or an insulated conductor. See *Coulomb's law, Electrical charge, Units.*

ELECTROSTATIC DEFLECTION. Control of the beam in a *cathode-ray tube* by the electric fields produced between two pairs of metal electrodes. See Fig. 18.

ELECTROSTATIC FIELD. A region in which a stationary charged particle would experience a force due to the distribution of *electrostatic charge.*

ELECTROSTATIC FOCUSING. Of an *electron beam,* a method of focusing the beam, usually with a focusing cylinder and an anode cylinder which direct a convergent field on the beam. Cf. *Electromagnetic focusing,* and see *Cathode-ray tube, Electron microscope.*

ELECTROSTATIC GENERATOR. A generator which depends on *electrostatic induction.*

ELECTROSTATIC INDUCTION. The process by which a body becomes charged when it is near a charged body but not touching it.

ELECTROSTATIC INSTRUMENT. An instrument, e.g. an *electrometer,* which depends on the forces of attraction or repulsion between charged bodies.

ELECTROSTATIC LENS. An arrangement of *electrodes* to produce a focusing effect on a beam of charged particles. See *Electrostatic focusing*.

ELECTROSTATIC LOUDSPEAKER. A loudspeaker in which the mechanical forces are produced by the action of *electrostatic fields*.

ELECTROSTATIC MACHINE. A device in which *electricity* is generated by friction or by spraying.

ELECTROSTATIC MEMORY (STORAGE). A device in which information is stored in the form of *electrostatic charge*. This usually takes the form of a special *cathode-ray tube* and associated circuits.

ELECTROSTATIC MICROPHONE. A *microphone* whose operation is based on variations of its *electrostatic capacitance*.

ELECTROSTATIC PRECIPITATION. The precipitation, by means of a unidirectional *electrostatic field* between electrodes, of solid or liquid particles held in suspension in a gas. The precipitated particles collect on the earthed electrode, which is usually the positive.

ELECTROSTATIC PRINTER. A printing device in which a pattern of electrostatic charges on the paper attracts a fine powder which is then fixed by heat. See *Xerography*.

ELECTROSTATIC SCREEN (SHIELD). A conducting screen placed around or between components which prevents or substantially reduces *electrostatic* coupling between components on opposite sides of the screen. See *Faraday cage*.

ELECTROSTATIC SEPARATOR. A device for separating fine powders of different *permittivities* by the different deflections which they undergo between highly charged electrodes.

ELECTROSTATIC SYSTEM OF UNITS. The system of absolute electrical units, based on the c.g.s. system, whose primary electrical unit is the unit quantity of *charge*. It involves the choice of the *permittivity* of free space as a fourth fundamental unit. Cf. *Electromagnetic system of units*. See *Units*.

ELECTROSTATIC VOLTMETER. An instrument for measuring high values of direct voltage by the electrostatic force which they produce between fixed and moving vanes.

ELECTROSTATICS. The science dealing with *electric charges* at rest.

ELECTROSTRICTION. See *Crystal electrostriction*.

ELECTROTHERAPY. The technique of treating disease by means of electricity.

ELECTROTHERMIC INSTRUMENT. An instrument whose operation depends on the heating effect of a current. See *Bolometer, Hot-wire instrument, Thermocouple*.

ELECTROTHERMICS. The science and technology of the direct transformation of electric energy and heat.

ELECTROTYPING. The production or reproduction of printing plates by *electroforming*.

ELECTROVALENCE. See *Valency*.

ELEMENT. A substance consisting entirely of atoms of the same *atomic number*. Just over 100 different elements have been identified.

E.M.F. Abbreviation for *electromotive force*.

EMISSION. The liberation of electrons or electromagnetic waves from a surface, due to heat, radiation, electrostatic field, bombardment, nuclear reaction, or other cause. See the following types of emission: *Counterfield, Electron, Field, Grid, Photoelectric, Schottky, Secondary, Temperature-limited, Thermionic*.

EMISSION CHARACTERISTIC. A relation, usually as a graph, between the emission and some factor, e.g. temperature, controlling the emission.

EMISSION EFFICIENCY. Of a *cathode,* the ratio of the electron emission in millamperes/sq. cm. to the cathode power dissipation density in watts/sq. cm.

EMISSIVE POWER, TOTAL. (1) The total energy emitted from unit area of the surface of a body at a specified temperature and in specified surroundings. (2) The rate at which the surface of a solid or liquid emits electrons when additional energy is given to the free electrons in the material by the action of heat or radiant energy, or by the impact of other electrons on the surface.

EMISSIVITY. See *Emissive power*.

EMITRON®. A television *camera tube,* almost identical in design with the *iconoscope,* which was one of the first to be used in high-definition television transmission.

EMITRON, CATHODE POTENTIAL STABILIZED (C.P.S. EMITRON). An improved form of *image orthicon* television *camera tube*, in which the optical image is projected on to a transparent photosensitive *mosaic*. A low-velocity electron beam scans the mosiac to stabilize it with the cathode potential and so prevent secondary emission.

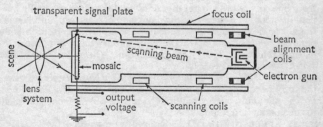

Fig. 50. Principle of C.P.S. Emitron television camera

EMITTER. (1) One of the three electrodes of a *transistor*. See also *Common-emitter connection, Majority emitter, Minority emitter*. (2) The part of a *cathode* which emits electrons.

EMITTER FOLLOWER. A circuit using a transistor in the *common-*

collector connection, having high input impedance and low output impedance.

EMITTER JUNCTION. Of a semiconductor device, usually a transistor: a junction through which the current carriers flow from a region in which they are *majority carriers* to one in which they are *minority carriers.*

EMMA. An electronic mask-making apparatus manufactured by Quest Automation Ltd. It converts engineers' rough drawings, which are coded by semi-skilled staff, directly into high-precision art work for printed-circuit boards, thick- and thin-film circuits, and *integrated circuits.*

EMPHASIZER. In *audio* engineering, a circuit which selects and amplifies particular frequencies or bands of frequency.

EMPIRE CLOTH. An insulating fabric made of cotton or linen cloth impregnated with linseed oil.

ENABLING PULSE. A *pulse* which opens an electrical *gate* which is normally closed.

ENCEPHALOGRAPH. See *Electroencephalograph.*

ENCODER. A device which applies a code or changes information into coded form. The commonest form is a *rotary encoder.*

END HATS. Metal structures fitted to the ends of the cathode in some *magnetrons* to minimize loss of efficiency due to *space-charge* effects.

END-FIRE AERIAL ARRAY. A linear array in which the direction of maximum radiation is along the axis of the array.

ENDORADIOSONDE. A miniature radio transmitter which may be swallowed or inserted into any of the body cavities for biological measurements and medical diagnosis.

ENERGY. The capacity for doing work. See *Erg, Units.*

ENERGY BAND. It is believed that each electron in the atoms of a solid can have a number of definite amounts of energy, and that change from one energy level to another does not take place gradually but in definite steps. The energy of the aggregate of electrons, and hence of the atoms of the solid of which they are a part, therefore falls into separate *bands* of energy with gaps, or *forbidden bands,* between. See *Conduction band, Energy level diagram, Forbidden band, Hole, Valence band,* and Fig. 51.

ENERGY GAP. Of a *semiconductor,* the *forbidden band* of energy between the top of the *valence band* and the bottom of the *conduction band*: i.e. the minimum energy required to raise a *carrier* from the valence to the conduction band.

ENERGY LEVEL DIAGRAM. A diagram representing, by a series of horizontal lines, the individual energy levels corresponding to the possible excited states of the atoms of an element. See Figure 51.

ENIAC. Electronic Numerical Integrator and Computer, a general purpose *digital computer* built by the University of Pennsylvania. It was the first large-scale digital computer to use electronic techniques and employs about 18,000 valves.

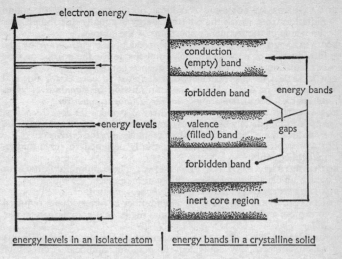

Fig. 51. Energy level diagram and energy bands

ENVELOPE DELAY. The time of propagation between two fixed points of the envelope of a *modulated wave*.

ENVELOPE DELAY DISTORTION. The distortion which occurs when the rate of change of *phase shift* with frequency of a system is not constant over the frequency range required for transmission.

EPISCOTIZER. A device used with a *photocell* to reduce the intensity of light on the sensitive surface. It consists of a rotating disc with alternate opaque and transparent sections.

EPITAXIAL TRANSISTOR. A *transistor* made from a *semiconductor* material in which the desired regular crystalline structure is determined by depositing the material on an appropriate monocrystalline support.

E-PLANE. Of *electromagnetic waves,* a plane containing the direction of maximum radiation.

EPOCH ANGLE. Of a *sinusoidal function,* the *phase angle* at time $t=0$ in the equation $x=A \cos (wt+\theta)$.

EPOXY RESIN. A type of synthetic resin, with high strength and low shrinkage, extensively used for embedding and encapsulating electronic components.

EPR. See *Electron paramagnetic resonance.*

EPUT METER. Trade name for a *counter* which measures the numbers of events per unit time.

EQUAL-ENERGY SOURCE. A source of electromagnetic or sound radiation from which the emitted energy is equally distributed over the whole of the frequency spectrum of the source.

EQUALIZATION. In electronics, the reduction of *distortion* by the introduction of networks which compensate for the particular type of distortion over the required frequency band.

EQUALIZER. An electrical network designed to correct for unequal *attenuation* or *phase shift* in the transmission of signals over wires and cables. See *Attenuation equalizer, Dialogue equalizer*.

EQUALIZING PULSE. In *television*, a pulse at twice the *line frequency*, applied immediately before and after the vertical *synchronizing pulse*, to reduce the effect of line frequency pulses on the *interlace*.

EQUATION SOLVER. A *computer* specially designed to solve mathematical equations.

EQUILIBRIUM REACTION POTENTIAL. The minimum voltage at which an electrochemical reaction can take place.

EQUIPOTENTIAL CATHODE. See *Cathode, unipotential*.

EQUIVALENT BINARY DIGITS. The number of binary digits required to express a given number of decimal digits or other characters. See *Binary code*.

EQUIVALENT CIRCUIT. An arrangement of simple circuit elements which is electrically equivalent to a more complicated circuit or device over a specified range.

EQUIVALENT DIODE. The imaginary *diode* consisting of the *cathode* of a multi-electrode valve and a virtual *anode* to which a composite controlling voltage is applied so that the resulting cathode current is the same as in the multi-electrode valve.

EQUIVALENT ELECTRONS. *Electrons* which occupy the same orbit in an atom.

EQUIVALENT NETWORK. One which may, under specified conditions, replace another *network* without material effect on the electrical performance.

EQUIVALENT RESISTANCE. Of an electric circuit, the value of total *resistance* which, if concentrated at one point, would dissipate the same power as the total of the various smaller resistances at different points in the circuit.

ERASE. (1) In *charge-storage tubes*, charge or discharge storage elements so as to remove information previously stored. (2) In *digital computers*, remove information from the store and leave the space available for new information.

ERASING HEAD. In *magnetic recording*, a device for obliterating any previous recordings on the magnetic recording media. See *Magnetic head*.

ERG. The unit of energy or work: the work done by a force of one dyne in moving its point of application through one centimetre.

ERGOMETER. An instrument for measuring the amount of energy produced or the work done by any device.

ERGONOMICS. The study of the relations between human beings and their working environment, and the engineering applications of that study.

ERIOMETER. An instrument for measuring very small diameters by *diffraction* methods.

ERNIE. Electronic Random Number Indicator Equipment. A machine designed and produced by the Post Office Research Station to find and print a long series (about 20,000 were required on the first draw) of nine-figure numbers which are random according to mathematical criteria. A *noise-source* is used for randomness, the noise arising from a *glow discharge* in an electronic tube, and a number is printed automatically every two and a half seconds.

ERROR SIGNAL. In an *automatic control system,* a signal representing the discrepancy between the desired and the actual performance, which is used to reduce that discrepancy.

ERROR-CORRECTING CODE. A *code* which improves the accuracy of reception of a coded message by supplying repeated symbols as a check.

ERROR-DETECTING CODE. See *Parity check*.

ESAKI DIODE. See *Tunnel diode*.

E-TRANSFORMER. In an *automatic control system,* an electric sensing device which produces an error voltage in response to linear motion.

ETTINGSHAUSEN EFFECT. A conductor carrying current in a magnetic field develops a transverse temperature gradient perpendicular to the field and the current.

EVERSHED MEGGER. A *megger* instrument with which very high *resistances* can be measured by the application of the *differential galvanometer* principle. The instrument is completely self-contained, current being produced by a built-in, hand-driven generator.

EWR. Abbreviation for *early warning radar*.

EXCESS CONDUCTION. In a *semiconductor,* conduction promoted by *excess electrons*.

EXCESS ELECTRON. An electron added to a *semiconductor* by, for example, a *donor impurity* which is not required to complete the bonding system of the semiconductor and is therefore available for conduction.

EXCESS-THREE CODE. In *computers,* a number *code* in which the decimal digit n is represented by the four-bit binary equivalent of $n + 3$. For example 6 is represented by 1001.

EXCITATION. (1) In general, the addition of energy to a system, transferring it from its ground state to an excited state. (2) In valves and tubes, a signal voltage applied to a control electrode. (3) The voltage applied to an oscillating crystal. (4) The r-f impulses applied to a tuned circuit.

EXCITATION ANODE. Of a mercury-pool *cathode* valve, an auxiliary anode used to maintain a *cathode spot*.

EXCITED ATOM. An atom in which orbital *electrons* are at energy levels higher than their normal level. See *Energy band*.

EXCITER. (1) A crystal or *self-excited oscillator* used to generate the carrier frequency of a transmitter. (2) The driving element of an aerial array. (3) The field excitation source in electric power generating machines.

EXCITING CURRENT. (1) Of a *transformer*, the primary current with no load on the secondary. (2) Of any electrical machine, the current flowing in the exciting circuit.

EXCITON. In a *semiconductor*, a bound hole-electron pair which results from an injection of *holes* into a crystal. See *Acceptor impurity*.

EXCITRON. A *rectifier* tube for large power used in electric power distribution. It consists of a mercury-pool *cathode* and a single anode surrounded by a grid.

EXPANDED SWEEP. A sweep electron of the beam of a *cathode-ray tube* in which the movement of the beam is speeded up during part of the sweep.

EXPANDED SWEEP GENERATOR. A circuit which produces a faster sweep voltage for a selected interval of the normal sweep of a cathode-ray tube presentation.

EXPANDER. A *transducer* which, for a given range of amplitude of input voltages, produces a larger amplitude range of output voltage. One type used in audio engineering employs the envelope of speech signals to expand their volume range. See *Compander*.

EXPANSION. In electronics, a process in which the effective amplification applied to a signal depends on the magnitude of the signal, being greater for large than for small signals. See *Automatic volume expansion*.

EXPLORER. The first U.S. earth satellite, placed into orbit on 31 January 1958.

EXPLORING COIL. A coil used to investigate *magnetic fields* by measuring the effects of changing *magnetic flux* linked with the coil.

EXPONENTIAL HORN. An acoustical horn whose cross-sectional area increases exponentially with axial distance from the throat.

EXPOSURE METER. A device, consisting of a *photoelectric cell* and a suitable indicating meter, which is used to determine the illumination of an object in photography.

EXTENSOMETER. An instrument for measuring extensions or deformations on a material by measuring the changes in the capacitance of a capacitor which result from the dimensional changes.

EXTERNAL FEEDBACK. See *Feedback, external*.

EXTERNAL MEMORY (STORAGE). Of a *computer*, a memory separate from the computer but available to it, e.g. a number of *magnetic tapes* in a separate cabinet.

EXTINCTION POTENTIAL. (1) Of a *gas-filled tube*, the voltage drop across the tube at which the *arc* will be distinguished as the anode voltage is gradually reduced. (2) Of a *gas discharge* tube, the minimum *anode potential* necessary to maintain a discharge.

EXTRACTION INSTRUCTION. In a *digital computer,* an instruction to form a new word by juxtaposing selected segments of given words.

EXTREMELY HIGH FREQUENCY (EHF). Frequencies between 30,000 and 300,000 megacycles per second. See *Frequency band.*

EXTRINSIC SEMICONDUCTOR. A *semiconductor* whose electrical properties depend on *impurities.*

F

FACSIMILE. The process of *scanning* any kind of fixed graphic material so that the image is converted into electric signals which may be used either locally or remotely to produce a recorded likeness of the original. See the following definitions.

FACSIMILE BANDWIDTH. The difference in cycles per second between the highest and lowest frequency components required for adequate transmission of the facsimile signals.

FACSIMILE BASEBAND. In a *carrier* wire or radio transmission system, the frequency band occupied by the signal before it modulates the carrier frequency to form the transmitted line or radio frequency.

FACSIMILE CONVERTER. A device which changes the type of *modulation* from transmitting to receiving or vice versa.

FACSIMILE DENSITY. A measure of the light-transmitting or light-reflecting properties of an area, expressed as the logarithm to base 10 of the ratio of incident to transmitted or reflected light.

FACSIMILE MODULATION. In facsimile transmission, the process by which the amplitude, frequency, or phase of the transmitted wave is varied with time in accordance with a signal.

FACSIMILE RECEIVER. An apparatus which translates the signals from the communications channel into a facsimile record of the original copy. Incoming signals are amplified and converted into a brightness-modulated light source which is then used to record the picture elements in sequence on a drum rotating in synchronism with the transmitting drum.

FACSIMILE SCANNING. Successive analysing of the facsimile densities of the subject copy according to a predetermined pattern. Normal scanning is from left to right and from top to bottom as when reading a page of print.

FACSIMILE TELEGRAPH. A telegraph system for the transmission of pictures.

FACSIMILE TRANSMITTER. The apparatus employed to translate the subject copy of a facsimile system into signals suitable for delivery to the communication channel. The details are picked up by an optical system as the copy rotates on a drum and differences in the brightness of reflected light converted into modulated electric signals by a *photocell,* and amplified.

FADER. A device for maintaining a constant level of electrical signals while one signal is faded out and another faded in.

FADING. Variation of radio field intensity caused by variations in the transmission medium. In radio broadcasting it is usually caused by destructive interference between two waves which reach the receiver by two different paths. Where all frequencies in the signal are attenuated by about the same amount the result is called amplitude fading. See also *Selective fading.*

FALL TIME. Of a *pulse,* the interval between the instants at which instantaneous amplitude falls from upper to lower limits, usually specified as from 90 per cent to 10 per cent of the peak amplitude of the pulse.

FALSE CURVATURE. The curvature of electron tracks in an ionization chamber in the absence of an applied magnetic field, caused by scattering collisions between electrons and gas atoms.

FAN-IN. The maximum number of inputs permitted to an *OR-gate* is sometimes called the fan-in of that gate.

FAN-OUT. The maximum number of inputs which may be driven from a given output stage is sometimes called the fan-out of that stage.

FARAD. The *capacitance* of a *capacitor* in which a charge of 1 *coulomb* produces a change of *potential difference* between the terminals of 1 volt. This is a very large unit and much smaller units – *microfarad* and *micromicrofarad* or *picofarad* – are used in practice. See *Units.*

FARADAY. The number of *coulombs* (96,500) required for an electrochemical reaction involving one chemical equivalent.

FARADAY CAGE. An *electrostatic screen,* made of wire mesh or network or a system of parallel conductors connected at one end to an earthed conductor, which shields objects inside it from the effects of an electrostatic field but is transparent to electromagnetic waves (except those whose wavelength is close to the dimensions of the mesh).

FARADAY CONSTANT. See *Physical constants.*

FARADAY DARK-SPACE. In a *gas discharge tube,* the relatively nonluminous region between the *negative glow* and the *positive column.*

FARADAY EFFECT. (1) The rotation of the plane of polarization of a plane polarized beam of light passing through certain transparent substances in the presence of a strong magnetic field. (2) Rotation of the plane of polarization of a plane polarized *microwave* beam passing through a *ferrite* which has a static magnetic field along the direction of propagation.

FARADAY TUBE. A unit of *magnetic induction* or *electric force.*

FARADAY'S LAW OF ELECTROMAGNETIC INDUCTION. The electromotive force induced in a circuit is proportional to the time rate of change of the *magnetic flux* linked with the circuit. See *Maxwell's equations.*

FARADISM. The treatment of disease by the use of an interrupted current to stimulate the nerves and muscles. Such current is obtained from an *induction coil.*

FARADMETER. An instrument for measuring capacitance, usually provided with a scale graduated in *microfarads.*

FAST TIME CONSTANT CIRCUIT. (1) An electric circuit consisting of *resistance* and *capacitance* which combine to give a short *time-*

constant for the discharge of the capacitor through the resistor. (2) In *radar,* a circuit of short time-constant used to emphasize signals of short duration.

FATHOMETER®. An *echo sounder* designed in the U.S.A.

FAULT CURRENT. The peak current which flows through a device under fault conditions, e.g. *arc backs* or short circuits.

FAWSHMOTRON. A *microwave* tube device based on FAst Wave Simple Harmonic MOTion.

FAX. Abbreviation for *facsimile.*

FCC. Abbreviation for Federal Communications Commission (U.S.).

FEEDBACK. (1) In general, for any system which converts energy and has an input and an output, the return of a fraction of the output to the input. (2) In particular, in an *amplifier,* the return of a portion of the output from any stage to the input of that stage, or of a preceding stage, so as to increase or reduce the amplification, according to the relative phase of the return. See the following definitions and especially *Feedback, negative* and *Feedback, positive.*

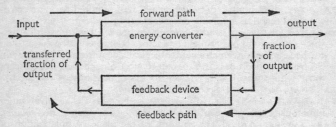

Fig. 52. The principle of feedback

FEEDBACK, ACOUSTIC. The return of a portion of an *audio* wave or vibration from the output of an audio system to the input or a preceding stage of the system. This may cause the form of oscillation known as *howling.*

FEEDBACK, CAPACITIVE. The application of *feedback* in a valve or transistor circuit by means of a *capacitance* which is common to output and input circuits.

FEEDBACK, EXTERNAL. In *magnetic amplifiers,* feedback in which the feedback voltage is derived from the rectified output current.

FEEDBACK, INDUCTIVE. *Feedback* by means of an inductor or inductive coupling.

FEEDBACK, MULTIPLE-LOOP. *Feedback* which may occur through two or more paths.

FEEDBACK, NEGATIVE. The process in which part of the power in the output circuit of an amplifying device reacts on the input circuit so as to reduce the initial power and hence the amplification. Cf. *Feedback, positive.*

FEEDBACK, POSITIVE. The process in which part of the power in the output circuit of an amplifying device reacts on the input circuit so as to reinforce the initial power and increase the amplification. The process is also described as *regenerative,* and if sufficient positive feedback is applied oscillation results.

FEEDBACK, SINGLE LOOP. *Feedback* which can occur through only one electrical path.

FEEDBACK, STABILIZED. *Feedback* employed to stabilize the gain of a transmission system or reduce noise and distortion in it.

FEEDBACK ADMITTANCE. Of a valve or tube: the short-circuit *transadmittance* from output electrode to input electrode.

FEEDBACK AMPLIFIER. An *amplifier* in which a portion of the output is fed back to the input. The portion returned may be a function of output voltage or current or a combination of both.

FEEDBACK CONTROL LOOP. In control systems, a closed transmission path which includes an active *transducer,* and which consists of a forward path, a feedback path, and one or more *mixing points,* so that it maintains a specified relation between the input and output signals of the loop. See *Closed-loop control, Feedback control system.*

FEEDBACK CONTROL SYSTEM. A control system which includes one or more *feedback control loops* and tends to maintain specified relations between the commands and the controlled signals. See *Automatic control system, Continuous control.*

FEEDBACK CONTROL WINDING. Of a *saturable reactor,* a *control winding* to which a feedback connection is made.

FEEDBACK FACTOR. In *negative feedback* (in an amplifier), the product of the fraction of the output voltage fed back and gain of the amplifier before feedback is applied.

FEEDBACK PATH. In a *feedback control loop,* the transmission path from the loop output signal to the loop feedback signal.

FEEDBACK REGULATOR. A *feedback control system* which tends to maintain a specified relation between certain system signals and other predetermined quantities. See *Servomechanism.*

FEEDTHROUGH. A conductor used to connect patterns on opposite sides of a *printed-circuit board.* Also, as an adjective, for components used in this manner.

FELICI BALANCE. A *bridge* for the determination of *mutual inductance* between transformer windings.

FEMTOAMPERE (fA). 10^{-15} *amperes.*

FERMI LEVEL. The value of the electron energy at which the *Fermi-Dirac distribution function* has the value one half.

FERMI-DIRAC DISTRIBUTION FUNCTION. In solid-state physics, the probability that an *electron* in a *semiconductor* will occupy a particular quantum state of energy when it has attained thermal equilibrium.

FERNICO®. An alloy of iron, nickel, and cobalt used in glass-metal seals.

FERRIMAGNETISM. A kind of magnetism found in *ferrites* in which the spins of adjacent ions in the presence of a magnetic field are in opposite senses. Similar to, but not identical with, *ferromagnetism*.

FERRISTOR®. A coil wound on a ferromagnetic core which acts as a *ferroresonant* amplifier.

FERRITE. One of a class of magnetic materials made from ceramic ferromagnetic compounds of the general formula $XF_2 O_4$ where X may be a metal such as nickel, cobalt, zinc. Cores made of these materials have very low *eddy-current loss* and are used for high frequency circuits and computer memories. See *Maser, solid-state*.

FERRITE BEAD MEMORY. A *magnetic memory* device in which a mixture of *ferrite* powders is fused directly on to the current-carrying wires of a memory matrix to form small beads.

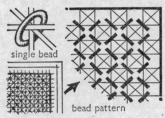

single bead

bead pattern

Fig. 53. Typical ferrite bead in a memory matrix

FERROCART CORE. A core made of finely powdered iron mixed with a binder, used at radio-frequencies.

FERRODYNAMIC INSTRUMENT. An *electrodynamic instrument* in which ferromagnetic material is used to increase the operating forces.

FERROELECTRICS. *Dielectric* materials, such as barium titanate and Rochelle salt, which exhibit electric properties analogous to certain magnetic properties. In particular they possess *hysteresis* in an alternating electric field, and have marked *piezoelectric* properties which makes them useful as detectors of vibrations.

FERROMAGNETIC. A *paramagnetic* material whose *permeability* is much greater than that of a vacuum and varies with the magnetizing force. Ferromagnetic materials are composed of iron, nickel, cobalt, or manganese, or alloys of these, with small additions of other elements. See *Magnetism*.

FERROMAGNETIC AMPLIFIER. A *parametric amplifier* depending on the non-linear behaviour of *ferroresonance* at high rf power levels.

FERROMAGNETIC ANISOTROPY. In some ferromagnetic crystals, the magnetization follows the crystal axes and the material thus exhibits different magnetic properties in different directions.

FERROMAGNETIC RESONANCE. A particular case of *paramagnetic resonance* characterized by the fact that the effective internal field is much stronger than the applied field, sometimes thousands of times stronger. It is supposed that ferromagnetic resonance involves unpaired spins of electrons very close together and therefore subject to large exchange forces.

FERROMAGNETICS. In *computers,* a general description of the techniques of storing information and the logical control of pulse sequences, using magnetically polarized materials as binary storage elements.

FERRO-MANGANESE. A *ferromagnetic* alloy of iron and manganese.

FERROMETER. An instrument for measuring the *magnetization* of a *ferromagnetic* with the application of alternating current.

FERRO-NICKEL. A *ferromagnetic* alloy of iron and nickel.

FERRORESONANCE. See *Ferromagnetic resonance*.

FERRORESONANT CIRCUIT. A *resonant* circuit in which one of the circuit elements is a *saturable reactor*.

FERROSPINEL. A crystalline material which satisfies the requirements for X in the formula for a *ferrite*.

FERROXCUBE®. A commercially available *ferrite*.

FET. Abbreviation for *Field Effect Transistor*.

FIBRE OPTICS. Assemblies of large numbers of very fine tubes made from a material with high refractive index – usually Perspex – each tube being coated with a material of still higher refractive index. Bunched together, these composite fibres act as flexible light guides with very low loss.

FIDELITY. The degree to which an electrical transmission system reproduces at its output the characteristics of a signal applied to its input. See *High fidelity*.

FIELD. (1) In general, the space in which any physical agency, e.g. electricity, magnetism, gravitation, exerts its influence. (2) In *computers,* a set of one or more symbols, not necessarily in the same *word,* treated as a unit of information. (3) In *television,* one of the two (or more) equal parts into which a frame is divided in *interlaced scanning*.

FIELD COIL. A coil used for magnetizing an electromagnet, such as the magnet of a generator.

FIELD CONTROL. The control of the voltage of a generator or the speed of a motor by varying the field current.

FIELD-CONTROL CONVERTER. A *synchronous converter* with a variable commutator voltage, which is controlled by varying the field current.

FIELD DENSITY. Of a specified cross-sectional area of an electric or magnetic field, the number of electric or magnetic lines of force passing through it.

FIELD EMISSION. The *emission* of electrons produced by sufficiently strong electric fields acting on unheated surfaces.

FIELD FRAME. See *Yoke*.

FIELD FREQUENCY (REPETITION RATE). In *television systems*, the product of picture frequency and the number of fields contained in one *frame*.

FIELD INTENSITY. See *Field strength*.

FIELD INTENSITY, MINIMUM. The minimum strength of an incident radio signal which will permit satisfactory communication. It depends primarily on the *noise level* at the receiver.

FIELD INTENSITY, RADIO. See *Field strength, radio*.

FIELD MAGNET. A magnet used to provide a *magnetic field*.

FIELD SCAN. In *television*, the vertical movement of the electron beam downwards across the face of the picture tube, in order to *scan* alternate lines of the picture.

FIELD STRENGTH. (1) The magnitude of a vector at any point of a vector field measured by the force exerted on a known charge or magnetic pole. (2) In *television*, the intensity, measured in microvolts per metre, of the carrier signal from a given transmitter which is present in a dipole aerial 1 metre long. (3) In *radar*, the intensity, in microvolts or decibels below a millivolt, of a signal received from a given transmitter.

FIELD STRENGTH, RADIO. The intensity of the electric or magnetic field at any point associated with the passage of radio waves. Usually expressed as the electric field intensity in microvolts or millivolts per metre.

FIELD-EFFECT TRANSISTOR (FET.) A *semiconductor* device in which the effect of an electric field, applied transversely to a wafer of the material, on the charge carriers of the material, is used to vary the effective conductance and hence to achieve current and voltage amplification. See *Fieldistor, Tecnetron*.

FIELD-ENHANCED PHOTOELECTRIC EMISSION. The increased *photoelectric emission* resulting from the action of a strong electric field on the emitter.

FIELD-ENHANCED SECONDARY EMISSION. The increased *secondary emission* resulting from the action of a strong electric field on the emitter.

FIELD-FREE EMISSION CURRENT. The electron current emitted by a *cathode* at whose surface the electric field is zero.

FIELDISTOR®. An early (1952) *field-effect transistor* made in the U.S.A. and used as an amplifying device with high impedance input and low impedance output.

FIELD-SEQUENTIAL COLOUR TELEVISION SYSTEM. A system of *colour television* in which the camera sees red, blue, and green images in turn through a rotating disc consisting of segments of colour filters in the correct sequence. A similar disc at the receiving end is rotated in synchronism and phase so that the viewer looks at the sequences of colour images through a corresponding set of filters and the eye synthesizes the complete coloured picture.

FIELD-STRENGTH METER. A calibrated radio receiver designed to measure *field strength*.

FIGURE OF MERIT (Q). A criterion, sometimes arbitrary, for comparison of the performance of similar devices. See *Q*.

FILAMENT. (1) A *cathode* of a thermionic valve, usually a wire or ribbon, which is heated by passing current through it. (2) An electrode used to heat the cathode but separate from it.

FILAMENT CURRENT. The current required to heat a *filament*.

FILAMENT GETTER. A *getter* used to maintain a high vacuum especially in vacuum valves, in the form of an auxiliary *filament,* and often made of zirconium.

FILAMENT REACTIVATION. Of *thoriated-tungsten* filaments, raising the temperature of the filament above normal for a short time in order to replenish the thorium at the surface from the material in the interior.

FILAMENT SATURATION. See *Temperature saturation*.

FILAMENT TRANSFORMER. A *transformer* designed only to supply power to the filaments of valves or tubes.

FILAMENT VOLTAGE. The voltage between the terminals of a *filament*.

FILAMENT WINDING. A separate winding on a *power transformer* to provide a low voltage, usually 6 or 12, for the valve filaments.

FILLED BAND. The *valence band* in the *energy level diagram* of a solid, which has its full complement of energy levels.

FILM RESISTOR. See *Metal-film resistor*.

FILMISTOR. A *film resistor*.

FILTER. In electronics, a large class of devices for the selective transmissions of different frequencies of electromagnetic, electromechanical, and electro-acoustic vibrations. See the following types of filter: *Band-elimination, Band-pass, Crystal, High-pass, Ladder, Lattice, Low-pass, Magnetostrictive, Mode, Rectifier, Resistance-capacitance, Ripple, Waveguide iris*.

General symbol.

FILTER ATTENUATION. Loss of signal power, usually expressed in *decibels,* through a filter, due to absorption, reflection, or radiation.

FILTER ATTENUATION BAND. A frequency band in which the *attenuation constant* of the filter is not zero.

FILTER DISCRIMINATION. The difference between the minimum *insertion loss* at any frequency in a *filter attenuation band* and the maximum insertion loss at any frequency in the operating range of the *filter transmission band*.

FILTER NETWORK. A *transducer* for separating waves on the basis of their frequency, usually composed of passive, linear, reactive elements, such as *capacitors* and *inductors*.

FILTER TRANSMISSION BAND (PASS BAND). A frequency band in which, if dissipation is disregarded, the *attenuation constant* is zero.

FINE TUNING. In electronic instruments, provision for greater sensitivity and accuracy in tuning.

FIRED TUBE. In *ATR* and *TR switches,* the condition of the tube during which a radio-frequency glow discharge exists at either the resonant gap or resonant window or both.

FIREFLASH. An air-to-air missile produced for the R.A.F.

FIRESTREAK. An air-to-air missile using *infra-red homing* produced for the R.A.F.

FIRING. In a *gas discharge tube,* the act of establishing the discharge.

FIRING POWER, MINIMUM. In *switching tubes,* the minimum rf power required to initiate a discharge in the tube for a specified ignitor current.

FIRING TIME, IGNITOR. In *switching tubes,* the time between the application of a direct voltage to the ignitor electrode and the beginning of the ignitor discharge.

FISH PAPER. A flexible insulating material, usually of varnished cambric, used to line metal chassis, separate coil windings from cores, etc.

FIXED CAPACITOR. An electrical *capacitor* which has no provision for varying the capacitance.

FIXED RESISTOR. A *resistor* with no provision for varying its resistance.

FIXED STORE. In *computers*: a store whose content cannot be changed automatically by a programme, but which may be changed by an alteration to the construction of the store.

FIXED-CYCLE OPERATION. Of a *computer,* a mode of operation in which each process is allocated a fixed time, in advance.

FIXED-POINT REPRESENTATION. In *computers,* a form of presentation of numerical quantities which employs a constant number of digits with the decimal, binary, or other point fixed in a given location. Modern computers employ *floating-point representation.*

FLASH. Subject to a large current for a short time, e.g. with thoriated filaments in order to replenish the thorium at the surface.

FLASH ARC. In a *thermionic valve,* a sudden violent increase in the emission from the cathode usually resulting in its destruction.

FLASHBACK VOLTAGE. In a *gas tube,* the inverse peak voltage at which ionization occurs.

FLASHED FILAMENTS. Carbon filaments which are *flashed* in a carbon atmosphere to produce uniform cross-section.

FLASHER. A device, usually mechanical or thermal, for the rapid and automatic on–off switching of electric lamps.

FLASHOVER. A *disruptive discharge* around or over the surface of a solid or liquid insulator.

FLASHOVER VOLTAGE. Of an impulse, the highest value of a voltage impulse which causes *flashover.*

FLEXIBLE RESISTOR. A *wire-wound resistor* resembling a flexible cable, made by winding nichrome wire around an asbestos cord and covering with a coloured braiding.

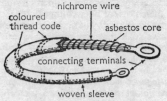

Fig. 54. Flexible resistor

FLEXOWRITER®. An automatic typewriter, often used at the input and output of a *digital computer,* which can read, reproduce, and punch punched paper tape and punched or tabulated card.

FLICKER. A subjective sensation resulting from periodic fluctuation in the intensity of light at rates of less than about twenty-five to thirty times a second. In *television* this sensation prevents complete continuity of the images.

FLICKER EFFECT. In a *vacuum valve,* small random variations in anode current thought to be due to positive *ions* from the cathode.

FLICKER NOISE. In valves and tubes, *noise* present at lower frequencies due to random fluctuations in the emission from the cathode, especially from an *oxide-coated cathode.*

FLIP-CHIP. Colloquial description of a technique used in *micro-electronics* where a transistor chip of crystalline silicon is mounted face down on a *thin-film circuit* using specially prepared mounting pads.

FLIP-FLOP. (1) A device having two stable states and two input terminals corresponding to these states. The device remains in either state until caused to change to the other by application of the appropriate signal. (2) Such a bistable device with an input which allows it to function as a single-stage *binary counter.* See *Magnetic flip-flop, Multivibrator.*

FLOATING. (1) Not connected to any source of potential. (2) A method of using storage batteries in which they are maintained in a constant state of charge by applying a constant voltage to their terminals.

FLOATING ACTION. In control systems, action in which there is a predetermined relation between the deviation or displacement of a final control element and its speed. See *Feedback control system.*

FLOATING BATTERY. A system of electrical supply in which a generator and storage battery are connected in parallel so that they share the load, and the battery accepts the full load if the generator fails.

FLOATING GRID. In a valve, a *grid* which has no connections to it and which usually becomes negatively charged with respect to the *cathode* as it accumulates electrons.

FLOATING POTENTIAL. In a valve, transistor, or other device, the direct *potential* of an electrode which is open-circuited with all other electrodes having their normal bias voltages.

FLOATING-POINT REPRESENTATION. In computers, a method of representing numbers such that any number x can be presented by two other numbers y and z such that $x = y(n)^2$ where z is an integer and n is usually 2 or 10. Thus a decimal number 1,230,000 can be shown as 1·23 6 by choosing n to be 10. y is called the mantissa or fraction and the integer z is called the exponent or characteristic.

FLOW CHART. In *computer* programming, a pictorial representation of the nature and sequence of operations to be carried out by the computer according to a programme.

FLOWMETER. An instrument for measuring the rate of flow of a fluid through a pipe.

FLUORESCENCE. Emission of electromagnetic energy as the result of absorption of energy from some other radiation, which lasts only as long as the stimulus responsible for it is continued. Also the electromagnetic radiation so produced. Cf. *Phosphorescence.*

FLUORESCENT LAMP. An electric discharge lamp in which *phosphors* transform the energy of the electric discharge into wavelengths which give greater luminosity.

FLUORESCENT SCREEN. (1) A sheet of suitable material which fluoresces visibly when roentgen rays, radium rays, or electrons impinge on it. (2) The inner face of a *cathode-ray tube* coated with a suitable fluorescent material.

FLUORIMETER. An instrument for measuring the intensity of fluorescent radiation or of the radiation which causes fluorescence.

FLUORINE (F). An extremely reactive gaseous element, Atomic No. 9. Combines to form complex compounds with carbon, such as *Teflon,* which are very high *resistivity* insulators.

FLUOROGRAPHY. Photography of an image on a fluorescent screen.

FLUOROSCOPE. An instrument consisting of a suitably mounted fluorescent screen, sometimes in conjunction with an X-ray tube, which produces visible images of the X-ray shadows of objects placed between the screen and an X-ray tube.

FLUX. A term used either to describe a flow of particles, *photons,* or *lines of force,* or as a synonym for the *flux density* of these entities. See *Displacement flux, Electric flux, Leakage flux, Magnetic flux.*

FLUX DENSITY. (1) Of electromagnetic radiation, the energy per unit time passing through a unit area of surface normal to the beam. (2) Of particles and *photons,* the number passing through a unit area of surface normal to the beam.

FLUX GATE. The direction-sensing element of a *gyro flux-gate compass* operated by the earth's magnetic field.

FLUX VALVE. See *Flux gate*.

FLUXGRAPH. An instrument for delineating the magnetic field around a coil automatically.

FLUX-GUIDE. For *induction heating,* suitably shaped magnetic material for directing the electromagnetic flux into desired paths.

FLUXMETER. An instrument for measuring *magnetic flux* calibrated directly in *maxwells* or *webers*. It may be simply a *moving-coil galvanometer* with negligible restoring torque in the suspension.

FLYBACK. (1) On a *cathode-ray* screen, the return of the spot (due to the electron beam) to its starting point after reaching the end of its trace. (2) The shorter of the two time intervals which make up a *sawtooth wave*.

FLYING-SPOT SCANNER. (1) A method of producing *video* signals from films or transparencies by scanning the photographic image with a flying spot of illumination generated by a special *cathode-ray tube* provided with magnetic deflection circuits. The modulated beam of light produced in this way is focused on a photocell and resulting electrical signals are amplified. (2) A device for scanning the object to be transmitted in a television system, using a very bright point source of light and a rotating disc suitably perforated as the sole illumination of the object.

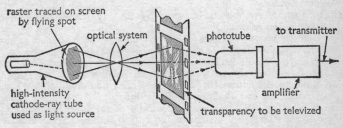

Fig. 55. Method of televising a film with a flying-spot scanner

FLYWHEEL EFFECT. In an *oscillator,* the effect of the electrical inertia of the circuit, analogous to the inertia of a flywheel, which results in maintaining oscillations during the intervals between the pulses of energy exciting the circuit.

FLYWHEEL TIME BASE. In television receivers, a method of controlling the *frame frequency* by the electrical inertia of the *scanning* circuits and not by synchronizing pulses.

FM. Abbreviation for *frequency modulation*.

FM RADIO. A system of radio transmission and reception in which the intelligence is transmitted by *frequency modulation* of a radio-frequency carrier.

FM RECEIVER. See *Frequency-modulated receiver*.

FM RECEIVER DEVIATION SENSITIVITY. The smallest *deviation* of frequency required to produce a specified output power.

FM TRANSMITTER. See *Frequency-modulation transmitter*.

FOCUSING COIL. An assembly for producing the magnetic field used to focus an *electron beam*, e.g. in a television picture tube or *electron microscope*. See *Electromagnetic focusing*.

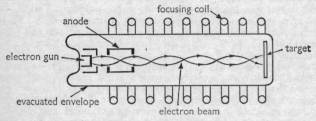

Fig. 56. Simplified diagram of action of focusing coil in a camera tube

FOCUSING ELECTRODE. An *electrode,* usually of a *cathode-ray tube,* to which a voltage is applied to control the cross-sectional area of the electron beam.

FOLDED HORN. An acoustic horn which is folded or convoluted so as to occupy less space.

FORBIDDEN BAND. A range of energy levels in a crystalline solid which is not attained by any electrons in the crystal. In the *energy level diagram* forbidden bands appear as gaps between *allowed bands*.

FORBIDDEN-COMBINATION CHECK. In *computers,* a checking routine, usually automatic, which reveals any non-permissible code expression, i.e. one which represents an error.

FORCE FACTOR. Of a *transducer,* the ratio of the force required to block the mechanical system to the corresponding current in the electrical system.

FORCED OSCILLATION. *Oscillation* in which periodic forces outside the oscillating system determine the period of the oscillations.

FORM FACTOR. Of an alternating quantity such as current or voltage, the ratio of the *effective value* of the quantity to its half-period average value.

FORM FACTOR, RECTIFIER. The ratio of the average unidirectional output current of a *rectifier* to the r.m.s. *value* of this current.

FORTRAN. FORmula TRANslation: a computer programming language – developed by I.B.M.

FORWARD CURRENT. Current resulting from the application of a *forward voltage*.

FORWARD PATH. In a *feedback control loop,* the transmission path from the loop actuating signal to the loop output signal.

FORWARD RECOVERY TIME. Of a *semi-conductor diode,* the time required for the *forward current* or voltage to reach a specified value after instantaneous application of a forward bias in a given circuit.

FORWARD TRANSFER FUNCTION. Of a *feedback control loop,* the *transfer function* of the *forward path.*

FORWARD VOLTAGE. In any circuit or device, voltage of that polarity which produces the larger current.

FORWARD WAVE. In *travelling-wave tubes* a wave whose group velocity is in the same direction as the motion of the electron stream.

FOSTER-SEELEY DISCRIMINATOR. See *Frequency discriminator.*

FOUR-ADDRESS. In *computer* programming, an *instruction code* in which the operation and *address* of each of four registers is included in each instruction.

FOURIER ANALYSIS. A mathematical method for determining the *harmonic* components of a complex periodic wave function. A *wave-analyser* embodies the physical application of the method.

FOURIER TRANSFORM. A mathematical expression relating the energy in a *transient* to that in a continuous energy spectrum of adjacent frequency components.

FOUR-POLE. Of an electric switch, having provision for connections to four separate circuits.

FOUR-TERMINAL NETWORK. A *network* with four accessible terminals.

FOUR-TERMINAL RESISTOR. A *resistance* standard provided with four terminals: two which connect the resistor to the current source and two to the measuring instrument. This arrangement eliminates errors due to incidental resistance or potentials at the contacts. See *Kelvin Bridge.*

FOUR-WIRE CIRCUIT. A two-way circuit with two paths arranged so that electric waves are transmitted in one direction only by one path and in the other direction only by the other path.

FRAME. (1) In *television,* the total area of the picture which is scanned while the picture signal is not blanked. (2) In communications using pulse trains, one cycle of a periodically recurring number of electric pulses. (3) In *facsimile* transmission, a rectangular area whose width is the available line and whose length is determined by the service requirements.

FRAME CONTROLS. In *television receivers,* controls which vary the free-running periods of the deflection oscillators. There are usually two: for vertical and horizontal adjustment respectively.

FRAME FREQUENCY. In *television,* the number of times per second that the *frame* is scanned.

FRAME PERIOD. The reciprocal of the *frame frequency.*

FRAMING SIGNAL. In *facsimile* transmission, a signal used to adjust the picture to a desired position in the direction of the progression.

FREE CHARGE. In *electrostatics,* a charge which is not bound by an equal or greater charge of opposite polarity.

FREE ELECTRON. An *electron* which is not permanently attached to a

151

particular atom and is free to move under the influence of an electric field. *Conductors* contain a high proportion of free electrons and *insulators* a very small proportion.

FREE FIELD. A field in which the effect of the boundaries are negligible for the region of interest.

FREE GRID (FLOATING GRID). A *grid* of a valve to which there is no electrical connection, and which then accumulates a charge and causes the valve to behave erratically.

FREE IMPEDANCE. Of a *transducer,* the *impedance* at the input of the transducer when the impedance of its load is made zero.

FREE OSCILLATION. *Oscillation* of any system under the influence of internal forces only, or of a constant force originating outside the system, or of both.

FREE SPACE CHARACTERISTIC IMPEDANCE. The ratio of the electric and magnetic intensities of an electromagnetic wave in space: numerically equal to 376·6 ohms.

FREE WAVE. An idealized wave in a medium free from boundary effects.

FREQUENCY. Of a periodic quantity in which time is the independent variable, the number of periods which occur in unit time. The c.g.s. unit is cycles per second (c/s); SI unit: *Hertz.* Symbol: f.

FREQUENCY AND TIME MEASUREMENT COUNTER. An electronic *counter* with added *frequency converters* which can measure frequency and time intervals outside the normal range of the counting circuits.

FREQUENCY BAND. A continuous range of frequencies within specified limits selected from a more extensive range of frequencies. A table of internationally agreed frequency bands is provided in Fig. 57.

Frequency 300,000 Mc/s		Wavelength 1 millimetre
30,000 Mc/s	**EHF** EXTREMELY HIGH FREQUENCY	1 centimetre
3,000 Mc/s	**SHF** SUPER HIGH FREQUENCY	10 centimetres
300 Mc/s	**UHF** ULTRA HIGH FREQUENCY	1 metre
30,000 Kc/s	**VHF** VERY HIGH FREQUENCY	10 metres
3,000 Kc/s	**HF** HIGH FREQUENCY	100 metres
300 Kc/s	**MF** MEDIUM FREQUENCY	1,000 metres
30 Kc/s	**LF** LOW FREQUENCY	10 kilometres
3 Kc/s	**VLF** VERY LOW FREQUENCY	100 kilometres

Fig. 57. Frequency-band designations

FREQUENCY BAND OF EMISSION. In communications, the frequency band required for a given type of transmission and speed of signalling.

FREQUENCY BRIDGE. An *ac bridge* whose balance is a function of the frequency of supply.

FREQUENCY CHANGER. (1) A multi-electrode valve which performs the two functions of generating the local oscillations and mixing them with the signal in *heterodyne* reception. The most popular form is a *triode-hexode* system in one envelope. See also

Block symbol.

Mixer. (2) A machine for converting ac at one frequency to ac at another frequency.

FREQUENCY COMPENSATION. Modification of an electronic circuit or device so as to improve its *frequency response*.

FREQUENCY CONTROL. The regulation of the frequency of an electronic system or electric generator within a narrow range.

FREQUENCY CONVERTER. See *Frequency changer*.

FREQUENCY DEPARTURE. The amount of variation of a *carrier* frequency from its assigned value.

FREQUENCY DEPARTURE METER. An instrument used in broadcasting stations to measure the drift of a transmitted carrier frequency.

FREQUENCY DEVIATION, FM. In *frequency modulation,* the peak difference between the instantaneous frequency of the modulated wave and the *carrier* frequency.

FREQUENCY DEVIATION, SYSTEM MAXIMUM. The greatest deviation specified in the operation of a *frequency modulation* system. For FM broadcast systems in the range 88 to 108 megacycles the maximum deviation is $\pm$ 75 kilocycles.

FREQUENCY DISCRIMINATOR. (1) A type of circuit used in *frequency modulation* systems which converts frequency modulated signals into amplitude modulated signals. The best known is the *Foster-Seeley discriminator* which includes a tuned *reactance valve* followed by a *limiter* to prevent amplitude variations appearing as noise in the audio output. (2) A device which responds only to a set of frequencies which have a particular characteristic in common, e.g. amplitude or period.

FREQUENCY DIVERSITY. A method of *diversity reception* or transmission in which different frequencies are transmitted carrying the same information.

FREQUENCY DIVIDER. A device for producing an output wave whose frequency is a proper fraction, usually a sub-multiple, of the input frequency.

Block symbol

FREQUENCY DIVISION MULTIPLEXING. The process of transmitting two or more signals over a common path by employing a different frequency band for each signal.

FREQUENCY DOUBLER. A device whose output frequency is twice its input frequency.

FREQUENCY DRIFT. A gradual change in the frequency of an oscillator or ac generator towards or away from the nominal frequency.

FREQUENCY METER. An instrument for measuring the frequency of an alternating quantity.

FREQUENCY MODULATED ALTIMETER. See *Absolute altimeter*.

FREQUENCY MODULATED (FM) RADAR. A form of *radar* in which the radiated wave is *frequency modulated* and the returning echo beats with the radiated wave, thus allowing the range to be measured.

FREQUENCY MODULATION (FM). A type of *modulation* in which the frequency of a sine wave carrier alters in synchronism with the amplitude of the modulating signal. In practice frequency variations are usually accompanied by phase variations and the term 'frequency modulation' is applied to the combined effects. See Fig. 3.

FREQUENCY MODULATION RECEIVER. A radio receiver designed to convert frequency modulated signals, usually into radio signals. The main advantage is the high signal-to-noise ratio possible even under conditions which make other types of receiver unsuitable. See Fig. 152.

FREQUENCY MONITOR. An instrument for the continuous indication of the departure of a frequency from its assigned value.

FREQUENCY MULTIPLIER. A device whose output frequency is designed to be a multiple of its input frequency. *Frequency doublers* and *triplers* are particular cases.

Block symbol.

FREQUENCY OVERLAP. In *colour television*, that part of the frequency band common to both *monochrome* and *chrominance channels*.

FREQUENCY PULLING. Of an *oscillator*, the change in the generated frequency caused by a change of load impedance.

FREQUENCY PUSHING. Of a *magnetron*: the rate of change of frequency with respect to anode current, usually expressed in MHz per ampere.

FREQUENCY RANGE. The range of frequencies for which a device may be considered useful with various circuits and operating conditions. See *Frequency band, Band width*.

FREQUENCY RECORD. A recording of various known frequencies at known amplitudes, usually for testing electroacoustic equipment.

FREQUENCY REGULATOR. A regulator for maintaining the frequency of a generator at a predetermined value, or for varying it in a predetermined manner.

FREQUENCY RESPONSE CHARACTERISTIC. Of a system or device, the variation with frequency of its transmission gain or loss.

FREQUENCY RUN. A series of tests at different frequencies made to determine the *frequency response characteristic*.

FREQUENCY SELECTIVITY. The degree to which an electric circuit or apparatus can differentiate between desired signals and other signals or interference at different frequencies.

FREQUENCY SHIFT TRANSMISSION. A type of *modulation* used in communications and radio *telemetering* systems in which the *carrier* is made to shift between two frequencies indicating mark (or 'on' pulse) and space (or 'off' pulse) respectively.

FREQUENCY SHIFT TRANSMITTER. A radio-telegraphy transmitter which transmits on two frequencies representing mark conditions and space conditions respectively.

FREQUENCY SPECTRUM OF ELECTROMAGNETIC WAVES. See chart in Fig. 46.

FREQUENCY STABILITY. The degree of constancy in the output frequency of devices such as valve oscillators and ac generators.

FREQUENCY STABILIZATION. The process of maintaining the frequency of any device or system so that it never departs by more than a specified amount from a reference frequency.

FREQUENCY STANDARD, PRIMARY. A very precise and stable *oscillator* which can be calibrated against national standard frequency radio transmissions.

FREQUENCY SWING. In *frequency modulation,* the peak difference between maximum and minimum values of the instantaneous frequency.

FREQUENCY TRIPLER. A device designed to deliver an output voltage at a frequency three times that of the input frequency.

FRESNEL. A unit of frequency equal to 10^{12} cycles per second.

FRICTIONAL ELECTRICITY. Static electricity produced by rubbing materials together.

FRINGE AREA. In radio or television broadcasting, the region round the transmitter in which satisfactory reception is not always certain.

FRONT PORCH. In a *television* signal, the period of time immediately preceding a synchronizing pulse during which the signal is held at *black level*.

FSD. Abbreviation for *full scale deflection*.

FSM. Abbreviation for *field-strength meter*.

FULL-SCALE DEFLECTION. Of an instrument: the maximum value of the measured quantity for which the instrument is scaled.

FULL-WAVE RECTIFICATION. *Rectification* in which both halves of the alternating current cycle are transmitted as unidirectional current. See Figure 58.

FULL-WAVE RECTIFIER. Any device, valve, or circuit which uses both positive and negative alternations in an alternating current to produce direct current. Cf. *Half-wave rectifier*.

FUNCTION GENERATOR. (1) A *signal generator* designed to produce test signals of various different waveforms over a wide range of low frequencies. (2) In *analogue computers,* an element which supplies the

Fig. 58. Full-wave rectification

value of a given function for a given value of the independent variable.

FUNCTION SWITCH. In *computers,* a network with a number of inputs and outputs so connected that its output signals are a representation of the input information in a code different from that of the input.

FUNCTION UNIT. In *computers,* a device which can store a functional relationship and release it continuously or in increments.

FUNCTIONAL DIAGRAM. A diagram representing the functional relationships between the different parts of a system.

FUNDAMENTAL. Of a complex vibration, the lowest frequency present.

FUNDAMENTAL COMPONENT. Of a *periodic quantity,* the *harmonic* component having the same *period* as the periodic quantity.

FUNDAMENTAL FREQUENCY. Of a *periodic quantity,* the frequency of a sinusoidal quantity having the same period as the periodic quantity.

FUNDAMENTAL TONE. The component of lowest pitch in a complex tone.

FUNDAMENTAL UNITS. Arbitrarily chosen units which serve as the basis of a system of *absolute units.*

FUSE. A device for protecting electric circuits in which some fusible material melts with a current overload and thus breaks the circuit.

FUSE CHARACTERISTIC. The relation between the current through the fuse and the time for the fuse to perform its interrupting function.

FUSE CURRENT RATING. The maximum current the fuse will carry under specified conditions without melting.

FUSE FREQUENCY RATING. The frequency at which a fuse is designed to operate.

FUSE VOLTAGE RATING. The voltage at which the fuse is designed to operate.

G

GAIN. An increase in signal power, usually expressed as the ratio of output power to input power in *decibels*.

GAIN CONTROL. A device for varying the amplification of an *amplifier*.

GALACTIC RADIO NOISE. Electrical noise which originates from sources outside the earth.

GALAXY NOISE. Electrical noise from interstellar space, mainly from certain areas in the Milky Way.

GALENA. A crystalline form of lead sulphide which acts as a *semiconductor* and has been used in *crystal rectifiers*.

GALL POTENTIOMETER. A coordinate *potentiometer* which includes two identical potential measuring networks, one in phase with the current and the other in *quadrature*.

GALLIUM (Ga). A metallic element, Atomic No. 31. Gallium arsenide has acquired importance as a *semiconductor*.

GALVANIC CELL. An *electrolytic cell* which produces electric energy by electrochemical action.

GALVANIC CURRENT. A *direct current* such as that produced by a galvanic cell.

GALVANISM. The therapeutic use of direct current.

GALVANOMAGNETIC EFFECT. See *Ettingshausen effect, Hall effect, Magnetoresistance,* and *Nernst effect*.

GALVANOMETER. An instrument for indicating or measuring small electric currents by a mechanical motion derived from electromagnetic or electrodynamic forces set up by the current. See *Ballistic, D'Arsonval, Differential, Mirror, Moving-magnet, Photoelectric, Tangent,* and *Vibration galvanometers*.

symbol

GALVANOMETER CONSTANT (FIGURE OF MERIT). The current necessary to produce a deflection of one division on the galvanometer scale.

GALVANOMETER SHUNT. A *resistor* connected in *parallel* across a *galvanometer* to reduce its sensitivity by a known amount and thus enable a large amount of current to be measured. Most *ammeters* are shunted galvanometers.

GALVANOTROPISM. The tendency of an organism to grow, turn, or move into a certain relation with an electric current.

GAMMA, PICTURE OR CAMERA TUBE. The exponent of the power law which approximately represents the relation between output and input magnitude of the tube.

GAMMA RAYS. Very high frequency radiation emitted from radioactive atoms, similar to the X-rays produced by a high voltage X-ray tube. See *Frequency spectrum of electromagnetic waves*.

GAMMA RAY SPECTROMETER. An instrument for determining the energy distribution of *gamma rays*.

GANGED TUNING. Simultaneous tuning of two or more circuits with a single mechanical control.

GAP ADMITTANCE, CIRCUIT. The *admittance* of the circuit at a gap between electrodes in the absence of an electron stream.

GAP ADMITTANCE, ELECTRONIC. The difference between the *gap admittance* with an electron stream traversing the gap and with no stream.

GAP CODING. In navigation, a technique of communication by so interrupting the transmission of a signal that the interruptions form a telegraphic message.

GAP FACTOR. In *travelling-wave tubes,* the ratio of the maximum energy gained in the accelerating gap to the maximum gap voltage.

GAP LENGTH. In *magnetic recording,* the distance between adjacent surfaces of the poles of a magnetic head.

GAS AMPLIFICATION. The ratio of the charge collected in a gas-filled *radiation counter tube* or ionization chamber, to the charge produced in the active volume by the preliminary ionizing event. In *gas phototubes* the gas amplification factor is the ratio of radiant or luminous sensitivities with and without *ionization*.

GAS BREAKDOWN. The phenomenon which occurs in a gas when the voltage across a *Townsend discharge* is increased to the point where the discharge is self-maintained. See *Gas discharge, Glow discharge, Thyratron.*

GAS CELL. A cell whose action depends on the absorption of gases by the electrodes.

GAS CURRENT. In a vacuum valve or tube, the *ionization* current flowing to a negatively biased electrode, composed of positive ions which result from collisions between electrons and molecules of the residual gas.

GAS DIODE. An electron valve containing two electrodes in an inert gas.

GAS DISCHARGE. The discharge of *electrons* which takes place when the potential difference between two cold electrodes in an *ionized* gas is made sufficiently high.

GAS DISCHARGE TUBE. An evacuated enclosure containing a gas at low pressure which permits the passage of electricity through the gas if sufficient voltage is applied. The tube is usually provided with metal electrodes but discharge can occur without electrodes by induced voltage.

GAS FOCUSING. In a *cathode-ray tube,* the focusing of the electron beam by the use of an inert gas inside the tube.

GAS MASER. A *maser* in which the microwave radiation interacts with the molecules of a gas, as e.g. in the *ammonia clock*.

GAS NOISE. *Noise* due to random ionization of gas molecules in a *gas tube* or a partially evacuated vacuum tube.

GAS PHOTOTUBE. A *phototube* in which a rarefied gas, such as argon, is introduced to counteract the *space charge* caused by electrons released by the action of light on the photosensitive cathode.

GAS RADIATION COUNTER. A *counter* in which the sample is introduced into the counter tube itself in the form of a gas.

GAS RATIO. The ratio of the *ion current* in a gas tube (valve) to the *electron current* which produces it.

GAS TUBE (VALVE). An electron *tube* or *valve* in which a contained gas or vapour plays a predominating role in the operation of the tube or valve.

Symbol for Gas tube

GAS-FILLED COUNTER. A radiation *counter* which detects radiation by the ionization of the gas in a gas-filled tube.

GAS-FLOW RADIATION COUNTER. A radiation *counter* in which a suitable gas flows slowly through the volume of the tube.

GASSING. The evolution of gases from one or more of the electrodes during *electrolysis*.

GATE. (1) In general, a device used for *gating*. (2) An electronic circuit with more than one input but only one output. (3) In *radar* or electronic control, a circuit or device designed to receive signals for only a selected fraction of the principal time interval.

Basic gate symbol

(4) An electric signal used to effect the passage of other signals through a circuit.

GATE CURRENT. Of a *magnetic amplifier*, the current in the *gate winding* of the *saturable reactor*.

GATE DETECTOR. A *detector* which operates only when a *gating pulse* is applied externally.

GATE IMPEDANCE. In a *magnetic amplifier*, the *impedance* of a *gate winding*.

GATE VOLTAGE. In a *magnetic amplifier*, the voltage across the terminals of the *gate winding*.

GATE WINDING. Of a *magnetic amplifier*, the *reactor* winding which produces the *gating* action.

GATED-BEAM TUBE. A *cathode-ray tube* with a special *accelerator* to control the beam width, used to replace a valve *gating* circuit, or as a *frequency discriminator*.

GATING. (1) The process of selecting those portions of a wave which occur during one or more time intervals, or which have magnitudes between selected limits. (2) Electronic switching using a *square wave* of desired time and duration.

GAUSITRON. A kind of mercury-arc valve *rectifier*.

GAUSS. The c.g.s. unit of *magnetic induction*. See *Units*.

GAUSSMETER. A *magnetometer* designed to measure the intensity of a magnetic field but not its direction.

GCA. Abbreviation for *Ground-controlled approach*.

GC/s. Abbreviation for Gigacycles (i.e., 10^9 cycles) per second.

GEE SYSTEM. A British radar navigation system operating on high frequencies (20–85 megacycles). Cf. *Loran.*

GEIGER COUNTER. A Geiger-Mueller *counter tube*, or such a tube together with its associated equipment. Geiger counters are *gas-filled counters.*

symbol

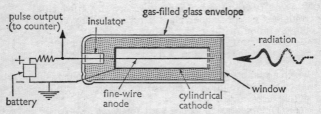

Fig. 59. Principle of Geiger counter tube

GEIGER-MUELLER THRESHOLD. Of a *counter tube*, the lowest applied voltage for which all pulses produced in the counter tube are substantially the same size, regardless of the size of the primary *ionizing event.*

GEISSLER TUBE. A special kind of discharge tube for showing the luminous effects of discharges through rarefied gases.

GENERATOR. (1) A machine for converting one form of energy into another. An electronic generator usually converts direct-current power into alternating-current power. (2) In *digital computers*, a programme which generates *coding.*

GEOMETRIC AMPLIFICATION FACTOR. Of a *triode* under static conditions: the ratio of the *anode* potential which produces a given electric field between *cathode* and *grid* to the grid potential which will produce the same field. It is always greater than 1.

GEOMETRIC CAPACITANCE. Of an insulated conductor in a vacuum, the *capacitance* determined by its geometrical shape.

GEOMETRIC DISTORTION. In television, any distortion in the *raster* due to imperfections of the *scanning.*

GEORGE. A large *digital computer*, which includes a store of 4,000,000 words on magnetic tape, built by the Argonne National Laboratory of America for its own use.

GERMAN SILVER (ELECTRUM). An alloy of copper, zinc, and nickel.

GERMANIUM (Ge). A dark grey, crystalline, metallic element. Atomic No. 32. It is one of the most important *semiconductors* and with suitable *doping* is widely used in *transistors.*

GETTER. A material with a strong chemical affinity for gases, which is vaporized inside a vacuum valve or tube, after the latter has been pumped and sealed, to remove residual gases. The most commonly used getter is metallic barium.

GEV. Abbreviation for giga-electron-volt, i.e. 10^9 electron volts.

GHOSTS (GHOST SIGNALS). In *television* and *radar*, unwanted images appearing on the screen of the cathode-ray tube or radar indicator usually caused by reflected rays from the transmitter or target reaching the receiver by paths other than the direct path.

GIGA-. Prefix denoting 1,000 million or 10^9.

GIGAHERTZ (Ghz). One thousand million cycles per second.

GILBERT. The unit of *magnetomotive force* in the c.g.s. electro-magnetic system. See *Units*.

GLASS VALVE (TUBE). An electron valve or tube which does not have a base of different material from the envelope, but in which the leads are brought out through metal-glass seals to reduce stray capacitances.

GLOSSMETER. A photoelectric instrument for determining the gloss factor of a polished surface, i.e. the ratio of reflected light in one direction to the total reflected light in all directions.

GLOVE BOX. A metal box with port openings sealed off by rubber gloves used for delicate manipulation in controlled atmospheres, e.g. transistor manufacture, and for handling radioactive isotopes.

GLOW DISCHARGE. A *discharge* of electricity through a gas in which the space potential near the cathode is much higher than the *ioniz-ation* potential of the gas. It is usually accompanied by a slight luminosity but not by much noise or heat.

GLOW DISCHARGE MICROPHONE. A *microphone* which, in place of a diaphragm, uses a *glow discharge* current which is modulated by the sound waves.

GLOW LAMP. A lamp in which the illuminating source is a non-rectifying glow-discharge. See *Neon lamp*.

GLOW SWITCH. An electron tube containing contacts which are operated thermally by a *glow discharge*.

GLOW-DISCHARGE COLD-CATHODE TUBE (VALVE). A tube or valve whose operation depends on the properties of a *glow dis-charge*.

GLUE-LINE HEATING. In *dielectric heating*, an arrangement of electrodes designed to give preferential heating to a thin film of material with a high *loss factor* between layers of material with a relatively low loss factor.

GOLD (Au). Metallic element, Atomic No. 79. Extensively used in electronics as a plating material where high resistance to corrosion and good electrical conductivity is required. Fine gold wires are used in gold-bonded *diodes* and *transistors*.

GOLD-LEAF ELECTROSCOPE. A sensitive instrument for the measure-ment of electric charge in which gold leaf is used as an indicator.

GOLDSCHMIDT ALTERNATOR. An alternator for generating radio

frequency by interaction of high frequency oscillating currents between rotor and stator.

GONIOMETER (RADIOGONIOMETER). A device for measuring angles or direction of a field, using a system of coils in which two coils are mounted at right angles and a third is pivoted to rotate between them.

GRADED-BASE TRANSISTOR. See *Drift transistor*.

GRADED FILTER. A multi-section *rectifier filter* designed to supply anode voltages to several amplifier stages in the most economical manner.

GRADIENT METER. A device for measuring the potential gradient of an electric field at the surface of a *conductor*.

GRAPHECON. A double-ended *storage tube* in which signals are recorded by the *conductivity* induced in the target by a high voltage writing beam. It is used for the integration and storage of radar information and as a translating mechanism.

GRAPHIC PANEL. In *automation* and *remote control* systems, a master control panel which uses coloured block diagrams to show the relation and functioning of the different parts of the control equipment. Controls and recording instruments may be mounted on the panel in their correct relative positions.

GRAPHICAL SYMBOLS. The graphical symbols illustrated in Fig. 60 (pp. 163–172) are mostly those recommended by the British Standards Institution for electronics, telecommunications, and allied subjects.

GRASS. A colloquial term for the pattern produced by random noise on the screen of a cathode-ray tube in certain types of radar display.

GRATING. An arrangement of alternate opaque and transparent, or reflecting and non-reflecting, elements which separates the incident electromagnetic or sound radiation into a frequency spectrum. Examples are *diffraction gratings* and wire screens.

GRATING REFLECTOR. An openwork metal structure designed to produce a reflecting surface for incident radiation of certain wavelengths.

GRAY CODE. A cyclically progressive code used in optical disc *digitizers*, in which only one bit changes between adjacent integers, thus avoiding the ambiguities of binary codes. See Fig. 37.

GREEN GUN. In a three-colour television *picture tube*, the *electron gun* used to excite the green phosphor.

GREY BODY. An imperfect *black body* which absorbs more radiation than it emits, but nevertheless emits equally at all wavelengths.

GRID. A valve or tube *electrode* with one or more openings for controlling the passage of electrons or ions. The term is frequently used to mean *control grid*.

GRID BASE. The value of *grid bias* at which *cut-off* occurs in a valve.

symbol

VALVE ELECTRODES AND VALVES

anode		cathode metallic and liquid	
anode luminescent		cathode photoelectric or radioactive	
grid		internal shield connected to cathode	
grid to which varying potential is applied		secondary emission electrode	
screen grid (if required to be specially distinguished)		cathode	
beam-forming electrode		cathode with heater	
filament directly-heated cathode heater			

VALVE EXAMPLES

diode triode

Fig. 60. Graphical symbols

163

electron
multiplier

half-wave
rectifying
valve

screen grid
variable mutual
conductance
indirectly-
heated

full-wave
rectifying
valve

photo-
electric
cell

pentode
indirectly-
heated

photo-
conductive
cell

Hexode

tetrode
indirectly
heated

diode
pentode

Beam power
valve

double
diode

diode
triode

Fig. 60. Graphical symbols

164

CATHODE RAY TUBES AND OTHER TYPES OF VALVE.

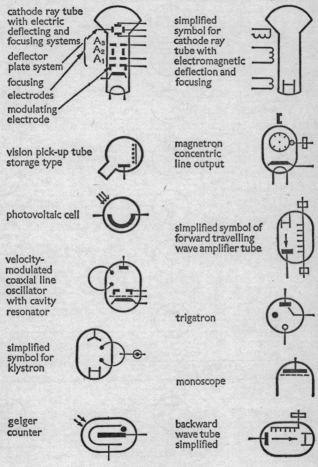

cathode ray tube with electric deflecting and focusing systems

deflector plate system

focusing electrodes

modulating electrode

simplified symbol for cathode ray tube with electromagnetic deflection and focusing

vision pick-up tube storage type

magnetron concentric line output

photovoltaic cell

simplified symbol of forward travelling wave amplifier tube

velocity-modulated coaxial line oscillator with cavity resonator

trigatron

simplified symbol for klystron

monoscope

geiger counter

backward wave tube simplified

Fig. 60. Graphical symbols

165

GAS-FILLED COLD-CATHODE DISCHARGE TUBES

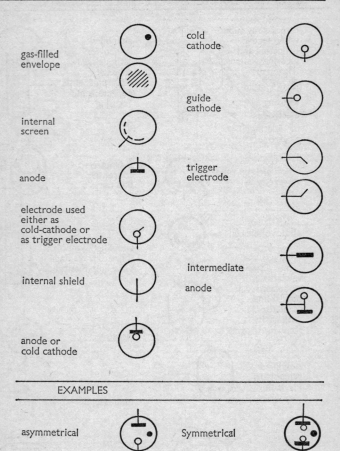

gas-filled envelope

cold cathode

internal screen

guide cathode

anode

trigger electrode

electrode used either as cold-cathode or as trigger electrode

internal shield

intermediate anode

anode or cold cathode

EXAMPLES

asymmetrical

Symmetrical

Fig. 60. Graphical symbols

166

Neon Trigatron

trigger tube
two trigger
electrodes multi-electrode

example: stabilovolt

typical multi-cathode tube
main cathodes brought out
separately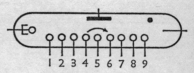

1 2 3 4 5 6 7 8 9

TRANSISTORS AND ALLIED SYMBOLS

point-contact or
junction triode
transistor with
n-type base
(arrow on emitter) p-type base

JUGFET

symbol for
jugfet with
n-type channel with p-type
channel

IGFET
symbols for depletion
type single-gate

n-channel p-channel

Fig. 60. Graphical symbols

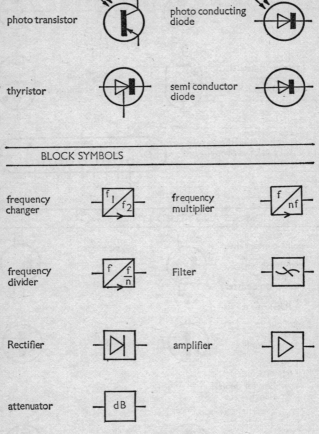

photo transistor

photo conducting diode

thyristor

semi conductor diode

BLOCK SYMBOLS

frequency changer

frequency multiplier

frequency divider

Filter

Rectifier

amplifier

attenuator

Fig. 60. Graphical symbols

RESISTORS, CAPACITORS, TRANSFORMERS AND MEASURING INSTRUMENTS.

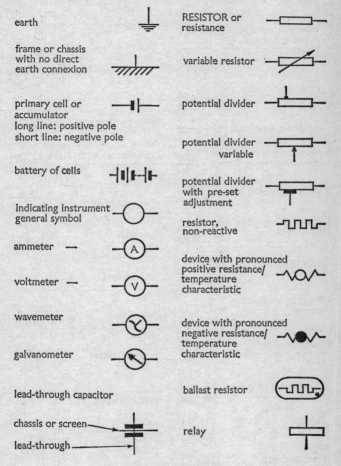

earth

RESISTOR or resistance

frame or chassis with no direct earth connexion

variable resistor

primary cell or accumulator
long line: positive pole
short line: negative pole

potential divider

potential divider variable

battery of cells

potential divider with pre-set adjustment

indicating instrument general symbol

resistor, non-reactive

ammeter

device with pronounced positive resistance/temperature characteristic

voltmeter

wavemeter

device with pronounced negative resistance/temperature characteristic

galvanometer

lead-through capacitor

ballast resistor

chassis or screen

relay

lead-through

Fig. 60. Graphical symbols

169

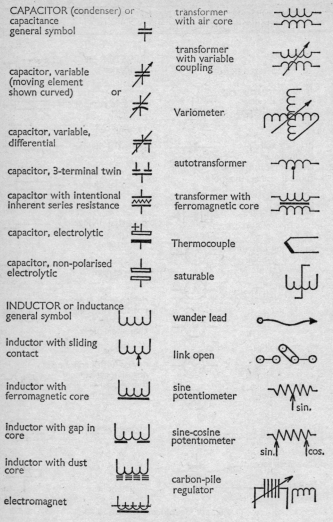

CAPACITOR (condenser) or capacitance general symbol	transformer with air core
capacitor, variable (moving element shown curved) or	transformer with variable coupling
	Variometer
capacitor, variable, differential	
capacitor, 3-terminal twin	autotransformer
capacitor with intentional inherent series resistance	transformer with ferromagnetic core
capacitor, electrolytic	Thermocouple
capacitor, non-polarised electrolytic	saturable
INDUCTOR or inductance general symbol	wander lead
inductor with sliding contact	link open
inductor with ferromagnetic core	sine potentiometer
inductor with gap in core	sine-cosine potentiometer
inductor with dust core	carbon-pile regulator
electromagnet	

Fig. 60. Graphical symbols

170

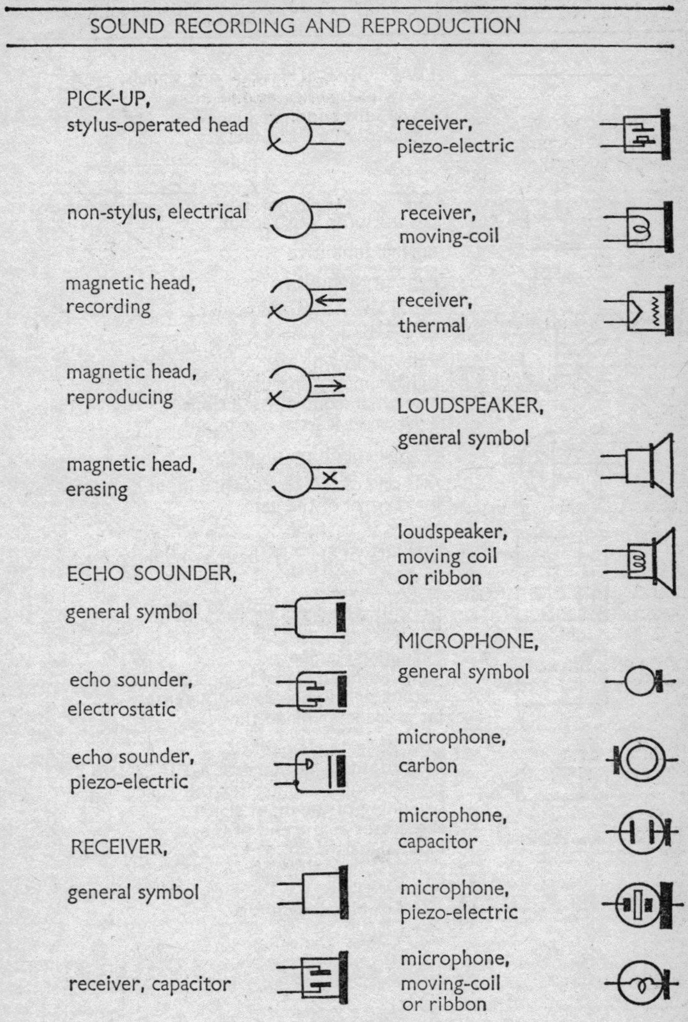

SOUND RECORDING AND REPRODUCTION

PICK-UP,
stylus-operated head

receiver,
piezo-electric

non-stylus, electrical

receiver,
moving-coil

magnetic head,
recording

receiver,
thermal

magnetic head,
reproducing

LOUDSPEAKER,
general symbol

magnetic head,
erasing

loudspeaker,
moving coil
or ribbon

ECHO SOUNDER,
general symbol

MICROPHONE,
general symbol

echo sounder,
electrostatic

echo sounder,
piezo-electric

microphone,
carbon

microphone,
capacitor

RECEIVER,
general symbol

microphone,
piezo-electric

receiver, capacitor

microphone,
moving-coil
or ribbon

Fig. 60 Graphical symbols

FUNCTION SYMBOLS FOR ELECTRONIC CIRCUITS

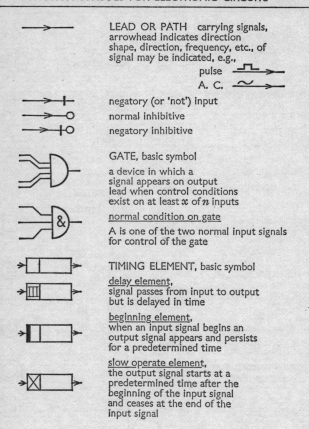

LEAD OR PATH carrying signals, arrowhead indicates direction shape, direction, frequency, etc., of signal may be indicated, e.g.,

pulse

A. C.

negatory (or 'not') input

normal inhibitive

negatory inhibitive

GATE, basic symbol

a device in which a signal appears on output lead when control conditions exist on at least x of n inputs

normal condition on gate

A is one of the two normal input signals for control of the gate

TIMING ELEMENT, basic symbol

delay element,
signal passes from input to output but is delayed in time

beginning element,
when an input signal begins an output signal appears and persists for a predetermined time

slow operate element,
the output signal starts at a predetermined time after the beginning of the input signal and ceases at the end of the input signal

Fig. 60. Graphical symbols

GRID BIAS. The negative potential applied to the *control grid* of a valve to fix the operating range with respect to the anode current-voltage characteristic.

GRID BIAS, AUTOMATIC. The *grid bias* provided by the difference in potential across resistance in the grid or cathode circuit caused by grid or cathode current or both.

GRID (BLOCKING) CAPACITOR. A small *capacitor* used to pass the signal current—in the grid circuit of a valve while blocking the direct anode current of the previous stage.

GRID CHARACTERISTIC. Of a valve, the graph of *grid current* versus *grid voltage*, with all other electrodes being maintained at constant voltage.

GRID CURRENT. Current flow between *cathode* and *grid* in a valve.

GRID EMISSION. *Electron* or *ion* emission from the *grid* of a valve.

GRID LEAK. A *resistor*, usually of a fairly high value, connected between the control grid and cathode of a valve to develop a *grid bias* voltage.

GRID LIMITING. The use of a high-value *resistor* in series with the *grid* of a valve to limit the range of positive volts on the grid.

GRID MODULATION. *Modulation* produced by the application of a modulating voltage to the control grid of a valve in which the *carrier* wave is present. Where the grid voltage contains externally generated pulses the process is called grid pulse modulation.

GRID NEUTRALIZATION. A method of *neutralizing* an amplifier in which the requisite phase shift of 180 degrees is provided by an inverting network in the grid circuit. See *Phase inverter, Rice neutralization*.

GRID RESISTOR. Any *resistor* in the grid circuit. See *Grid leak, Grid limiting, Grid-suppressor resistor*.

GRID RETURN. The external conducting path by which the grid current returns to the cathode.

GRID SWING. The maximum excursion of the *control grid* voltage from the *bias* voltage, either negatively or positively, when a signal is applied to the grid.

GRID VOLTAGE. The voltage between *grid* and *cathode* of a valve, measured at the valve terminals.

GRID-DIP METER. A portable instrument for measuring radio-frequency which uses an oscillator and a grid-current meter. The oscillator is electromagnetically coupled to the measured circuit and resonance is indicated by a sharp dip in the reading of the grid-current meter.

GRID-DRIVE CHARACTERISTIC. A graph of the relation between electrical or light output and control electrode voltage measured from *cut-off*.

GRID-GLOW TUBE. A cold-cathode *gas discharge* tube employing electrostatic starting, which maintains a constant voltage from anode to cathode for a wide range of anode current and is used for voltage regulation.

GRID-LEAK DETECTION. The use of electron current to the grid in a valve for *rectification*. The voltage associated with the current flowing through a high resistance in the grid circuit appears in an amplified form in the anode circuit. Now largely replaced by the *diode detector*.

GRID-POOL TUBE. A gas discharge tube with a solid or liquid pool-type cathode, e.g. a *mercury-pool cathode*.

GRID-PULSE MODULATION. See *Grid modulation*.

GRID-SUPPRESSOR RESISTOR. A resistor connected between the control grid and tuned circuit of an rf amplifier to suppress *parasitic oscillations*.

GROUND (EARTH). A conducting connexion between an electric circuit or equipment and earth or to a body at zero *potential* with respect to the earth.

GROUND ABSORPTION. The loss of energy in the transmission of radio waves due to absorption in the ground.

GROUND CAPACITANCE. The *capacitance* between an electric circuit or equipment and earth, or to some conducting body which is at zero potential to earth.

GROUND CLUTTER. The pattern produced on the screen of a radar indicator due to unwanted *ground return*.

GROUND CONTROL. Guidance of an aircraft or missile by a person on the ground; also the electronic equipment used for this purpose.

GROUND CURRENT. See *Earth current*.

GROUND ELECTRODE. See *Earth electrode*.

GROUND PLATE. See *Earth plate*.

GROUND POSITION INDICATOR (GPI). A dead-reckoning computer for continuous indication of the position of an aircraft with respect to some fixed point on the ground.

GROUND REFLECTION. In *radar* transmission, the wave which reaches the target after reflection from the earth's surface.

GROUND RETURN. In *radar*, the sum of received echoes due to reflection from the surface of the earth and from objects thereon.

GROUND WAVE. A radio wave which is propagated over the earth and is normally influenced by the presence of the ground and *troposphere*. It has two components: the *space wave* and the *surface wave*.

GROUND-CONTROL INTERCEPTION (GCI). A combination of radar and radio systems used for the control of *guided missiles* and *interceptors*.

GROUND-CONTROLLED APPROACH (GCA). A ground radar system providing information to enable aircraft approaches to be directed by radio.

GROUNDED. *Earthed*.

GROUNDED-ANODE AMPLIFIER. See *Cathode follower*.

GROUNDED-CATHODE AMPLIFIER. The conventional valve *amplifier* circuit with a *cathode* at ground potential at the operating

174

frequency and with the input applied between the *control grid* and ground and output load between *anode* and ground.

GROUNDED-GRID AMPLIFIER. A valve *amplifier* in which the *control grid* is at ground potential at the operating frequency, with input applied between *cathode* and ground, and output load connected between *anode* and ground.

GROUND-REFLECTED WAVE. The component of the *ground wave* which is reflected from the ground.

GROUP FREQUENCY. The frequency corresponding to the *group velocity* of a propagated wave in a *transmission line* or *waveguide*.

GROUP OPERATION. Of a multiple switch or circuit-breaker, operation of all poles by one operating mechanism.

GROUP VELOCITY. The velocity of propagation of the envelope of a wave occupying a frequency band over which the *envelope delay* is approximately constant.

GROWN JUNCTION. In *semiconductors*: a junction produced by changing the types and amounts of impurities added to the semiconductor crystal during its growth from the melt.

GROWN-DIFFUSED TRANSISTOR. A *junction transistor* in which the final junctions are formed by diffusion of impurities near the grown junction.

G-STRING. See *SWTL*

GUARD BAND. A *frequency band* between two neighbouring channels which is left vacant to give a margin of safety against mutual interference.

GUARD RING. An auxiliary electrode in a *counter tube* or *ionization chamber* used to control potential gradients, reduce insulator leakage, and define the sensitive volume.

GUARD-RING CAPACITOR. A standard *capacitor* with circular parallel plates surrounded by a ring held at the potential of one of the plates to reduce the *edge effect*.

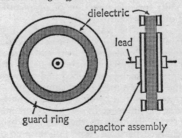

Fig. 61. Diagrammatic top view and section of guard-ring capacitor

GUARD-WELL CAPACITOR. A primary standard *capacitor*, fixed or variable for values of capacitance below 0·1 *micromicrofarad*.

The *guard ring* forms a well in which is mounted a Pyrex disc for the accurate location of the electrode assembly.

GUDDEN-POHL EFFECT. The momentary illumination produced when an electric field is applied to a phosphor previously excited by ultra-violet radiation.

GUIDE FIELD. Of a *betatron* or *synchrotron*, the magnetic flux provided to hold the particles in a stable circular orbit during the accelerating period.

GUIDED MISSILE. A missile which is directed to its target, either by preset or self-reacting devices inside it, or by radio command from outside.

GUIDED ROCKET. A *guided missile* with rocket propulsion.

GUIDED WAVE. A *wave* whose energy is concentrated substantially within the boundaries between material of different properties and which is propagated along the path so defined. See *Waveguide*.

GUILLEMIN EFFECT. The tendency of a bent rod of magnetostrictive material to straighten in a longitudinal magnetic field. See *Magnetostriction*.

GUILLEMIN LINE. A network used to generate a nearly square pulse having a steep rise and fall.

GULP. In *data processing*: several *bytes*, forming part of a *word*.

GUN. See *Electron gun*.

GUNN EFFECT. The effect observed by the English scientist J. B. Gunn in 1963 that the application of a steady electric field above a certain threshold level to low resistivity gallium arsenide caused the *charge carriers* to break up into domains. The rate at which these domains moved along the potential gradient was determined by the *carrier mobility*. The effect is used in some solid-state micro-wave oscillators.

GUTTA-PERCHA. An insulating material with properties similar to those of natural rubber.

GYRATOR. A waveguide component in which a *ferrite* section gives zero phase shift for one direction of propagation and 180° phase shift for the other direction.

GYRO FLUX-GATE COMPASS. A compass in which a triangular *flux-gate* stabilized by a gyroscope, senses the horizontal component of the earth's magnetic field and consequently changes in the heading of an aircraft in which it is fixed. These changes appear as current change which is amplified and converted into mechanical motion on the dial of a master indicator.

GYRO FREQUENCY. Natural frequency of a charged particle, e.g. an *electron*, under the influence of a constant magnetic field.

GYROMAGNETIC. Pertaining to the magnetic properties of rotating electrically charged particles, especially atomic particles.

GYROMAGNETIC EFFECT. See *Barnett effect*.

GYROSYN®. An aircraft compass system in which directional readings are obtained from a gyroscope under the long-term control of either a single flux-valve, or a pair of flux-valves, situated in the wing tips.

H

HALF-ADDER. In *computers*, a circuit with two input and two output channels for *binary* signals. Two half-adders can be used to form a binary *adder*.

HALF-CELL. A single electrode and that part of the electrolyte with which it is in contact.

HALF-DUPLEX OPERATION. Of a telegraph system, *duplex operation* limited to working in either direction but not in both directions simultaneously.

HALF-LIFE. The time required for the radioactive decay of one half of the atoms in a radioactive substance.

HALF-POWER POINT. The point on a response characteristic which represents half the power intensity of that corresponding to the point of maximum power.

HALF-WAVE RECTIFICATION. *Rectification* in which only one half of the alternating current cycle is transmitted as unidirectional current.

Fig. 62. Half-wave rectification

HALF-WAVE RECTIFIER. A *rectifier* employing *half-wave rectification*.

HALF-WAVE VOLTAGE DOUBLER. A voltage doubler in which load current is drawn from the ac source during only half the cycle.

HALL ANGLE. An electric angle associated with the *Hall effect*.

HALL COEFFICIENT. A coefficient associated with the *Hall effect* where this results in a transverse potential gradient in a *conductor*.

HALL EFFECT. If a thin sheet of metal or semiconductor with an electric current flowing along its length is placed in a magnetic field at right angles to the direction of the current, an electromotive force is developed in a direction at right angles to both the current and the magnetic field. See *Suhl effect*.

HALL MOBILITY. The mobility of *electrons* or *holes* in a *semi-conductor*, numerically equal to the drift velocity of the charge *carrier* under a field of 1 volt/cm.

177

HALLWACHS EFFECT. When a negatively charged body is irradiated in a vacuum it becomes discharged.

HAND COUNTER. A radiation *counter*, usually of the *Geiger* type designed to measure the extent of radioactive contamination of the hands.

HANDSET. The standard telephone set, in which the transmitter and receiver are contained in a single unit designed for the hand.

HARCO. Hyperbolic ARea COverage. An improved version of the Decca navigator.

HARD VALVE. A high vacuum electronic valve.

HARDWARE. A popular expression for the physical parts making up any electronic equipment, particularly, as in *computers*, when it is desired to distinguish them from the *software*.

HARMODOTRON. A device which uses a tightly-bunched, high energy, electron beam to excite harmonics in a resonator for the generation of frequencies of more than a thousand megacycles per second.

HARMONIC. An oscillation of a periodically varying quantity having a frequency which is an integral multiple of the *fundamental frequency*.

HARMONIC, Nth. The harmonic whose frequency is n times the *fundamental* frequency.

HARMONIC ANALYSER. A device (usually mechanical) for measuring the amplitude and phase of the harmonic components of a periodic function from its graph. See *Wave analyser*.

HARMONIC DISTORTION. Of a *transducer*, the production of harmonic frequencies at the output of the transducer when a *sinusoidal* voltage is applied at the input.

HARMONIC GENERATOR. An oscillating device, controlled by a tuning fork, which is capable of producing a large range of odd and even *harmonics* of the fundamental tuning fork frequency.

HARMONIC SERIES. An acoustic series in which each basic frequency is an integral multiple of a fundamental frequency.

HARRIS FLOW. The flow of *electrons* in a cylindrical beam in which the divergence due to *space charge* is counteracted by the application of a radial electric field.

HARTLEY. A unit of information content, equivalent to the designation of one of ten equally likely values or states of anything used to store or convey information. One hartley equals (log of 10 to base 2) times one *bit*, i.e. 3·223 bits.

HARTLEY OSCILLATOR. An *oscillator* in which a parallel tuned *tank circuit* is connected between grid and anode of a valve; or between base and collector of a transistor.

HARTLEY'S LAW. A law relating the information in a message, its time of transmission, and the bandwidth of the channel used.

HARTREE DIAGRAM. For *travelling-wave magnetrons*, a diagram showing the relation between the different modes of oscillation and the anode voltage for various values of magnetic field flux density.

HAY BRIDGE. An *ac bridge* for the measurement of *inductance* in terms of *capacitance*, *resistance*, and frequency.

HAZELTINE NEUTRALIZATION. A method of anode *neutralization* used in single-stage power amplifiers.

H-BEND. In *waveguides*, a smooth change in the direction of the axis which is maintained perpendicular to the plane of polarization.

HEAD. A device which reads, records or erases information in a storage medium.

HEAD AMPLIFIER. An *audio-frequency* amplifier which amplifies the sound signals in a motion picture projector.

HEADING MARKER. A radial trace on the *plan position indicator* screen of a mobile radar station to indicate the heading of the vehicle.

HEARING LOSS. Of a human ear at a specified frequency, the ratio, in *decibels*, of the *threshold* of audibility for that ear to the normal threshold.

HEAT COIL. A protective device which opens a circuit when the current in that circuit causes the temperature to rise above a specified value.

HEAT DETECTOR. In a burglar alarm system, a temperature sensitive device which operates an alarm in the event of an attack by heat or burning.

HEATER. Generally, any *resistance* used to supply heat. In particular the heating element in an *indirectly heated cathode* of a valve. See also *Induction heating*.

HEATER VOLTAGE. The voltage between the terminals of a *heater*.

HEATING DEPTH. In *dielectric heating*, the depth below the surface of a material to which effective dielectric heating can be confined when *applicator* electrodes are applied to one surface only.

HEATING ELEMENT. The complete resistor, including the structure on which the resistance wire is wound, used in electric fires, ovens, kettles, etc.

HEATING PATTERN. In *induction* and *dielectric heating*, the distribution of temperature in the load or charge.

HEATING TIME. This usually refers to *cathode heating time*.

HEAVISIDE LAYER. A region of the *ionosphere* about 55 to 85 miles from the earth's surface which maintains its ionized condition night and day.

HEAVISIDE-CAMPBELL BRIDGE. A *bridge* for the comparison of *self* and *mutual inductance*.

HEAVISIDE-LORENTZ ELECTROMAGNETIC SYSTEM OF UNITS. A rationalized system of units used in electromagnetic theory devised so that the same constants appear in corresponding electrical and magnetic laws. See *Units*.

HEAVY CURRENT. Colloquial for large current.

HEIGHT CONTROL. (1) In *television* receivers, the control which is used to adjust the vertical *sweep* and hence the vertical height of the picture. (2) In some radar receivers, the control used to match the amplitudes of two echoes which determine the height of the target.

HEIL TUBE. An early version of the *klystron*.

HELICON. A semiconductor device which amplifies or oscillates by an action similar to that of a travelling wave tube without the use of *p-n junctions*.

HELIPOT®. A helically wound, continuously variable resistor which is widely used in *servomechanisms*.

HELIUM (He). Inert, gaseous element, Atomic No. 2.

HELIXYN®. An analogue *transducer* developed by Associated Electrical Industries Ltd.

HELMHOLTZ COIL. A kind of *variometer*.

HELMHOLTZ RESONATOR. A *cavity resonator* which opens to the outside medium through a single small hole.

HENRY. The practical electromagnetic unit of *inductance*, the inductance of a circuit which produces a counter electromotive force of 1 volt when the current flowing through it is changing at the rate of 1 ampere per second. See *Units*.

HEPTODE. A valve with seven electrodes: anode, cathode, control electrode, and four additional electrodes which are usually grids.

HERTZ. Unit of frequency used on the Continent equal to one cycle per second. See SI units.

HERTZIAN WAVES. Radio waves.

HETERODYNE. In a radio receiver, the process of combining in a non-linear device a received wave with a locally generated wave of slightly different *carrier* frequency. The two combining frequencies produce sum and difference frequencies called *beats*. See *Superheterodyne*.

HETERODYNE DETECTION. *Detection* using the *heterodyne* principle. See *Beat frequency oscillator, Superheterodyne*.

HETERODYNE FREQUENCY. A separate frequency equal to the sum or difference of the two frequencies of a *heterodyne* circuit.

HETERODYNE INTERFERENCE (WHISTLE). A high audio note in a *superheterodyne* receiver caused by the beating of the carrier of two transmissions whose frequencies are close together.

HETERODYNE WAVEMETER. A calibrated variable frequency oscillator whose output is *heterodyned* with the frequency of the circuit to be measured.

HEUSLER ALLOYS. A group of strongly *ferromagnetic* alloys of manganese, copper, and aluminium which do not contain iron, cobalt, or nickel.

HEXODE. A six-electrode valve containing: anode, cathode, control electrode and three additional electrodes which are usually grids.

HF. Abbreviation for *High Frequency*.

HIERARCHIC CONTROL. A method of controlling large or complex systems in which each process has its own controller and a central controller governs the whole process. See e.g. ARCH.

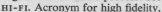
symbol

HI-FI. Acronym for high fidelity.

HIGH FIDELITY AMPLIFIER. An *amplifier* designed to reproduce an input signal with an exceptionally high degree of accuracy.

HIGH FIDELITY RECEPTION. The reception and reproduction of all *audio frequencies* without serious distortion.

HIGH FREQUENCY (H-F). Radio frequencies between about 3,000 and 30,000 kilocycles per second. See *Frequency band*.

HIGH RECOMBINATION-RATE CONTACT. A contact between *semiconductors* or between a metal and a semiconductor, at which the densities of *charge carriers* are maintained substantially independent of current density.

HIGH TENSION. Synonym for high *voltage* or high *potential*.

HIGH-FREQUENCY CARRIER TELEGRAPHY. *Carrier telegraphy* in which the frequency of the *carrier* currents is above the range transmitted over a *voice-frequency* telephone channel.

HIGH-FREQUENCY TREATMENT. The treatment of diseases by the use of high-frequency, intermittent, and isolated trains of heavily damped oscillations of high voltage and relatively low current.

HIGH-FREQUENCY RESISTANCE. See *Effective resistance*.

HIGH-LEVEL DIFFUSION LENGTH. In a *semiconductor* or *transistor*, the average distance through which a *hole* diffuses between generation and recombination.

HIGH-LEVEL FIRING TIME. In *switching tubes*, e.g. *TR switch*, the time required to establish a radio-frequency discharge in the tube after the application of radio-frequency power.

HIGH-LEVEL MODULATION. *Modulation* at any point in a system where the power level approximates to that at the output of the system.

HIGH-LEVEL R-F SIGNAL. A radio-frequency signal of sufficient power to fire a *switching tube*.

HIGH-PASS FILTER. A wave *filter* with a single transmission band from some critical or cut-off frequency (not zero) up to infinite frequency.

HIGH-RESISTANCE VOLTMETER. A *voltmeter* having a resistance greater than 1,000 ohms per volt and drawing negligible current.

HIGH-VACUUM VALVE (TUBE). A *valve* or *tube* evacuated to a degree of vacuum where its characteristics are essentially unaffected by gaseous *ionization*.

181

HIGH-VELOCITY SCANNING. The *scanning* of a target with electrons of velocity sufficient to produce a *secondary-emission* ratio greater than 1.

HILL-AND-DALE RECORDING. See *Recording, vertical*.

HIPAR. HIgh Power Acquisition Radar.

HIPERCO. A magnetic alloy of iron, cobalt, and chromium.

HIPERNIK. A range of alloys of iron and nickel.

HIPERSIL. A magnetic alloy of iron and silicon, in which the grains of material are oriented in a particular direction.

HITTORF DARK-SPACE. The *cathode dark-space*.

HITTORF PRINCIPLE. Discharge between electrodes in a gas at a given pressure will not always occur between closest points of the electrodes if these do not represent the points of minimum ignition potential.

H-NETWORK (H-PAD, H-SECTION). A *network* of five *impedance* branches connected in the form of an 'H'.

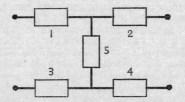

Fig. 63. Typical H-network using resistors

HODOSCOPE. An array of radiation detectors used to trace the path of a *cosmic-ray* particle.

HOFFMAN ELECTROMETER. An extremely sensitive form of *quadrant electrometer*.

HOLD. (1) In *charge-storage* tubes, maintain storage elements at equilibrium potential by electron bombardment. (2) In *computers*, retain the information contained in one storage device after copying it into another storage device.

HOLDING ANODE. An auxiliary anode to insure continuity of ionization in *mercury-arc rectifiers*, and to stabilize the *cathode-spot* in *ignitrons*.

HOLDING BEAM. A diffuse beam of electrons for regenerating the charges retained on the dielectric surface of an *electrostatic memory* or *storage tube*.

HOLE. In a *semiconductor*, a mobile vacancy in the *valence band* which acts as a positive electronic charge with a positive mass. It is mathematically equivalent to the *positron*.

HOLE CONDUCTION. A mechanism of conduction in an *extrinsic semiconductor* in which *holes* are, in effect, propagated through the structure under the influence of an electric field by a continuous process of exchange with adjacent electrons. The apparent movement of such holes is equivalent to a movement of positive charges in the same direction.

HOLE CURRENT. In a *semiconductor*, the current associated with the formation and transfer of *holes* which behave as positive charge carriers, and drift in the same direction in the presence of an electric field.

HOLE DENSITY. The density of *holes* in an otherwise full band in a *semiconductor*.

HOLE INJECTION. The injection of *holes* in a *semiconductor* which can be effected by application of a sharp conducting point in contact with an *n-type* semiconductor.

HOLE TRAP. An impurity in a *semiconductor* which can release electrons to the *conduction* or *valence bands* and can therefore trap a *hole*. See *Donor impurity*.

HOLLOW-CATHODE TUBE. A *gas discharge tube* which emits radiation in the form of *cathode glow* from a hollow cathode closed at one end.

HOLOGRAPHY. A technique which employs a *laser* for the formation of visible, and apparently 3-dimensional, images. A hologram is made by illuminating an object with a laser beam in such a way that both the direct light from the beam and the light reflected from the object are recorded on photographic film. Multiple images of the original object can then be observed by viewing the film with light of a single wavelength.

HOMEOSTASIS. The dynamic condition of a system whose input and output are evenly balanced and which therefore gives the appearance of a steady state.

HOMING. Following a course directed towards a point by maintaining some navigational coordinate (not altitude) constant.

HOMING GUIDANCE. Of a *guided missile*, a technique of automatic steering towards a target by recognition of some distinguishing characteristic of the target.

HOMODYNE RECEPTION. A system of reception using a locally generated voltage of *carrier* frequency.

HOMOPOLAR GENERATOR. A dc generator in which the voltage generated in the active conductors maintains the same direction with respect to those conductors.

HOMOPOLAR MAGNET. A magnet whose pole pieces are concentric.

HONEYCOMB COIL. A coil in which the turns are wound in crisscross fashion to reduce the *distributed capacitance*.

HOOK TRANSISTOR. See *Transistor, hook*.

HOOK-UP. Colloquial for any electric or electronic circuit, especially of a temporary nature.

HOPPER. In *data processing*: a device which holds cards and makes them available to a feed mechanism.

HORIZON SENSOR. A *thermistor* device which detects the thermal discontinuity between earth and space, used to establish a stable vertical reference level for missiles and satellites.

HORIZONTAL BLANKING. Of a *cathode-ray tube*, *blanking* applied to the tube to eliminate the horizontal trace during *flyback*.

HORIZONTAL CENTRAL CONTROL. Of a *cathode-ray tube*, a control which shifts the starting point of the horizontal *sweep* in a horizontal direction with respect to the tube face.

HORIZONTAL HOLD CONTROL. In *television*, the control which varies the free-running period of the horizontal deflection oscillator.

HORIZONTAL POLARIZATION. The *polarization* of a wave in which the electric-field vector is parallel to the horizon. Horizontally polarized aerials, e.g. some television aerials, have their *dipoles* arranged horizontally.

HORIZONTAL RESOLUTION. In *television*, the number of picture elements resolved along the *scanning* lines.

HORIZONTAL SYNCHRONIZING PULSE. See *Line synchronizing pulse*.

HORN FEED. A radar feed in the shape of a horn.

HORN LOUDSPEAKER. A loudspeaker in which the radiating element is coupled to the air by means of a horn.

HORSEPOWER. British unit of power equal to 746 *watts*. See *Units*.

HORSESHOE MAGNET. A bar magnet bent to the shape of a horseshoe so that its N. and S. poles are close together.

HOT CATHODE. A *thermionic cathode*.

HOT CATHODE STEPPING TUBE. A multi-electrode hot cathode tube where the electron beam has two or more stable positions and can be made to step in sequence when a suitably shaped signal is applied to an input cathode.

HOT SPOT. A small portion of an electrode in a valve or tube whose temperature is higher than the mean temperature of the electrode.

HOT-WIRE GAUGE. See *Pirani gauge*.

HOT-WIRE INSTRUMENT. An instrument, such as an *ammeter*, whose operation depends on the expansion of a wire or thin strip which is carrying a current.

HOWL. A high-pitched audio tone produced by a receiver in which there is some undesirable acoustic or electrical *feedback*.

HOWLER. (1) A valve *oscillator* designed to attract attention, e.g. to a telephone whose receiver has been left off, or to warn an operator that radar signals are appearing on the screen. (2) An audio oscillator or hummer used as a null-point indicator for certain *bridges*.

H-PARAMETERS. See *Transistor parameters*.

H-PLANE. Of *electromagnetic waves*, a plane containing the direction of maximum radiation. The magnetic vector lies in the plane but the electric vector is normal to it.

HUFF-DUFF. From **HF-DF**: High Frequency Direction Finder.

HUM. In *audio-frequency* systems, a low-pitched droning noise, usually at frequencies which are harmonics of the mains supply frequency, and usually originating from the mains supply.

HUM MODULATION. *Modulation* of a radio-frequency or detected signal by *hum* resulting in noise or distortion.

HUMMER. A *microphone* used to detect the passage of current in conductivity measurements.

HUNTING. A condition of instability resulting from over-correction by a control device in which the controlled magnitude fluctuates about the required value without reaching it.

HYBRID COMPUTER. A computer in which both analogue and digital techniques are used.

HYBRID ELECTROMAGNETIC WAVE. An *electromagnetic wave* which has both transverse and longitudinal components of displacement.

HYBRID JUNCTION. A *waveguide* transducer designed to be connected to four branches of a circuit so as to render these branches *conjugate* in pairs.

HYBRID RING JUNCTION. A *hybrid junction* in the form of a ring, colloquially termed a *rat-race*.

HYBRID SET. Two or more *transformers* connected to form a network having four pairs of terminals to which four impedances may be connected so that the branches containing them may be conjugate in pairs.

HYDROGEN (H). Atomic No. 1 and the lightest of all gases.

HYDROGEN THYRATRON. A *thyratron* containing hydrogen in place of mercury, used to provide high peak currents at high anode voltages in radar circuits.

HYDROPHONE. An electro-acoustic *transducer* which responds to water-borne sound waves and delivers essentially equivalent electric waves.

HYGROMETER, ELECTRICAL. An instrument for indicating the humidity of the environment by electrical means.

HYPERBOLIC NAVIGATION. A technique for determining position by measuring the difference in the distance of the navigating receiver from two or more fixed transmitters. See *Decca, Gee, Loran*.

HYSTERESIS. The lagging or retardation of an effect behind a change in the mechanism or force causing the effect. See *Counter-tube, Dielectric, Ferroelectric, Magnetic, Rotational,* and *Magnetic hysteresis*. See also Fig. 64.

HYSTERESIS DISTORTION. Distortion due to a non-linear hysteresis characteristic of one or more magnetic or ferroelectric elements in a system.

HYSTERESIS FACTOR. The increase in effective resistance of a coil produced by *hysteresis* of its magnetic core when a current of one ampere at specified frequency is flowing.

HYSTERESIS LOOP. For a magnetic material, the closed figure formed by plotting the values of the *magnetic flux density* (*B*) against the *magnetizing force* (*H*) when the latter is taken through a complete cycle. The *hysteresis loss* is proportional to the area of this loop.

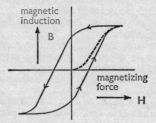

Fig. 64. Hysteresis loop

HYSTERESIS LOSS. See *Hysteresis loop*.

HYSTERESIS LOSS, ELECTRICAL. In a *dielectric* material, internal forces produced by a varying electrical field and usually accompanied by heat.

HYSTERESIS LOSS, INCREMENTAL. The *hysteresis loss* when a magnetic material is subjected to a pulsating magnetizing force.

HYSTERESIS METER. An instrument for measuring magnetic hysteresis.

HYSTERESIS MOTOR. A synchronous motor which starts by virtue of the hysteresis losses induced in its hardened steel secondary member by the revolving field of the primary.

HYSTERESIS OF A VALVE OSCILLATOR. Anomalous behaviour which occurs in valve oscillators, e.g. *reflex klystrons*, under certain conditions whereby multiple values of the output power or frequency are produced for given values of an operating *parameter*.

HZ. Abbreviation for Hertz.

I

IAS COMPUTER. A large electronic *digital computer* built by the U.S. Institute for Advanced Study, which is the American prototype for the modern high speed digital computer.

IATRON. A storage cathode-ray tube.

ICBM. Abbreviation for Inter-Continental Ballistic Missile.

ICONOSCOPE. A television *camera tube* in which a high-velocity electron beam scans a photoactive *mosaic* which has the capacity for electrical *storage*.

ICONOTRON. A type of *image iconoscope*.

ICW. Abbreviation for *interrupted continuous wave*.

IDEAL CRYSTAL. A crystal which is perfectly periodic in structure and therefore contains no foreign atoms or other imperfections.

IDEAL NOISE DIODE. A diode which has an infinite internal impedance and in which the current exhibits full *shot-noise* fluctuations.

IDEAL TRANSDUCER. A hypothetical linear passive *transducer* which transfers the maximum possible power from a specified source to a specified load.

IDEAL TRANSFORMER. A hypothetical *transformer* for inductively coupling two separate circuits, which has no losses of capacitance, infinite reactance, and coefficient of coupling = 1.

IDIOCHROMATIC CRYSTAL. An *intrinsic crystal*.

IFF. Abbreviation for identification, friend or foe. *Radar* equipment for automatic identification of ships, aircraft, etc.

I-F STRIP. The chassis of an *intermediate-frequency amplifier*.

IGFET. Abbreviation for Insulated Gate Field Effect Transistor.

IGNITION COIL. An *induction coil* which forms part of the ignition system for internal combustion engines.

symbols for depletion type single-gate

n-channel

p-channel

IGNITION INTERFERENCE. Interference in radio or television systems due to *ignition noise*, i.e. electrical noise mainly due to the ignition sparks.

IGNITION NOISE. See *Ignition interference*.

IGNITOR. The electrode of an *ignitron* used to initiate current flow in the main arc, made of a semiconductor such as boron carbide.

IGNITOR ELECTRODE. In *switching tubes*, an electrode used to initiate and sustain the *ignitor* discharge.

IGNITOR ROD. The electrode used to ignite the arc in *gas discharge* rectifiers.

IGNITOR VOLTAGE DROP. In *switching tubes*, the voltage between cathode and anode of the *ignitor* discharge at a specified ignitor current.

IGNITRON. A type of *mercury-arc rectifier* with only one anode. The arc is started at each cycle by an ignitor which dips into a pool of mercury forming the cathode.

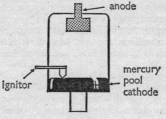

Fig. 65. Ignitron

IGNITRON RECTIFIER. A rectifier using an *ignitron* and usually a *thyratron* to control the firing time.

IGNORE. In *computer* terminology, a character indicating that no action should be taken.

ILS. Abbreviation for Instrument Landing System.

IMAGE ADMITTANCE. The reciprocal of *image impedance*.

IMAGE CONVERTER TUBE. An electron tube which reproduces on its fluorescent screen an image of an irradiation pattern incident on its photosensitive surface.

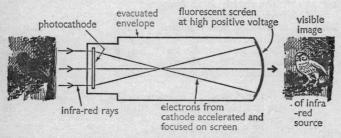

Fig. 66. Simplified diagram of an image converter tube and typical application

IMAGE DISSECTOR TUBE. A *camera tube* in which an electron image produced by a photo-emitting surface is focused in the plane of a defining aperture and is scanned past that aperture.

188

IMAGE FORCE. The force on a *charge* due to the charges which it may induce on conductors or dielectrics near it.

IMAGE FREQUENCY. In *heterodyne frequency* circuits, an undesired input frequency capable of producing the same beat frequency as the *sideband* selected.

IMAGE ICONOSCOPE. An *iconoscope* in which a translucent *photocathode* and an arrangement of accelerating and decelerating coils are used to intensify the electron image.

IMAGE IMPEDANCE. Of a *transducer*, the *impedance* which will simultaneously terminate all inputs and outputs so that at each input and output the impedances in both directions are equal.

IMAGE INTERFERENCE. Interference in superheterodyne receivers, caused by transmitters which differ in frequency by twice the value of the *intermediate frequency*.

IMAGE ORTHICON. A *camera tube* in which an electron image is produced by a photo-emitting surface and focused on a separate storage target, which is then scanned on its opposite side by a low-velocity electron beam. See Fig. 99.

IMAGE PHASE CONSTANT. The imaginary part of the *transfer constant*.

IMAGE RATIO. In a *heterodyne* receiver, the ratio of the *image frequency* signal input at the aerial to the desired signal input, for identical outputs.

IMAGE SIGNAL. In a *superheterodyne receiver*, a signal whose frequency differs from the received signal by twice the *intermediate frequency*.

IMAGE TUBE. See *Image converter tube*.

IMPACT FLUORESCENCE. *Fluorescence* caused by particle bombardment.

IMPACT IONIZATION. The loss of orbital electrons by an atom of a crystal lattice which has undergone a high-energy collision.

IMPEDANCE. In general, the ratio of a quantity with the nature of a force to a related quantity with the nature of a velocity. Electrical impedance is the ratio of the *R.M.S. electromotive force* in a circuit to the R.M.S. *current* which it produces in the circuit. Symbol: Z. Practical unit: *ohm*. $Z = R + jX$, where R is the *resistance* and X the *reactance* and $j = \sqrt{-1}$. See *Units*.

IMPEDANCE CHARACTERISTIC. The ratio of voltage to current at every point along a transmission line on which there are no standing waves.

IMPEDANCE COUPLING. The use of a tuned circuit or impedance coil to couple two circuits.

IMPEDANCE MATCHING. For the transfer of maximum power from a generator to a load, the load *impedance* is made the *conjugate* of the generator impedance. For the condition of no *reflection* the load impedance is made equal to the generator impedance.

IMPEDANCE, NORMALIZED. Any impedance of a system divided by its characteristic impedance.

IMPEDOMETER. A device consisting of a short *transmission line*, to measure *impedance* at very high frequencies.

IMPEDOR. A circuit element used mainly for its *impedance*. Sometimes referred to simply as an impedance.

IMPERFECT DIELECTRIC. A *dielectric* in which part of the energy required to establish an electric field in the dielectric is not returned to the electric system when the field is removed but is converted into heat.

IMPERFECTION. Of a crystalline solid: any deviation in structure from that of an *ideal crystal.*

IMPREGNATED CATHODE. A *cathode* made of porous tungsten impregnated with a barium compound, so that the barium evaporated by emission at the surface is continuously replenished from within.

IMPULSE EXCITATION. A method of producing oscillations in an electric circuit in which the duration of the impressed voltage is short compared with the duration of the current produced. Also called *shock excitation.*

IMPULSE GENERATOR. A device for generating very short, high voltage pulses, usually by charging capacitors in parallel, and discharging them in series. See *Surge generator.*

IMPULSE INERTIA. A property of an insulator which requires that the shorter the time of application of a voltage, the higher must the voltage be to cause *disruptive discharge.*

IMPULSE TRANSMISSION. A form of signalling which uses impulses to reduce the effects of low-frequency interference.

IMPURITIES. (1) In semiconductor elements: foreign atoms. (2) In semiconductor compounds: either an excess or deficiency of atoms belonging to the compound.

IMPURITY DIFFUSION. The formation of a region of *p-type* or *n-type* conductivity in a semiconductor crystal by diffusing into the crystal *acceptor* or *donor* impurity atoms.

IMPURITY LEVELS. In a *semiconductor*, energy levels due to the presence of impurity atoms which are different from the normal energy levels of the material.

IN PHASE. The condition which exists when two waves of the same frequency pass through their maximum and minimum values with the same polarity at the same instant.

IN QUADRATURE. The condition which exists when two waves of the same frequency differ in phase by 90°, i.e. one wave is at its maximum value when the other is passing through zero and vice-versa. See Fig. 118.

INCIDENT WAVE. A wave, travelling through a medium, which impinges on a discontinuity or a medium of different characteristics.

INCREMENTAL INDUCTION. At a point in a magnetic material where it is subjected to a polarizing magnetic force, one half the algebraic difference of the maximum and minimum values of the *magnetic induction* reached at the point during a cycle.

INCREMENTAL PERMEABILITY. The *permeability* of a material to a small alternating magnetic field superimposed on a larger constant magnetic field.

INDICATING INSTRUMENT. An instrument in which only the present value of the quantity measured is indicated visually.

INDICATOR GATE. A voltage of rectangular waveform applied to the grid or cathode circuit of an indicator *cathode-ray tube* to sensitize or desensitize it as required.

INDICATOR TUBE. An *electron-beam tube* in which information is conveyed by varying the cross section of the beam at a luminescent target. A common example is the *electron-ray tube*.

INDIRECT RAY. The part of a radio or radar signal reflected by the ground or by an *ionized* layer of the atmosphere.

INDIRECTLY HEATED CATHODE. The cathode of a thermionic valve in which an independent *heater* supplies the heat necessary, and is usually electrically insulated from the cathode.

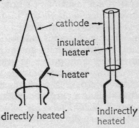

Fig. 67. Indirectly heated and directly heated cathodes compared

INDIUM (In). Soft, silvery, metallic element with low melting point, Atomic No. 49. Used in manufacture of *transistors*.

INDUCED CURRENT. Current resulting from an *induced e.m.f.*

INDUCED ELECTROMOTIVE FORCE (E.M.F.). An *electromotive force* in a circuit produced by a change of *magnetic flux* in the circuit.

INDUCED POLARIZATION. *Polarization* induced in a *dielectric* by an electric field, as distinct from polarization due to permanent *dipoles* in the dielectric.

INDUCTANCE (1). That property of a circuit, which when carrying a current is characterized by the formation of a magnetic field and the storage of magnetic energy. (2) The magnitude of this property. Symbol L. Practical unit: Henry. See also *Electromagnetic induction, Faraday's Law, Lenz's Law.*

INDUCTANCE, DISTRIBUTED. The inductance between the turns of a coil or along the length of a transmission line.

INDUCTANCE, LUMPED. The total distributed inductance along a length of transmission line, considered as acting at one point on that length of line.

INDUCTION BALANCE. A *bridge* to measure the *inductance* of a coil or device.

INDUCTION COIL (SPARK COIL). A *transformer* in which the *ferromagnetic* core is open and the current in the primary winding is interrupted periodically.

INDUCTION COMPASS. A compass whose indication depends on the current generated in a coil which revolves in the earth's magnetic field.

INDUCTION FURNACE. A transformer of electrical energy to heat by *electromagnetic induction*.

INDUCTION GENERATOR. An *induction machine* driven above synchronous speed by an external source of mechanical power.

INDUCTION HEATING. The heating by virtue of its internal losses of a nominally conducting material in a varying electromagnetic field. Contrast *Dielectric heating*.

INDUCTION INSTRUMENT. An instrument whose operation depends on the reaction between *magnetic fluxes* in fixed windings and electric currents due to *induction* in moving parts.

INDUCTION MACHINE. An asynchronous alternating current machine in which magnetic and electric circuits are interlinked and rotate with respect to each other, and power is transferred from one circuit to another by *electromagnetic induction*.

INDUCTION METER. An alternating current meter which incorporates a voltage coil as well as a current-carrying coil.

INDUCTION MOTOR. An *induction machine* which converts electric power delivered to its primary circuit into mechanical power.

INDUCTION OUTPUT VALVE. A valve in which the voltage of the output electrode is produced by *inductive coupling* and not by the normal method of collecting electrons.

INDUCTION SATURATION. The maximum intrinsic *electromagnetic induction* possible in a material.

INDUCTIVE. Of a circuit or winding, having a *self-inductance* which, for the purpose, is appreciable compared with its *resistance*.

INDUCTIVE INTERFERENCE. *Interference* caused by electric supply systems to communication systems.

INDUCTIVE LOAD. A *load* whose *inductive reactance* exceeds its *capacitive reactance*. See *Lagging load*.

INDUCTIVE NEUTRALIZATION. A method of neutralizing an amplifier in which the feedback *susceptance* due to capacitance between circuit elements is cancelled by the equal and opposite susceptance of an *inductor*.

INDUCTIVE REACTANCE. The *reactance* presented by an *inductance* to an alternating current or direct pulsed current.

INDUCTIVELY COUPLED CIRCUIT. A two-mesh network in which the two meshes have only a *mutual inductance* in common.

INDUCTOMETER. An instrument for measuring *inductance*.

INDUCTOR. A piece of apparatus used primarily because it possesses the property of *inductance*.

INDUCTOR MICROPHONE. A *microphone* in which the moving element is a straight-line conductor.

INDUCTOSYN®. A resolver designed to indicate angular position very accurately in applications such as digital to analogue conversion, control of machine tools, and servomechanisms.

INERTIA SWITCH. A switch operated by an abrupt change in its velocity.

INERTIAL CONTROL. A system of automatic control used for *guided missiles* and rockets based on the forces of inertia and independent of information obtained from outside the missile.

INFINITY. In *computer* terminology, a number greater than the largest number which the computer can store in any *register*.

INFORMATION CONTENT. Of a message, is expressed as the minimum number of *bits* or *hartleys* required to transmit the message with specified accuracy over a noiseless medium.

INFORMATION RETRIEVAL. In *data processing*: the study of methods and procedures for recovering specific information from stored data.

INFORMATION WORD. See *Machine word*.

INFRA-BLACK REGION. In *television*, the *blacker-than-black level* of the transmitted signal.

INFRA-RED. The range of invisible *electromagnetic* radiations which extends from the visible red region to the *microwave* region. See Fig. 46.

INFRA-RED DETECTOR. A thermal device for observing and measuring infra-red radiation. See *Bolometer*, *Thermopile*, *Photocell*.

INFRA-RED HOMING. A *homing-guidance* system which depends on the infra-red radiation emitted by the target.

INFRA-RED IMAGE CONVERTOR. An electron tube which converts an invisible scene illuminated with infra-red into a visible image on a fluorescent screen.

INFRA-RED MASER. An optical *maser* which employs an infra-red frequency for stimulation, and radiates or detects signals of millimetre wavelengths. See *Laser*.

INFRASONIC FREQUENCY. See *Subsonic frequency*.

INHERITED ERROR. In a computer, the error present in the initial values supplied to the machine.

INHIBITING INPUT. A *gate* input which prevents any output which might otherwise occur.

INITIAL IONIZING EVENT. An *ionizing event* which initiates a *pulse* in a *radiation detector*.

INJECTION. (1) Generally, the technique of applying signals to an electronic circuit. (2) See *Hole injection*.

INJECTION EFFICIENCY. In *transistors*, the fraction of the current flowing across the *emitter junction* which is contributed by the *minority carriers*.

INJECTION GRID. A *grid* in a multi-grid valve to which signals are applied to modify the electron stream.

INPUT. (1) The current, voltage, power, or driving force applied to a circuit or device; or the terminals where these may be applied. (2) Of computers, information which is transferred from outer external storage elements to the internal storage of the computer.

INPUT GAP. In *microwave* valves and tubes, an *interaction space* used to initiate a variation in an electron stream.

INPUT IMPEDANCE. The *impedance* presented by a device to a source.

INPUT-OUTPUT (I/O). A general term for the equipment used to communicate with a computer and the data involved in the communication.

INSCRIBER. An input *transcriber*.

INSERTION GAIN. The gain resulting from the insertion of a *transducer* in a transmission system expressed as the ratio of the power delivered to that part of the system following the transducer to the power delivered to the same part before insertion of the transducer.

INSERTION LOSS. The reciprocal of *insertion gain*.

INSTANTANEOUS AUTOMATIC GAIN CONTROL. In *radar*, a quick-acting *automatic gain control* which responds to variations in mean *clutter* level thereby reducing the clutter.

INSTANTANEOUS FREQUENCY. The time rate of change of the angle of a wave which varies with time.

INSTANTANEOUS POWER. At the points of entry of an electric circuit into a region, the rate at which electric energy is being transmitted by the circuit into the region.

INSTANTANEOUS SAMPLING. The process of obtaining a sequence of instantaneous values of a wave.

INSTANTANEOUS VALUE. The value, at a given instant, of any variable which alters with time.

INSTRUCTION. Information which, suitably coded and introduced as a unit into a *digital computer*, causes it to perform one or more of its operations.

INSTRUCTION CODE. An artificial *language* to formulate the instructions which can be carried out by a *digital computer*.

INSTRUMENT ACCURACY. Of an instrument or meter, a number or quantity which defines its limits of error.

INSTRUMENT APPROACH. An aircraft landing approach made

without visual reference to the ground by the use of instruments in the aircraft and electronic devices or communication systems on the ground.

INSTRUMENT DAMPING. A means of preventing an instrument pointer from oscillating about its correct reading when a sudden change of input is applied.

INSTRUMENT LANDING SYSTEM. A radio-navigational aid which provides the aircraft with the directional, longitudinal, and vertical guidance necessary for landing.

INSTRUMENT RATING. The designation, by the manufacture, of the limits between which it can operate. This does not necessarily correspond to the full-scale marking.

INSTRUMENT SENSITIVITY. The ratio of the magnitude of the response of an instrument to the magnitude of the quantity measured. It may be expressed directly, e.g. divisions per volt, or indirectly, e.g. ohms per volt for a stated deflection.

INSTRUMENT SHUNT. A special type of *resistor* designed to be connected in *parallel* with the instrument so as to extend its current range.

INSTRUMENT TRANSFORMER. A *transformer* for use with a measuring instrument, relay, or control device, designed to maintain precisely under specified conditions a given relationship as regards magnitude and phase-displacement between the *primary* and *secondary* currents or voltages.

INSULATED. Separated from other conducting surfaces by a *dielectric* offering a high resistance to the passage of current and to *disruptive discharge*.

INSULATOR. A material of such low *conductivity* that the flow of current through it can usually be neglected.

INTEGRATED CIRCUIT. The combination of the equivalent of a number of discrete components, such as transistors, resistors and capacitors into a device, resulting in miniaturization and increased reliability. See Figures 68b and 68c.

INTEGRATING INSTRUMENT. An instrument which sums the quantity of current, voltage, etc., which it measures over a period.

INTEGRATOR. (1) A *transducer*, circuit, or network whose output waveform is substantially the time integral of its input waveform. See *Capacitance integrator*, *Miller integrator*. (2) In some *digital*

Epitaxial layer

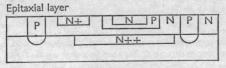

Fig. 68a. Integrated circuit construction

195

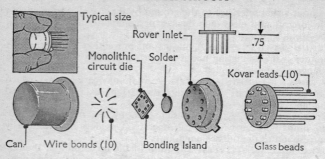

Fig. 68b. Integrated circuit – Exploded view of a typical package

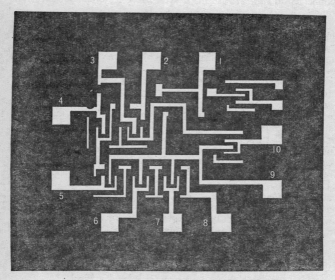

Fig. 68c. Appearance of a typical integrated circuit acting as a high-gain operational amplifier and formed from a single chip of silicon crystal about 0·2 inches square

computers, a device for obtaining a numerical approximation to the mathematical process of integration.

INTENSIFIER ELECTRODE. In a *cathode-ray* tube, an electrode to which a potential is applied to produce acceleration of the electron beam after it has been deflated.

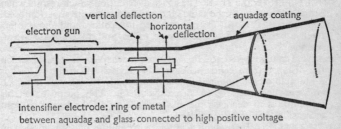

Fig. 69. Diagram of intensifier electrode in cathode-ray tube

INTENSITY LEVEL (OF A SOUND). Expressed in *decibels*, ten times the logarithm to the base 10 of the ratio of the intensity of this sound to a reference intensity (which must be stated).

INTENSITY MODULATION. Control of the brilliance of the trace on the screen of a *cathode-ray tube* in accordance with the magnitude of a signal.

INTERACTION FACTOR. For a *transducer*, the factor in the equation for the received current which allows for the effect of multiple reflections at the transducer terminals.

INTERACTION GAP. An *interaction space* in a microwave tube.

INTERACTION IMPEDANCE. Of *travelling-wave valves*, a measure of the radio-frequency field strength in the electron stream for a given power in the interaction circuit.

INTERACTION SPACE. The region of a valve or tube, roughly corresponding to the inter-electrode space, in which electrons interact with an alternating *magnetic field*.

INTERBASE CURRENT. In *junction-type tetrode transistors*, current flowing between the two *base* connectors.

INTERCEPTOR MISSILE. A *guided missile* used to intercept aircraft or guided missiles.

INTERFACE. Generally, an imaginary plane through which data are converted from one form to another.

INTERFERENCE. In any system or circuit which depends on electric currents or electromagnetic waves for the transmission, reception, or storage of information, the undesired transfer of energy from other sources which affects or disturbs the information.

INTERFERENCE FADING. *Fading* produced because different wave

197

components of a signal travel by slightly different paths to reach the receiver.

INTERFERENCE INVERTER. In a *television receiver*, a circuit which reverses the drive to the grid of the *cathode-ray tube* during a large pulse of *interference*.

INTERFERENCE PATTERN. The space distribution which results when progressive waves of the same kind and frequency are superimposed.

INTERFEROMETER. An instrument used to produce and show interference between two or more wave trains and to compare wavelengths.

INTERLACED SCANNING. In *television*, a *scanning* process in which the distance from centre to centre of successively scanned *lines* is a multiple of the nominal line width (usually twice), and in which adjacent lines belong to successive fields. The purpose of this technique is to decrease the effect of *flicker* on the image without increasing the signal *band width*.

INTERLEAVING. In *colour television*, a technique of transmission in which *chrominance* and *luminance* signals occupy the same *video* band of frequencies.

INTERLOCK. A device designed to make the operation of one piece of apparatus dependent on the fulfilment of predetermined conditions by another piece of apparatus.

INTERMEDIATE FREQUENCY (I-F). In *superheterodyne* reception, a frequency resulting from the combination of the received *modulated carrier* frequency and the *local oscillator* frequency.

INTERMEDIATE FREQUENCY AMPLIFIER (I-F AMPLIFIER). A fixed-tuned *amplifier* between the first and second detectors in a *superheterodyne* receiver.

INTERMEDIATE FREQUENCY STRIP. See *I-f strip*.

INTERMEDIATE FREQUENCY TRANSFORMER. A *transformer* designed for use in an *intermediate frequency amplifier*.

INTERMEDIATE-FREQUENCY HARMONIC-INTERFERENCE. See *Image interference*.

INTERMEDIATE-FREQUENCY RESPONSE RATIO. In a *superheterodyne receiver*, the ratio for the intermediate frequency signal input at the aerial to the desired signal input for identical outputs.

INTERMODULATION. The *modulation* of the components of a complex wave by each other, producing waves of frequencies equal to the sums and differences of integral multiples of the components of the original complex wave.

INTERMODULATING DISTORTION. Distortion which results from *intermodulation*.

INTERNAL CORRECTION VOLTAGE. Of a valve or tube, the voltage added to the *composite controlling voltage* as the voltage equivalent of effects such as those produced by initial electron velocity and contact potential.

INTERNAL MEMORY. Of a *computer*, the *memory* circuits and equipment which are under the direct control of the computer.

INTERNATIONAL ELECTRICAL SYSTEM OF UNITS. An international system for expressing the magnitudes of electrical quantities in terms of four fundamental units: the international ampere, international ohm, the centimetre, and second. It has now been discarded in favour of the *mksa system*.

INTERPRETER. In *digital computers*, a routine which translates a programme stored in an arbitrary code into machine language and performs the operations as they are translated.

INTERROGATOR. An electronic device for transmitting pulses which challenge or interrogate a *transponder*; also the radar or radio *beacon* used for this purpose.

INTERROGATOR-RESPONSER. An electronic device which combines an *interrogator* for sending pulses to a *transponder* and a *responder* which receives and displays the answering pulses.

INTERRUPTED CONTINUOUS WAVES (ICW). *Continuous* waves which are interrupted at a constant *audio-frequency* rate.

INTERRUPTER. A device which periodically interrupts the flow of a continuous current to produce *pulses*.

INTERRUPTION. In *data processing*: a suspension of the operation of a sequence of instructions followed by starting another sequence, or reverting to the one suspended.

INTRINSIC CONDUCTION. In a *semiconductor*, conduction associated with the drift in opposite directions of *electrons* and *holes* under the influence of an electric field.

INTRINSIC CRYSTAL. A crystal which does not depend on impurities for its *photoelectric* properties.

INTRINSIC MOBILITY. The mobility of electrons in an *intrinsic semiconductor*. See *Hall mobility*.

INTRINSIC SEMICONDUCTOR. A pure semiconductor in which the electron and hole densities are equal under conditions of thermal equilibrium.

INVERSE CURRENT, ELECTRODE. See *Electrode inverse current*.

INVERSE LIMITER. A device used generally to remove the low-level portions of signals from an output wave.

INVERSE NETWORKS. A pair of two-terminal networks the product of whose *impedances* is independent of frequency.

INVERSE PEAK VOLTAGE. The peak value of the instantaneous *inverse voltage* across a valve during the half of the cycle for which it is not conducting.

INVERSE VOLTAGE. The voltage impressed across a valve or tube during the half-cycle when the anode is negatively charged.

INVERTER. (1) Any device which converts dc into ac, especially a rotating machine designed for this purpose. (2) An *amplifier* which inverts *polarity*.

ION. An atom or group of atoms possessing an *electric charge*.

ION BEAM. A beam of *ions*, produced by subjecting the ions to very large forces as in a high-voltage *accelerator*.

ION BURN. In a *cathode-ray tube*, a localized permanent deactivation of the *phosphor* due to bombardment of the screen by heavy negative ions. Compare *Ion spot*.

ION DENSITY. The number of *ion* pairs per unit volume.

ION EXCHANGE. The reversible interchange of *ions* between a solution and a solid, e.g. between water and a resin as in an ion-exchanger for the purification of water.

ION IMPLANTATION. A technique used in the fabrication of integrated circuits and transistors, either in conjunction with or as an alternative to diffusion. E.g. in some field-effect transistors beams of phosphorus or boron ions are implanted in the silicon at energies of 10–140 KeV.

ION MIGRATION. A movement of *ions* produced in a *semiconductor*, electrolyte, etc. by applying a voltage between electrodes.

ION MOBILITY. The velocity of an *ion* moving through a gas under the influence of unit electric field.

ION SOURCE. A device for producing positive or negative *ions*, e.g. for *accelerators*.

ION SPOT. (1) On a *cathode-ray tube* screen, an area of localized deterioration of luminescence (not necessarily of the *phosphor*) caused by bombardment with negative ions. (2) In *camera* or *image tubes*, the spurious signal resulting from alteration of the target or cathode by ions.

ION TRAP. A device to prevent the *ions* present in *cathode-ray tubes* impinging on the *phosphor* coating and so causing blemishes.

IONIC CONDUCTION. In a *semiconductor*: the continuous movement of charges within a substance due to the displacement of ions in a crystal lattice, the movement being maintained by a continuous contribution of external energy.

IONIC POTENTIAL. The ratio of ionic charge to radius.

IONIC SEMICONDUCTOR. A solid in which the electrical conductivity due to the flow of ions predominates over that due to the motion of electrons and holes.

IONIC-HEATED CATHODE. A *hot cathode* heated primarily by ionic bombardment of the emitting surface.

IONIZATION. Any process by which a neutral atom or molecule gains or loses *electrons* and is left with a net charge.

IONIZATION, PRIMARY. In *counter tubes*, the total *ionization* produced by incident radiation without *gas amplification*.

IONIZATION CHAMBER. A device for detecting *ionizing radiation* by measuring the *ionization* produced in a volume of gas.

IONIZATION COUNTER. An *ionization chamber*, which has no internal amplification due to multiplication of *ions*, used to count ionizing particles.

IONIZATION CURRENT. The current resulting from the movement of *electric charges* in an *ionized* medium, under the influence of an electric field. See also *Gas current*.

IONIZATION GAUGE. A high-vacuum gauge in which the pressure of the gas in the gauge is measured by its degree of *ionization*.

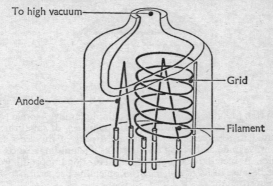

Fig. 70. Principle of ionization gauge

IONIZATION GAUGE VALVE (TUBE). A valve designed to measure low gas pressures using the relation between gas pressure and *ionization current*.

IONIZATION POTENTIAL. (1) For a particular atom, the energy per unit charge required to remove an *electron* from the atom to an infinite distance; usually expressed in volts. (2) In a gas-filled valve or tube, the potential at which *ionization* begins; usually slightly lower than the *firing* potential at which ionization is complete.

IONIZATION TIME. Of a *gas discharge* tube, the time interval between the initiation of conditions for conduction and the commencement of conduction for a given value of voltage drop.

IONIZING EVENT. Any process by which an *ion* or group of ions is produced, e.g. the passage of a charged particle.

IONOSPHERE. That part of the earth's outer atmosphere where *ions* and free electrons are normally present in sufficient quantities to affect the propagation of *radio waves*. See Figure 71.

IONOSPHERE E-REGION. The region of the *ionosphere* about 55 to 100 miles above the surface of the earth.

IONOSPHERE F-REGION. The region of the *ionosphere* about 85 to 220 miles above the surface of the earth.

IONOSPHERIC WAVE. A radio wave propagated by reflection from the ionosphere. Sometimes called the *sky wave*.

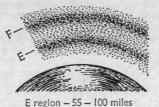

E region – 55 – 100 miles
F region – 85 – 220 miles

Fig. 71. Ionosphere

IRIDIUM (Ir). Metallic element of the platinum family, Atomic No. 77.

IRIS. In a *waveguide*: a conducting plate or plates, of small thickness compared to a wavelength, occupying a part of the cross-section of the waveguide and acting effectively as a shunt impedance.

IRON (Fe). A metallic element, Atomic No. 26. Important in electronics because of its *ferromagnetism*, shielding properties, low cost, and high strength.

IRON LOSS. In electrical machinery and devices such as *transformers* the power dissipated in *eddy currents* and *hysteresis* in the iron or magnetic material in the magnetic circuit.

ISLAND EFFECT. The restriction of emission from the *cathode* of a valve to certain small areas on the surface when the grid voltage is less than a minimum value.

ISOCLINE. A graphical method for the determination of the performance of an *oscillator* when the *negative resistance* characteristic of the oscillator is accurately known. See *Cyclogram*.

ISOCON. A development by English Electric of the *image orthicon*: a television camera tube in which the electron scanning beam is steered along a helical path by two pairs of electrodes. Because the design of the tube converts noise into black spots instead of white as in conventional tubes, greatly improved performance is claimed.

ISOELECTRONIC. Referring to groups of atoms having similar electrical qualities, e.g. similar distributions of *electrons* in their outer shells.

ISOLANTITE®. An insulation suitable for high frequencies.

ISOLATION DIODE. A *diode* used in a circuit to pass signals in one direction but to block signals and voltages in the other.

ISOLATION TRANSFORMER. A *transformer* used to isolate electrically any equipment from its power supply, i.e. to derive the power from the source with no continuous wire connection.

ISOLATOR FERRITE. A microwave device which allows the energy to pass through in one direction with very little loss while absorbing power in the reverse direction.

ISOPERMS. Magnetic alloys containing nickel, iron, and cobalt used in making *inductors* for high frequencies.

ISOTOPES. All atoms which are isotopes of a particular element have the same number of *protons* and *electrons*, but varying numbers of *neutrons* in their nuclei so that their masses are different. As they have the same extra-nuclear electron structure they are chemically indistinguishable.

ISOTOPIC DIELECTRIC. A *dielectric* material which has only one dielectric constant, which is therefore independent of the direction of the applied field.

ITERATIVE IMPEDANCE. Of a *transducer*, that impedance which, when connected to one pair of terminals, produces a like impedance at the other pair of terminals.

ITERATIVE OPERATION. In digital computers: repeating a cycle of operations so that they approach closer and closer to the desired result.

I-TYPE SEMICONDUCTOR. See *Intrinsic semiconductor*.

J

JACK. A connecting device to which wires of a circuit may be attached, which is arranged for insertion of a plug.

JACK BOX. A box with connections or switches for changing circuits, especially in aircraft communication systems.

JAM, CARD. A pile-up of cards in a *data-processing* machine.

JAMMER. Any electronic transmitting set used for *jamming*.

JAMMING. The intentional transmission of interfering radio signals in order to disturb the reception of other signals.

JANET. A system of radio communications in which VHF radio signals are given increased range by forward scattering them from ionized meteor trails.

JAR. A unit of *capacitance* formerly used in the British Navy.

J-CARRIER SYSTEM. A broad-band *carrier telephone* system which provides twelve telephone channels and uses frequencies to 140 kc/s.

JIM CREEK TRANSMITTER. The most powerful radio transmitter in the world. In use by the U.S. Navy at Jim Creek, Washington, it has a power of one million watts.

JITTER. Short term instability, either in amplitude or phase, of a signal, especially a signal on a *cathode-ray tube*.

JITTER SCOPE. An oscilloscope for measuring *jitter*.

JITTERS. The distortion in a received *television* or *facsimile* picture due to momentary errors in synchronism between scanner and recorder.

JOHNSON NOISE. *Thermal noise*.

JOHNSON-LARK-HOROWITZ EFFECT. In a metal or degenerate *semiconductor*, the resistivity which is attributed to electron scattering by impurity atoms.

JOSEPHSON JUNCTION. A device based on the behaviour, discovered by Josephson in 1962, of a superconductor-insulator – superconductor junction in which the insulating layer is extremely thin: about 10–15 Angstroms. A direct current can flow with no voltage across the junction: i.e. the insulator behaves as if it were a super-conductor. With an alternating voltage high frequency current flows, the frequency of oscillation depending on the applied voltage.

the Josephson junction

JOSHI EFFECT. In a *gas discharge*, the effect of light on the discharge current.

JOULE. A unit of work or energy in the *mksa system*, in which it is the work done by a force of one newton acting through a distance of 1 metre. Also equal to 10^7 ergs. See *Units*.

JOULE EFFECT. A conductor becomes heated by the passage through it of an electric current due to the *resistance* of the conductor.

JOULE MAGNETOSTRICTION. Positive *magnetostriction*. The length of a magnetostrictive rod increases under the influence of a longitudinal magnetic field.

JOULE METER. An integrating *wattmeter* calibrated in *joules*.

JOULE'S LAW. The rate at which heat is produced in an electric circuit is proportional to the square of the current.

JUGFET. Abbreviation for junction gate field effect transistor.

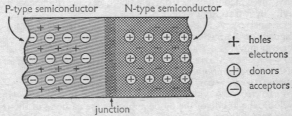

symbol for jugfet with n-type channel

with p-type channel

JUMP. (1) In *computer* technology, cause the next instruction to be selected from a specified *storage* location. (2) The *transition* of an orbital electron from one energy level to another.

JUNCTION. (1) A transition region between layers of *semiconductors* with different electrical characteristics. (2) A contact between two different metals or materials as in a *rectifier* or *thermocouple*. (3) A connection between two or more conductors or sections of transmission line.

P-type semiconductor   N-type semiconductor

+ holes
− electrons
⊕ donors
⊖ acceptors

junction

Fig. 72. Diagrammatic representation of a P-N junction at equilibrium

JUNCTION, RECTIFIER. See *Rectifier, junction*.

JUNCTION, TRANSISTOR. See *Transistor, junction*.

JUNCTION BOX. An enclosed distribution panel for connecting or branching electric circuits without using permanent splices.

JUNCTION COUPLING. In *coaxial* line *cavity resonators*, *coupling* by direct connexion to the coaxial conductor.

K

K BAND. A *frequency band* extending from 11,000 to 33,000 Mc/s.

KC. Abbreviation for *kilocycle*(s).

KC/S. Abbreviation for kilocycles per second.

KEEP-ALIVE ARC. In *mercury-arc rectifiers*, an auxiliary arc current which provides a *cathode spot* to facilitate the establishment of the main arc.

KEEP-ALIVE CIRCUIT. In a *TR* or *ATR switch*, a circuit to produce residual *ionization* in order to reduce the initiation time of the main discharge.

KEEP-ALIVE ELECTRODE. An auxiliary electrode which carries the current in a *keep-alive circuit*.

KEEPER. A piece of *ferromagnetic* material placed against the extremities of a permanent magnet to complete the magnetic circuit.

K-ELECTRON. One of the two *electrons* in the orbits which are closest to the *nucleus*.

KELL FACTOR. In *television*: the relation between the number of scanning lines in the picture and the bandwidth of the video signal.

KELVIN BALANCE. An *electrodynamic ammeter, voltmeter,* or *wattmeter*, in which the electromagnetic forces are balanced against gravity by a travelling weight.

KELVIN BRIDGE. A *bridge* embodying two similar sets of ratio coils, designed for the measurement of low values of resistance.

KELVIN EFFECT. See *Thomson effect*.

KELVIN-VARLEY SLIDE. A device for reducing the effect of *contact resistance* in *potentiometers*.

KENNELLY-HEAVISIDE LAYER. See *Heaviside layer*.

KENOPLIOTRON. A valve containing a *diode* and a *triode* in one envelope, the anode of the diode acting as the cathode of the triode.

KENOTRON. A high-vacuum, high-voltage, low-current *diode*, used for *rectification* in high-voltage systems.

KERR CELL. A device using the *Kerr effect* to modulate a beam of light.

KERR EFFECT. (1) If a beam of light is passed through certain liquid dielectrics, the plane of polarization of the light may be rotated by applying an electric potential to the liquid. See *Electro-optical effect*. (2) Plane-polarized light can be changed to elliptically-polarized light by reflecting it from the highly polished pole of a powerful electromagnet. See *Magneto-optical effect*.

KEV. Abbreviation for Kilo-electron-volt = 1,000 electron-volts.

KEY CLICK. The noise created by *transient* signals generated by opening or closing a key switch.

KEY PUNCH. A keyboard-operated device for punching holes in a card to represent data.

KEYING. In telegraph or continuous-wave transmission: forming signals by abrupt *modulation* of a current, e.g. by interrupting it or by suddenly changing its amplitude, frequency, etc.

KEYSTONE DISTORTION. Distortion of a rectangular pattern into a keystone-shaped or trapezoidal pattern. In television systems this distortion usually originates in a *camera tube* where a plane target area is not normal to the electron beam.

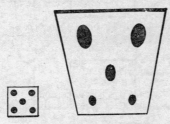

Fig. 73. Keystone distortion

KILOCYCLE (KC). One thousand *cycles* (per second).

KILOVOLT (KV). One thousand *volts*.

KILOVOLT-AMPERE (KVA). One thousand volt-amperes, a unit of apparent electric power.

KILOWATT (KW). One thousand watts, a unit of electric power.

KILOWATT-HOUR (KWHR). One thousand watts acting for a period of one hour: a unit of work.

KINESCOPE. A television *picture tube*.

KIRCHHOFF'S LAWS. (1) The algebraic sum of the currents flowing towards any point is zero. (2) The algebraic sum of the products of current and resistance in each of the conductors forming a closed path in a network is equal to the algebraic sum of the electromotive forces in the path.

KLYDONOGRAPH. A surge voltage recorder.

KLYSTRON. An electron valve (tube) in which an *electron beam* is *velocity modulated* to generate or amplify electromagnetic energy of very high frequency in the *microwave* region. In a typical klystron, *bunching* occurs in an input resonant cavity or *buncher* and the resultant intensity-modulated beam is converted into high frequency oscillations in an output cavity or *catcher*. The cavities are electrically and mechanically coupled.

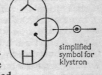

simplified symbol for klystron

KLYSTRON, FREQUENCY MULTIPLIER. A *klystron* in which the resonant frequency of the output cavity can be adjusted to a desired multiple of the resonant frequency of the input cavity.

KLYSTRON, REFLEX. See *Reflex klystron*.

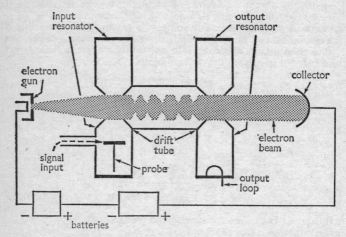

Fig. 74. Principle of a two-cavity klystron amplifier

KLYSTRON AMPLIFIER. An *amplifier* using a *klystron* together with its associated tuned cavities.

KLYSTRON OSCILLATOR. See *Velocity-modulated tube*.

KNEE. A region of maximum curvature on a curve or characteristic.

KOCH RESISTANCE. Of a vacuum *phototube* or *photocell*, the resistance when light is incident on the active surface.

KONEL METAL. An alloy of iron, nickel, cobalt, and titanium used in the fabrication of cylinders, on which the emitting compound is sprayed, for *indirectly heated cathodes*.

KOVAR. An alloy of cobalt, iron, and nickel which has a thermal coefficient of expansion nearly the same as a range of glass materials used in vacuum technique. It is therefore widely used in glass-to-metal seals for components such as valves and transistors.

KRYPTON (Kr). Inert gaseous element, Atomic No. 36.

KV. Abbreviation for *kilovolt*.

KVA. Abbreviation for kilovolt-ampere.

KVP. Abbreviation for kilovolts, peak; usually referring to the peak voltage across an *X-ray* tube.

KYMOGRAPH. An instrument for recording periodic movements, especially in muscles.

L

L BAND. A *frequency band* extending from 390 to 1,550 Mc/s.

L ELECTRON. An electron from the second shell outward from the atom nucleus.

LABYRINTH LOUDSPEAKER. A loudspeaker in a housing provided with air chambers to reduce acoustic *standing waves*.

LADAR. LAser Detection And Ranging.

LADDER FILTER. A network of series and shunt *impedances* in alternate succession.

LADDIC. A magnetic device designed for *logical operations*.

LAG. (1) The interval of time, or the angle in electrical degrees, by which a particular phase in one wave follows a similar phase in another wave. (2) The time elapsing between the operation of a transmitting device and the response of a receiving device. (3) In a control system the delay in response to a correcting signal. (4) In *camera tubes*, persistence of the electrical image for a few frames. See *Delay*.

LAGGING CURRENT. An *alternating current* the phase of which is behind the electromotive force giving rise to it. In a pure *inductance* the phase is 90 degrees or one quarter of a complete cycle behind.

LAGGING LOAD. A *reactive* load in which the phase of the current is behind the phase of the voltage at the terminals. See *Inductive load*.

LAMBERT (L). A unit of *luminance*. The average luminance of a surface emitting or reflecting light at the rate of one *lumen* per sq. cm. For practical purposes the *millilambert* (0·001 lambert) is used.

LAMINATION. A thin stamping of *ferromagnetic* material used in the assembly of magnetic cores, for *transformers*, etc.

LANGEVIN ION. An *ion* moving under the influence of an electric field in a gas.

LANGUAGE. In electronic *computer* terminology, a system consisting of a well-defined set of characters or symbols, and rules for combining these to form words or other expressions. It is usual to assign specific meanings to such words or expressions in order to supply information or data to a group of people or machines, but a language may be studied independently of any meanings.

LANGUAGE, ALGORITHMIC. An *arithmetic* language designed for precise presentation of numerical procedures to a computer in standard form. See *Algol*.

LAPPING. A fine grinding operation used in electronics to provide a smooth surface for wafers of crystalline *semiconductors*, for the final adjustment of *quartz crystals* to determine their fundamental frequency, and for similar operations.

209

LARYNGOPHONE. A *throat microphone*.

LASA. Larger Aperture Seismic Array. A complex array of instruments, including 21 separate clusters of seismographs and associated instruments, built for the U.S. Department of Defence to detect distant underground nuclear explosions.

LASER. Light Amplification by Stimulated Emission of Radiation. An optical *maser* first made in the U.S.A. in 1960, using a small crystal of ruby which generated intermittent but very pure and concentrated beams of infra-red light. A gas laser produced by Bell Telephone Laboratories at the end of 1960 was the first laser to operate continuously. It used a low-energy discharge in a mixture of helium and neon gases to generate a continuous, very intense, and very sharply defined beam of infra-red radiation.

General symbol

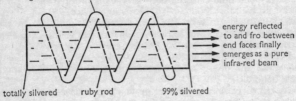

light from helical flash tube pumps (stimulates) ruby atoms

energy reflected to and fro between end faces finally emerges as a pure infra-red beam

totally silvered   ruby rod   99% silvered

silvered end surfaces polished flat and exactly parallel

Fig. 75. Principle of ruby laser

LATENCY. In *digital computers*, the delay in the delivery of information to the *arithmetical unit* after it is called for from the *memory*.

LATERAL RECORDING. A mechanical *recording* in which the modulation of the groove is perpendicular to the motion, and parallel to the surface, of the recording medium.

LATERAL REPRODUCER. A reproducing device, e.g. a *pick-up*, used in *lateral recording*.

LATTICE FILTER. A *filter* whose elements are arranged in the form of a *bridge* network.

LATTICE NETWORK. A network in which the input and output terminals are connected to a junction point of two or more components. See Figure 76.

LATTICE PARAMETER (LATTICE SPACING). The lengths of the edges of a unit cell in a crystal.

LAURITSEN ELECTROSCOPE. A robust but sensitive *electroscope* in which the sensitive element is a metallized quartz fibre.

LAWNMOWER. Colloquially in radar terminology: a kind of pre-

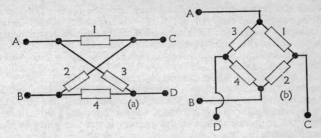

Fig. 76. Two ways of drawing a lattice network with resistive elements ((b) is also known as a 'bridge' network)

amplifier used in front of the radar receiver input, which cuts down the level of the '*grass*' on the indicator screen.

LAWRENCE TUBE. A single gun *colour television tube* in which colour switching occurs at the screen. See *Chromatron*.

LC. Of a *tuned circuit*: the product of *inductance* and *capacitance*. It is a constant for a given frequency.

LEAD (Pb). Metallic element, Atomic No. 82. Mainly used in electronics in *solders*, and for storage batteries.

LEAD-ACID CELL (ACCUMULATOR). A storage battery in which the negative plates are sponge lead, the positive plates are lead dioxide, and the electrolyte is dilute sulphuric acid.

LEAD-IN. Of an *aerial*: the cable which connects the elevated part of the aerial to the transmitter or receiver.

LEAD-IN GROOVE. A blank spiral at the beginning of a gramophone record with a greater pitch than that of the recorded grooves.

LEADING CURRENT. An *alternating current* the phase of which is in advance of the impressed electromotive force giving rise to the current. In a pure *capacitance* it is 90 degrees in advance.

LEADING EDGE, PULSE. See *Pulse leading-edge*.

LEADING LOAD. A *reactive* load in which the current wave is in advance of the voltage wave. See *Capacitive load*.

LEAD-LAG NETWORK. A *resistance-capacitance* network used in some circuits for *series stabilization*.

LEAD-OUT GROOVE. A blank spiral at the end of a gramophone record which facilitates automatic stopping or changing of records.

LEAKAGE. Electrical loss due to poor *insulation*. See also the terms following.

LEAKAGE CURRENT. A stray current of relatively small value which flows through or across the surface of an insulator when a voltage is impressed on the insulator.

LEAKAGE FLUX. In electrical machines and *transformers* in which there is a magnetic circuit, *flux* whose circuit is outside the useful flux circuits. Lost flux.

LEAKAGE POWER. In *TR* and *pre-TR tubes*, the radio-frequency power transmitted through a tube which has fired.

LEAKAGE REACTANCE. In *alternators* and *transformers*, *reactance* due to leakage flux.

LEAKANCE. The reciprocal of insulation *resistance*.

LEAPFROG TEST. In *computers*, a programme used to test the internal operation of the computer. The name derives from the observed jump on a monitoring oscilloscope when the testing routine transfers itself from one section of the memory to another. See *Crippled leapfrog test*.

LEARNING, MACHINE. The capability of a device to improve its performance, based on its past performance.

LECHER LINE. A section of open-wire *transmission line* consisting of two parallel wires on which *standing waves* are set up. It is used for measurement of wavelength and for impedance matching.

LECLANCHE CELL. A *primary cell* with carbon and zinc electrodes immersed in an electrolyte of sal ammoniac plus a depolarizer.

LEDUC CURRENT. An interrupted direct current, each pulse of which is of approximately the same strength and in the same direction.

LENARD RAYS. *Cathode rays* which have passed outside a *Lenard tube*.

LENARD TUBE. A *discharge* tube with a thin window opposite the cathode to allow the cathode rays to pass outside.

LENZ'S LAW. For currents induced by motion in a magnetic field, the induced currents have such a direction that their reaction tends to oppose the motion which produces them.

LEO®. A large *digital computing* and *data-processing* system developed by J. Lyons and Co. Ltd., and taken over by English-Electric–Leo-Marconi.

LEPTON. A particle of small mass, especially an *electron*, *positron*, or *neutrino*.

LEVEL COMPENSATOR. A device or circuit used to compensate automatically for the effects of variations in amplitude of a received signal.

LEVEL METER. An instrument for indicating the absolute level of electro-acoustical or transmission systems. These instruments usually have an internal calibrating system for standardizing gain.

LEYDEN JAR. An early form of fixed *capacitor*.

LF. Abbreviation for *low frequency*.

LIFETIME. Of a *semiconductor* charge *carrier*, the mean time between the birth and death of the carrier.

LIGHT EFFICIENCY. Of an electric lamp, the ratio of the total luminous flux to the total power input expressed in *lumens* per watt.

LIGHT PEN. A device for producing a visible image on the screen of a cathode-ray tube by movements similar to those of writing with a pen on paper. In conjunction with a computer such a device becomes an extremely powerful design tool.

LIGHT VALVE. A device whose light transmission can be made to vary in accordance with an externally applied electrical quantity such as current or voltage, electric or magnetic field, etc., or an electron beam. See *Kerr cell*.

LIGHTHOUSE VALVE. An *ultra high frequency* valve with parallel plane electrodes encased in three stepped cylinders suggestive of a lighthouse. The purpose of this design is to reduce the electron *transit time* to a minimum and thus raise the limit of the frequency of oscillation and amplification.

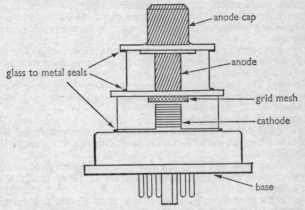

Fig. 77. Simplified construction of lighthouse valve

LIGHTNING ARRESTER. A device with the property of reducing the voltage of a surge applied to its terminals.

LIGHTNING CONDUCTOR. The part of a protective system designed to lead the current of a lightning discharge from an air terminal to ground.

LIMITED SIGNAL. In *radar*, a signal which is intentionally limited in amplitude by the *dynamic range* of the system.

LIMITED STABILITY. Characteristic of a system which is stable when the input signal falls within a certain range and unstable if the signal falls outside this range.

LIMITER. Generally, a device in which some characteristic of the output is automatically prevented from exceeding a certain value. In particular, a *transducer* in which the amplitude of the output is substantially linear with respect to the input up to a predetermined value, and substantially constant thereafter. See *Audio-frequency peak limiter, Clipper, Current limiter, Inverse limiter, Peak limiter, Schmitt limiter, Volume limiter.*

LIMITING. The action performed by a *limiter*. See also *Double limiting, Grid limiting.*

LINCOMPEX. LINked COMPressor and EXpander: a new type of terminal equipment developed by the Post Office. It is used on high frequency *radiotelephone* circuits to improve two-way communications when much noise and fading are present.

LINDEMANN ELECTROMETER. A highly sensitive, low-capacity *electrometer*, in which a metallized quartz needle is suspended by a metallized quartz fibre between two vertical plates.

LINE BIAS. The effect on the length of teletypewriter signals of the electrical characteristics of the transmission line.

LINE BROADCASTING. The transmission by means of *line communication* of a programme of sound, vision, or facsimile for general reception.

LINE COMMUNICATION (WIRE COMMUNICATION U.S.A.). The use for communication of a physical path, such as wire or waveguide between the terminals.

LINE DROP. The voltage drop between any two points on a *transmission line* caused by resistance, reactance, or leakage.

LINE EQUALIZER. See *Equalizer.*

LINE FLYBACK. See *Flyback.*

LINE FREQUENCY. In television, the number of times per second that a fixed vertical line in the picture is crossed in one direction by the *scanning spot.*

LINE OF ELECTRIC FLUX. A line drawn in an *electric field* so that its direction at every point is the direction of the *electric flux* at that point.

LINE OF ELECTRIC FORCE. A line drawn in an *electric field* so that its direction at every point is the direction of the electric force at that point.

LINE OF MAGNETIC FLUX. A line drawn in a *magnetic field* so that its direction is the direction of the *magnetic flux* at that point.

LINE OF MAGNETIC FORCE. A line drawn in a *magnetic field* so that its direction at every point is the direction of the magnetic force at that point.

LINE PAD. A line *attenuator* used in radio transmission.

LINE PRINTING. In *data processing*: the printing of an entire line of characters at once.

LINE REFLECTION. In a *transmission line*, the reflection of signal energy due to some discontinuity in the line.

LINE STRETCHER. A mechanical device sometimes used in transmission lines to adjust their length for tuning.

LINE SYNCHRONIZING PULSE. In *television*, pulses transmitted at the end of each line to synchronize the start of the *scan* for the next line.

LINEAR. Used to describe the relation between two varying quantities when one varies in direct proportion to the other, so that the graph representing this relation is a straight line.

LINEAR AMPLIFIER. An amplifier in which there is a linear relation between instantaneous anode current and grid voltage.

LINEAR DETECTION. *Detection* in which the output voltage is proportional to the input voltage, over the useful range of the detecting device.

LINEAR DISTORTION. Any form of *distortion* which is independent of the amplitude of the signal.

LINEAR MODULATION. *Modulation* in which the change in the modulated characteristic of the *carrier* signal is proportional to the value of the modulating signal over the range of the *audio-frequency* band.

LINEAR PROGRAMMING. A technique for solving certain kinds of problems which involve many variables, to give a best value or set of values. Problems using this technique may be programmed on a computer.

LINEAR RECTIFICATION. *Rectification* in which the changes in the direct current output are proportional to the changes in the alternating current input.

LINEAR SCAN. (1) An electron beam in a *cathode-ray tube* which is swept across the tube face with constant velocity, e.g. by a *sawtooth waveform* on the *deflector plates*. See *Sweep*. (2) A *radar scan* projected and fixed in a straight line, used in *sector scanning* to increase the intensity of echoes. Also a radar beam which moves with a constant angular velocity.

LINEAR SWEEP. *Sweep* of a *cathode-ray* beam by application of a *sawtooth waveform* to the horizontal deflection plates of a *cathode-ray tube*. See *Sweep*, *Time base*.

LINEAR TIME BASE OSCILLATOR. A *relaxation oscillator* used to generate the *sawtooth waveform* for *linear sweep*.

LINEARITY CONTROL. See *Distribution Control*.

LINE-SEQUENTIAL COLOUR TELEVISION. A system of *colour television* in which each of the red, blue, and green cameras is connected to the transmission system for the period of one *scanning line* in sequence.

LINGERING PERIOD. The time spent by an *electron* which has reached the state of maximum energy in an atom, before it changes to a lower state of energy.

LINK. (1) In *communications*, a channel or circuit designed to be connected in tandem with other channels or circuits. (2) In automatic switching, a path between two units of switching apparatus which are part of a central control system.

LINKAGE. (1) Of *magnetic flux*, a measure of the product of the number of *lines of magnetic flux* and the number of turns of a coil or circuit through which they pass. See *Maxwell*. (2) In computer programming: coding which connects two separately coded routines.

LIN-LOG RECEIVER. A *radar receiver* having linear response for small signals and logarithmic for large signals.

LIP MICROPHONE. A *microphone* for use in contact with the lips.

LIQUID RHEOSTAT. A *rheostat* employing a liquid-metal contact as the resistance element.

LISSAJOUS FIGURES. On a *cathode-ray oscilloscope*, geometric patterns described by the trace of the electron beam when it is simultaneously displaced by two *sinusoidal* voltages applied to vertical and horizontal deflecting plates. The patterns can be used for accurate frequency matching. See Fig. 78.

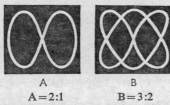

A   B

A = 2:1   B = 3:2

Fig. 78. Typical Lissajous figures corresponding to frequency ratios

LITHIUM (Li). The lightest metallic element, Atomic No. 3.

LITZ WIRE. A conductor composed of a number of fine, separately insulated strands, woven so that each strand behaves in substantially the same way when a high frequency current is passed through the wire.

L-NETWORK. A network of two *impedances* in series.

LOAD. (1) Of a machine or apparatus, the power which it delivers. (2) Any device which receives signal power from a *transducer*. (3) The power-absorbing device in definition (1). (4) In *dielectric* and *induction heating*, the material to be heated.

LOAD CELL. A *transducer* used for measuring weight in which the weight causes a strain in a supporting member and the strain is measured electronically.

LOAD CHARACTERISTIC. Of a valve, the dynamic relation between the instantaneous values of two variables, e.g. electrode voltage and current, with all direct-supply voltages to the electrodes maintained constant.

LOAD CIRCUIT EFFICIENCY. Of a valve circuit, the ratio between the useful power delivered by the load circuit to the load, and the input power to the anode circuit.

LOAD COIL. In *induction heating*, an electric conductor which, when energized with alternating current, is adapted to deliver energy to the *load*.

LOAD CURVE. A curve of power versus time showing the value of a specified *load* for each unit of time.

LOAD FACTOR. Of electrical power, the ratio of the average load over a specified period to the peak load in that period.

LOAD IMPEDANCE. The *impedance* presented by the *load* to the load circuit.

LOAD IMPEDANCE DIAGRAM. Of an *oscillator*, a chart showing the performance of the oscillator with respect to variations in the *load impedance*.

LOAD LEADS. In *induction* and *dielectric heating*, the connections or transmission lines between the power source or generator and the load, *load coil*, or *applicator*.

LOAD LINE. A line on the family of characteristic curves of a valve or transistor showing the graphical relation between the voltage and current established by the load for the particular circuit being considered.

LOAD MATCHING. In *dielectric* and *induction heating*, the process of adjusting the load circuit impedance to produce the desired energy transfer from source to load.

LOAD REGULATOR. A circuit or device which maintains the load at a predetermined value or varies it in a predetermined way.

LOAD TRANSFER SWITCH. A switch to connect a generator or power source optionally to one or another load circuit.

LOBE. One of the sections of the radiation pattern of an aerial.

LOBE SWITCHING. In *radar*, a form of *scanning* in which the direction of maximum radiation or reception is switched in turn to each of a number of preferred directions, e.g. first to one side and then to the other of a target.

LOCAL CONTROL. A method of radio-transmitter control in which the control functions are performed directly at the transmitter, and not by *remote control*.

LOCAL OSCILLATOR. An *oscillator* in a *superheterodyne* receiver whose output is mixed with the received signal to produce a sum or difference frequency equal to the *intermediate frequency* of the receiver.

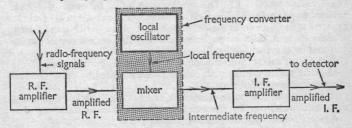

Fig. 79. Functional diagram of a local oscillator in a superheterodyne receiver

LOCAL-OSCILLATOR VALVE. The valve which provides the local *heterodyne* frequency for the *mixer valve* in a heterodyne *conversion transducer*.

LOCATION. In *digital computers*, a storage register.

LOCK-IN AMPLIFIER. A detector which responds only when the input signal is synchronous with the frequency of a control signal locally generated. It is usually employed as a null indicator in bridge circuits.

LOCK-OUT. In *data processing*: an inhibition of any other reference to a particular part of an equipment during an operation which uses that part.

LOCKING. Controlling the frequency of an oscillating circuit by an applied signal of constant frequency.

LOCKING RELAY. A *relay* which renders some other relay or device inoperative under predetermined conditions.

LOCKING-IN. Shifting and holding one or both of the frequencies of two coupled oscillating systems so that the two frequencies have the ratio of two integral numbers.

LOCKING-ON. Automatic following of a target by a radar aerial.

LOGARITHMIC DECREMENT. Of an *underdamped harmonic system*, the natural logarithm of the ratio of the excursions during two successive oscillations, with no external forces applied to maintain the oscillations.

LOGARITHMIC HORN. See *Exponential horn*.

LOGARITHMIC RESISTOR. A *variable resistor* designed so that the resistance it introduces into a circuit is directly or inversely proportional to the movement of the variable contact.

LOGIC. A system of computation based on the concepts of AND, EITHER-OR, NEITHER-NOR, etc. The use of only two conditions lends itself to being represented by bistable devices such as relays and diodes.

LOGIC ELEMENT. See *Logical element*.

LOGICAL COMPARISON. In computers, the operation of comparing two numbers.

LOGICAL DESIGN. (1) The basic planning of a *computer* or *data-processing* system. (2) The synthesizing of a network of *logical elements* to perform a specified function. (3) The result of (1) and (2).

LOGICAL DIAGRAM. In *logical design*, a diagram showing the *logical elements* and their interconnections, but not necessarily constructional or engineering details.

LOGICAL ELEMENT. In a *computer* or *data-processing* system, the smallest building blocks which can be represented by mathematical operators in symbolic logic.

LOGICAL OPERATION. In *computers*, any non-arithmetical operation.

LOGICAL SYMBOL. In *computers*, a graphical representation of a *logical element*. See *Graphical symbols*.

LOKTAL BASE. A valve base which incorporates a locking device.

LONG COUNTER. A proportional counter with a flat response to neutron flux.

LONG WAVE. A radio wave of wavelength between 1,000 and 10,000 metres.

LONGITUDINAL WAVE. A wave in which the direction of displacement at each point in the medium is the same as the direction of propagation, e.g. sound waves. Contrast *Transverse wave*.

LONG-LINE EFFECT. An *oscillator*, which is tightly-coupled to a load through a transmission line long in comparison to the wavelength of the oscillation, may sometimes jump from its desired frequency to a nearby undesired frequency.

LONG-PERSISTENCE SCREEN. A *cathode-ray tube* screen in which the use of *phosphorescent* compounds as well as *fluorescent* compounds allows the trace to persist for several seconds.

LONG-TAILED PAIR. Two matched thermionic valves connected so that they have a common *cathode bias* resistor and equal *anode loads*. With this arrangement the anodes remain at the same potential unless signals at the two *grids* differ in *phase*, if they are ac, or magnitude, if they are dc.

LOOP. (1) In *digital computers*, the repetition of a group of instructions in a routine. (2) A communications circuit between two private subscribers or between subscribers and the local switching centre.

LOOP ACTUATING SIGNAL. In control systems, the signal derived by mixing the *loop input signal* and the *loop feedback signal*.

LOOP ERROR. In control systems, the desired value minus the actual value of the *loop output signal*.

LOOP ERROR SIGNAL. In control systems, the *loop actuating signal* in those cases where it is the *loop error*.

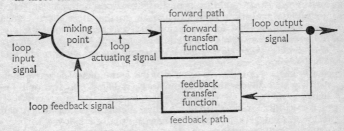

Fig. 80. Loop signals in a single-loop feedback system

LOOP FEEDBACK SIGNAL. In control systems, the signal derived as a function of the *loop output signal* and fed back to the mixing point for control purposes.

LOOP GAIN. The product of the values of gain acting on a signal which passes round a closed path.

LOOP INPUT SIGNAL. In control systems, an external signal applied to a *feedback control loop*.

LOOP OUTPUT SIGNAL. In control systems, the controlled signal extracted from a *feedback* control.

LOOP TEST. A method of locating a fault in the insulation of a conductor when the conductor can be connected to form part of a closed loop.

LOOP TRANSFER FUNCTION. The transfer function of the transmission path formed by opening and properly terminating a *feedback* loop.

LOOSE COUPLING. Any degree of coupling less than the *critical coupling*.

LORAN. A long-range system of radio-navigation of the *hyperbolic* type, in which position lines are determined by measuring the difference in the time of arrival of synchronized pulses.

LOSS FACTOR. The ratio of the average power loss to the peak load power loss during a specified period of time.

LOSSEV EFFECT. In a *semiconductor junction*: radiation due to recombination of charge carriers injected into a p-n or p-i-n junction which is biased in the forward direction.

LOSSY. Of an insulating material, dissipating more electrical energy than the normal for that class of material.

LOSSY LINE. A *transmission line* designed to have a high degree of attenuation.

LOUDNESS LEVEL. The sound pressure level in *decibels*, relative to 0·0002 *microbar*, of a simple tone of frequency 1,000 cycles, which is judged by the listeners to be equivalent in loudness to the sound under measurement.

LOUDSPEAKER. An electro-acoustic *transducer* normally intended to radiate acoustic power into the air so that it is effective at a distance. See the following entries: *Electrostatic loudspeaker, Labyrinth loudspeaker, Crystal loudspeaker, Magnetic armature loudspeaker, Magnetostriction, Permanent-magnet loudspeaker,* and entries below. See Figure 81.

LOUDSPEAKER DIVIDING NETWORK. A frequency-selective network which divides the spectrum to be radiated into two or more parts.

LOUDSPEAKER MICROPHONE. A dynamic loudspeaker which can also be used as a microphone, and is usually employed in intercommunication systems.

LOUDSPEAKER SYSTEM. A combination of one or more *loudspeakers* and associated baffles, horns, and dividing networks arranged to work together as the coupling means between the driving electric circuit and the acoustic medium.

LOVOTRON®. A modification of the *Trigatron* which has an improved performance under low voltage conditions with single pulses.

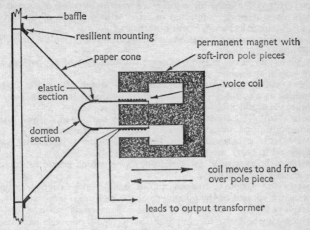

Fig. 81. Cross-section of typical permanent-magnet moving-coil loudspeaker

LOW FREQUENCY (LF). Radio frequencies between 30 and 300 kilocycles per second. See *Frequency band.*

LOW FREQUENCY COMPENSATION. In *amplifiers* at low signal frequencies compensation for the reactance of coupling capacitors to avoid distortion resulting from signal attenuation and phase shift.

LOW TENSION (LT). The low voltage which is used to supply heater current for a valve.

LOW-LEVEL MODULATION. *Modulation* produced at a point in a system when the power level is low compared with the power level at the output of the system.

LOW-LEVEL R-F SIGNAL. In *switching tubes*, a radio-frequency signal of insufficient power to cause the tube to become fired.

LOW-LOSS LINE. An electric *transmission line* with low series resistance and shunt conductance and therefore low energy dissipation per unit length.

LOW-PASS FILTER. A wave filter having a single transmission band extending from zero frequency up to some cut-off frequency which is not infinite.

LOW-VELOCITY SCANNING. The *scanning* of a target with electrons of velocity less than the minimum velocity required to give a *secondary-emission ratio* of unity.

LOWER SIDEBAND. The *sideband* containing all frequencies below the frequency of the carrier which has been *amplitude modulated.*

L-SECTION. Of a *filter* or *network*, a circuit element of two components, one being connected in *series* with one side of the circuit and the other in *parallel* with both sides.

LSI. Abbreviation for Large Scale Integration: referring to *integrated circuits*.

L.T. BATTERY. Low tension battery: the battery which provides the power for the *heater* supplies of a valve.

L-TYPE CATHODE. An *oxide cathode* in which the oxide is not applied as a coating, but is contained in a hollow reservoir.

LUMEN. The unit of *luminous flux*.

LUMINANCE. The *luminous intensity* of any surface in a given direction per unit of projected area viewed from that direction.

LUMINANCE CHANNEL. In a *colour television* system, any path designed to carry the luminance *signal*.

LUMINANCE FLICKER. In television, *flicker* which is due to fluctuations of *luminance* only.

LUMINANCE SIGNAL. In television, a signal wave designed to have exclusive control of the *luminance* of the picture.

LUMINESCENCE. A property of a class of materials called *phosphors* which causes them to emit light as a result of some excitation. (Incandescence is not included.) There are two kinds of luminescence: *fluorescence* and *phosphorescence*. See also *Cathodoluminescence, Cathodophosphorescence, Electroluminescence*.

LUMINOPHORE. A luminescent material.

LUMINOSITY. The ratio of *luminous flux* to the corresponding radiant flux at a given wavelength, expressed as *lumens* per watt.

LUMINOUS EFFICIENCY. Of radiant energy, the ratio of the *luminous flux* to the radiant flux, expressed as lumens per watt. Synonymous with *luminosity* for energy of a single wavelength.

LUMINOUS FLUX. The total visible light energy emitted by a source in unit time. The unit is the *lumen*.

LUMINOUS INTENSITY. The ratio of the *luminous flux* emitted in any direction by a light source, in an infinitesimal solid angle containing this direction, to the solid angle. Unit: *Candela*.

LUMISTER®. A class of semiconductor devices based on the *electroluminescent* and photoconductive effects, used in control circuits, and servo amplifiers.

LUMPED. Effectively concentrated at a single point.

LUMPED PARAMETER. Any circuit *parameter* which may be treated for circuit analysis as a single *inductance, capacitance, resistance* etc., throughout the frequency range under consideration.

LUX. The practical unit of illuminance in the metric system.

LUXMETER. An optical instrument for measuring illuminance, calibrated in lux units.

M

mA. Abbreviation for *milliampere*.

MAC. Acronym for 'Machine-Aided Cognition' or 'Multiple-Access Computer'; the nickname for a project started at the Massachusetts Institute of Technology in 1959, which developed into the first time-sharing computer systems in the U.S.

MACHINE CODE. A series of instructions in a form which a *digital computer* can accept directly.

MACHINE EQUATIONS. In *analogue computers,* equations which represent the relations between the voltages in the computer and the variable quantities which they simulate.

MACHINE LANGUAGE. The *language* occurring in a *computer.*

MACHINE UNITS. In dc *analogue computers,* arbitrary units of 100 volts selected to represent units of the variable which is simulated. The scale factor is then the number of these 100-volt units equal to one unit of the simulated variable.

MACHINE VARIABLE. In *analogue computers,* the voltage representing a mathematical variable in the problem presented to the computer.

MACHINE WORD. The number of characters a given *computer* can handle in one operation. See *Word.*

MADISTOR®. A magnetic semiconductor device used as a storage component.

MAGAMP. Abbreviation for *magnetic amplifier.*

MAGIC EYE. Colloquial for *electron-ray tube.*

MAGIC T. See *Hybrid junction.*

MAGNADUR®. A *ferrite* material used for making permanent magnets, especially for the focusing units of television picture tubes.

MAGNESIL®. A magnetic grain-oriented alloy of iron and silicon, used to make cores for *magnetic amplifiers.*

MAGNESIUM (Mg). A light metallic element, Atomic No. 12, extensively used in the construction of electronic components, often as an alloy with aluminium.

MAGNESTAT®. A *magnetic amplifier.*

MAGNESYN®. A small electromagnetic *transducer* which transmits a position signal. See *Selsyn.*

MAGNET. A body which produces a *magnetic field* external to itself.

MAGNET GAP. The space between the poles of a magnet, especially where a device such as a *magnetron* is mounted in the gap.

MAGNETIC ALLOYS. Alloys which exhibit *ferromagnetic* properties. See *Alnico, Ferrites, Heusler alloys, Mumetal, Permalloy.*

MAGNETIC AMPLIFICATION. *Amplification* of an electric signal based on the non-linear characteristics of *saturable reactors.*

MAGNETIC AMPLIFIER. An *amplifier* in which one or more *saturable reactors* is used in combination with other circuit elements to obtain a *power gain*, i.e. a small input power is used to control a large output power.

MAGNETIC ARMATURE. An assembly of *ferromagnetic* material near the pole pieces of a magnet, arranged so that it can move in relation to them.

MAGNETIC ARMATURE LOUDSPEAKER. A *loudspeaker* whose operation involves the vibration of a *magnetic armature*.

MAGNETIC ARMATURE MICROPHONE. A *magnetic microphone* in which the diaphragm causes an armature to move the coil in a magnetic field.

MAGNETIC BALANCE. An instrument for measuring the strength of a magnetic field by balancing the force exerted on a conductor carrying current in the field.

MAGNETIC BIASING. In *magnetic recording*, the superimposition, during recording, of an additional magnetic field on the signal magnetic field, in order to obtain a nearly linear relation between the signal amplitude and the remanent flux density in the recording medium.

MAGNETIC BOTTLE. A magnetic field used to confine the volume of a plasma stream in the application of *pinch effect* to a tube.

MAGNETIC CIRCUIT. A closed path of *magnetic flux*, the path having the direction of the *magnetic induction* at every point.

MAGNETIC COMPASS. A compass based on the tendency of a pivoted magnetized needle to align itself along the direction of the earth's magnetic field.

MAGNETIC CONTACTOR. A *contactor* actuated magnetically.

MAGNETIC CONTROLLER. An electric controller in which all the basic functions are controlled by electromagnets.

MAGNETIC CORE. Generally, any core of magnetic material. In a special sense, such a core is capable of assuming and retaining one of two or more conditions of magnetization and thus capable of providing *gating*, *storage*, or *switching* functions. For these applications the core takes the form of a small ring and is pulsed or polarized by currents through a toroidal winding round the core.

MAGNETIC CORE, SATURABLE. See *Saturable reactor*.

MAGNETIC CROSS-FIELD MODULATOR (REACTOR). A *magnetic core* in the form of a hollow toroid enclosing an annular winding and wound externally with a toroidal winding. When an alternating current is applied to the annular winding and a direct current bias to the toroidal winding a voltage develops in the toroidal winding whose frequency is double that of the signal.

MAGNETIC CUTTER. In *recording*, a cutter in which the motions of the recording stylus are produced by magnetic fields.

MAGNETIC DELAY LINE. A *delay-line memory*, composed of metallic sections in which the velocity of propagation is small compared with the speed of light and wave patterns containing information can be recirculated.

MAGNETIC DEFLECTION. *Deflection* of an electron beam produced by a magnetic field.

MAGNETIC DISPLACEMENT. *Magnetic induction*.

MAGNETIC DRUM. A rotating cylinder coated with magnetic material. Information is stored on the coating in the form of magnetized *dipoles*, the orientation or polarity of which is used to store *binary* information in *computers* or *data-handling systems*.

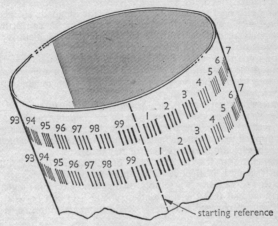

Fig. 82. Method of storing words on a magnetic drum. The two tracks shown each carry 100 words

MAGNETIC FIELD. The space in the neighbourhood of an electric current, or of a permanent magnet, throughout which the forces due to the current or magnet can be detected.

MAGNETIC FIELD INTENSITY. Synonym for *magnetizing force*.

MAGNETIC FLIP-FLOP. A *bistable multivibrator* employing one or more *magnetic amplifiers*.

MAGNETIC FLUX. A phenomenon produced in the medium surrounding electric currents or magnets. The amount of magnetic flux through any area is measured by the quantity of electricity caused by flow in an electric circuit of given resistance bounding the area when this circuit is removed from the magnetic field. It is the surface integral of the *magnetic flux density* or *magnetic induction* over any surface having the same boundary as the area. Unit in the *mksa* and *SI systems* is the *weber*; in the c.g.s. e.m.u. system the *maxwell*. One maxwell$=10^{-8}$ webers. See *Units*.

MAGNETIC FLUX DENSITY (MAGNETIC INDUCTION). At a point, the amount of *magnetic flux* per sq. cm. over a small area at the point. The direction of the flux is at right angles to this area. Symbol: *B*. Unit in the *mksa system*: *weber*/sq.m.; in the c.g.s. e.m.u. system *gauss*; in the SI system: *tesla*. See *Units*.

MAGNETIC FOCUSING. See *Electromagnetic focusing*.

MAGNETIC FORCE. See *Magnetizing force*.

MAGNETIC GATE. A *gate* circuit used in *magnetic amplifiers*.

MAGNETIC HEAD. In a *magnetic recorder*, a *transducer* which converts electrical to magnetic variations for magnetic storage, e.g. on *magnetic tape* (magnetic recording head); or for converting such a magnetic record back into electrical variations (magnetic reproducing head); or for erasing the record.

MAGNETIC HYSTERESIS. The property of a magnetic material which causes the *magnetic flux density* or induction, for a given *magnetizing force*, to depend on previous conditions of magnetization. See *hysteresis*.

MAGNETIC INDUCTION. See *Magnetic flux density*.

MAGNETIC INTENSITY. Synonym for *magnetizing force*.

MAGNETIC LAG. The delay between the time a *magnetizing force* is applied to a magnetic material and the time the *magnetic induction* reaches its equilibrium value for this force.

MAGNETIC LEAKAGE. That part of the *magnetic flux* following a path which is ineffective for the purpose desired.

MAGNETIC LENS. An assembly of magnets, electromagnets, or coils arranged so that the resulting magnetic field can be used to focus a beam of charged particles. See *Electromagnetic focusing*. See Figure 83.

MAGNETIC LOUDSPEAKER. A *loudspeaker* in which acoustic waves are caused by mechanical forces resulting from magnetic reactions.

MAGNETIC MEMORY. A device which uses the ability of certain magnetic materials to retain their state of magnetization indefinitely, to store information. See *Magnetic drum*, *Magnetic core*, *Magnetic memory plate*, *Memory*.

MAGNETIC MEMORY PLATE. A thin plate of *ferrite* pierced with holes in the form of a rectangular grid. Wires threaded through the holes allow pulses of information to be stored by magnetic induction round each hole independently of the others, and to be read out as required. Company memory matrices with a large storage capacity are made up by stacking these plates which can also have circuits printed directly on them.

MAGNETIC MICROPHONE. A *microphone* whose electric output is generated by the relative motion of a magnetic field and a coil or conductor located in the field.

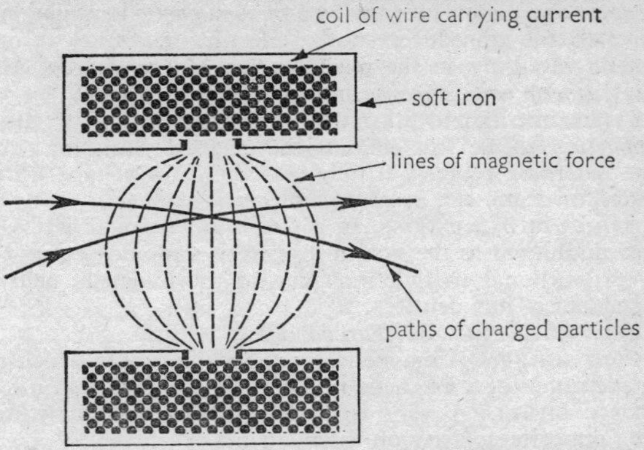

Fig. 83. Schematic diagram of one type of magnetic lens

MAGNETIC MODULATOR. A *modulator* in which modulation is effected by *magnetic amplification* elements. See *Magnettor*.

MAGNETIC MOMENT. Of a magnet or magnetic *dipole*, the product of the pole strength and the distance between the poles.

MAGNETIC PENDULUM. A small magnet, suspended from its centre by a wire from a fixed support, used as a torsion pendulum.

MAGNETIC POLARIZATION. The difference between the *magnetic induction* at any point in a medium and the induction which would exist there if the point were in a vacuum.

MAGNETIC POLE. Of a magnet, those parts of the magnet towards which the external *magnetizing force* tends to converge (south pole) or from which it tends to diverge (north pole).

MAGNETIC POLE, UNIT. A pole of such strength that, when situated one centimetre away from a similar pole in a vacuum, the force on each is one dyne.

MAGNETIC POTENTIOMETER. A *potentiometer*, consisting of a thin strip of non-magnetic material wound with many turns of fine wire connected to a *fluxmeter*, used to determine the magnetic potential difference between two points in a magnetic circuit.

MAGNETIC POWDER-COATED TAPE. A *magnetic tape* consisting of a non-magnetic base on which powdered *ferromagnetic material* is uniformly coated.

MAGNETIC PRINTING (MAGNETIC TRANSFER). The permanent transfer of a recorded signal from a section of a magnetic recording medium to another section of the same or a different medium when these sections are brought together.

MAGNETIC RECORDER. Equipment incorporating an electromagnetic *transducer* and means for moving a *ferromagnetic* recording medium relative to the transducer so that electric signals are recorded as magnetic variations in the medium. See *Magnetic head*, *Magnetic tape*, *Magnetic wire*, *Tape recorder*, and the next entry.

MAGNETIC RECORDING MEDIUM. A magnetizable material used in a *magnetic recorder* for retaining the magnetic variations imparted in the recording process. It may take the form of tape, wire, disc, cylinder, or drum, etc. See *Magnetic memory*.

MAGNETIC RESIDUAL LOSS. In a *ferromagnetic* material: a loss of energy additional to the normal hysteresis and eddy current losses. It is proportional to the frequency of the magnetic field and is independent of flux density.

MAGNETIC ROTATION. See *Faraday effect*.

MAGNETIC SCREEN. A magnetically permeable screen which reduces the penetration of a magnetic field into an assigned region.

MAGNETIC SHELL. A very thin sheet of ferromagnetic material having opposite polarity on the two sides.

MAGNETIC SHIELDING. A screen made of high *permeability* material used to isolate a device from unwanted magnetic fields.

MAGNETIC SHIFT REGISTER. A *shift register* in which the pattern of settings of a row of magnetic cores is shifted one step along the row by each new pulse.

MAGNETIC STORM. A disturbance in the earth's magnetic field, associated with abnormal solar activity, which can affect both radio and wire transmission.

MAGNETIC TAPE. A magnetic recording medium, of width at least ten times greater than its thickness, which may be entirely of magnetic material or coated with it. See Figure 84.

MAGNETIC THIN FILM. A layer of magnetic material usually less than one micron thick, which may be used for *storage* or *logic elements* in computers.

MAGNETIC TRANSITION TEMPERATURE. See *Curie point*.

MAGNETIC WIRE. A magnetic recording medium, approximately circular in cross section.

MAGNETICALLY DAMPED METER. A meter in which damping is achieved by moving a metal vane through a magnetic field.

MAGNETISM. It is now thought that magnetism is due to the unbalanced spins of electrons in the atoms of magnetic materials.

MAGNETIZATION CURVE. A curve showing the *ferromagnetic* characteristic of a material in which the *magnetic flux density*, B, is usually plotted against the magnetizing force, H.

MAGNETIZE. Induce magnetic properties in; convert into a magnet.

MAGNETIZING FORCE (MAGNETIC FIELD STRENGTH). The 'force' which produces or is associated with *magnetic flux density*. It is equal to the *magnetomotive force* per centimetre measured along the line of force. Symbol: H. Unit: *Oersted*.

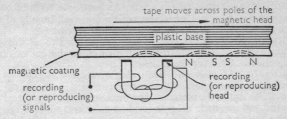

Fig. 84. Principle of magnetic tape recording and reproduction

MAGNETO. An *alternator* with permanent magnets, especially one used in conjunction with an *induction coil* to provide a high voltage for the ignition system of an internal combustion engine.

MAGNETOCALORIC EFFECT. See *Thermomagnetic effect.*

MAGNETOMETER. An instrument for measuring the intensity or direction of a magnetic field. See *Fluxmeter, Gaussmeter.*

MAGNETOMOTIVE FORCE (MMF). The magnetic analogue of *electromotive force.* It can be defined as the line integral of the *magnetizing force* around a closed path in a magnetic field; or as the work done in moving a unit *magnetic pole* around a closed magnetic circuit. See *Ampere-turn, Gilbert, Units.*

MAGNETO-OPTICAL EFFECT. The phenomenon observed when a beam of plane-polarized light is reflected from the polished pole face of a very strong electromagnet: the light beam becomes slightly elliptically polarized. See *Kerr effect.*

MAGNETOPHONE. A *magnetic recorder.*

MAGNETORESISTANCE. If a magnetic material is carrying electric current in the presence of a magnetic field, the *resistivity* of the material increases when the field is parallel to the current flow and decreases when it is at right angles.

MAGNETOSTATIC FIELD. A stationary *magnetic field.*

MAGNETOSTRICTION. The change in the dimensions of a *ferromagnetic* material when it is placed in a magnetic field. See *Guillemin effect, Wiedemann effect.*

MAGNETOSTRICTION, CONVERSE. The change in magnetic properties observed in *ferromagnetic* materials which are subjected to strain or pressure.

MAGNETOSTRICTION LOUDSPEAKER. A *loudspeaker* in which the mechanical displacement is derived from the *magnetostriction* of a magnetic rod, in the field of the loudspeaker coils.

MAGNETOSTRICTION MICROPHONE. A *microphone* whose operation depends on the generation of an electromotive force by the deformation of a material undergoing *magnetostriction.*

MAGNETOSTRICTION OSCILLATOR. An *oscillator* whose frequency is controlled by a rod of ferromagnetic material, usually of nickel or nickel alloy, with pronounced *magnetostriction*.

MAGNETOSTRICTIVE FILTER. A *filter* network which uses magnetostrictive bars or rods, together with their energizing coils, as part of the network. See *Magnetostriction*.

MAGNETRON. An electron valve (tube) in which electrons interact with the electric field of a circuit element in the presence of crossed, steady, electric and magnetic fields to produce an alternating current power output. The device consists of heater, cathode, and anode (usually with a number of radial segments) enclosed in a highly evacuated container which is positioned in the gap of an external magnet. It is capable of producing high output power at high frequencies in the *microwave* region. The output circuit is either a *coaxial line* or a *waveguide*.

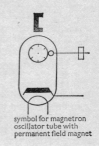

symbol for magnetron oscillator tube with permanent field magnet

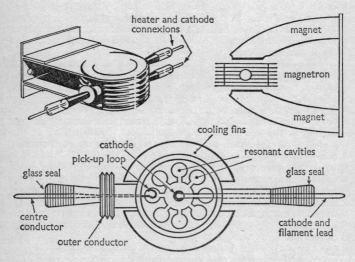

Fig. 85. Magnetron

MAGNETRON, INTERDIGITAL. A *magnetron* having axial anode segments around the cathode, alternate segments being connected together at one end, remaining segments connected at the opposite end.

MAGNETRON, MULTICAVITY. A *magnetron* in which the anode includes several cavities.

MAGNETRON, PACKAGED. An integral assembly comprising a *magnetron*, its magnetic circuit, and an output matching device.

MAGNETRON, RECTIFIER. An early form of magnetron.

MAGNETRON, RISING SUN. A *multicavity magnetron* in which resonators of two different frequencies are arranged alternately for the purpose of *mode separation*.

MAGNETRON, SPLIT-ANODE. A *magnetron* with an anode divided into two segments, usually by slots parallel to its axis.

MAGNETRON, TRAVELLING-WAVE. See *Travelling-wave magnetron*.

MAGNETRON ARCING. Internal breakdown between anode and cathode of a *magnetron*, especially in high power magnetrons.

Fig. 86. Anode of rising sun magnetron

MAGNETRON CRITICAL FIELD. The smallest (theoretical) value of steady *magnetic flux density* at steady anode voltage which would prevent an electron, emitted from the cathode at zero velocity, from reaching the anode.

MAGNETRON CRITICAL VOLTAGE. The highest (theoretical) voltage on the anode, at steady *magnetic flux density*, at which electrons emitted from the cathode would fail to reach the anode.

MAGNETRON MODE. In multi-anode *travelling-wave magnetrons*, modes of operation are defined which depend on the rotational velocity of the electron cloud as it passes the anode openings.

MAGNETRON OSCILLATOR. An *oscillator* in which the output from a magnetron is used to excite and sustain oscillations in a tuned circuit.

MAGNETRON PULLING. A frequency shift in a *magnetron* caused by faults in its output circuit.

MAGNETRON PUSHING. A frequency shift in a *magnetron* caused by faulty operation of the *modulator*.

MAGNETRON TRAVELLING WAVE OSCILLATIONS. Oscillations sustained by the interaction between the *space-charge* cloud of a *magnetron* and a travelling electromagnetic field whose *phase velocity* is nearly the same as the mean velocity of the cloud.

MAGNETTOR (MAGNETIC MODULATOR, FREQUENCY AMPLIFIER). A low-level signal converter which uses a *saturable reactor* to amplify direct current or low frequency alternating signals, or acts as *frequency multiplier*.

MAGNISTOR®. A solid-state device used in computers, consisting of a high permeability low-retentivity *ferrite* combined with a high retentivity ferrite.

MAGSLIP®. A *synchro* developed in Britain.

MAIN ANODE. Of a mercury-pool cathode tube, the anode which conducts load current (as distinct from auxiliary or exciting anodes).

MAIN BANG. In *radar*, the transmitter pulse as displayed on the screen.

MAIN GAP. In a *cold-cathode glow charge* tube, the conduction path between a principal *cathode* and a principal *anode*.

MAINTAINING VOLTAGE. Of a *glow discharge* valve or tube: the voltage between two specified electrodes carrying a glow discharge at a specified current.

MAJOR CYCLE. In a *memory* with consecutive access to *storage* positions, the time between successive appearances of a storage position, e.g. the time of rotation of a *magnetic-drum* memory.

MAJOR LOBE. Of an aerial radiation pattern, the *lobe* containing the direction of maximum radiation or reception.

MAJORITY CARRIER. In *semiconductors*, the type of *carrier* constituting more than half the total number of carriers.

MAJORITY CARRIER CONTACT. In *semiconductors*, an electrical contact in which the ratio of current to applied voltage is substantially independent of the polarity of the voltage for *majority carriers* but not for *minority carriers*.

MAJORITY EMITTER. Of a *transistor*, an electrode from which a flow of *majority carriers* enters the inter-electrode region.

MALLARD. An extensive technical plan for tactical voice communications between all British, United States, Canadian, and Australian armed services. Research and development is costed at £45 million and is expected to continue until the end of 1974.

MANGANESE (Mn). Metallic element, Atomic No. 25. Used in *manganin* and *Heusler alloys*.

MANGANIN. An alloy of manganese, copper, and nickel, with a high coefficient of *resistivity* and a low temperature coefficient, used in the manufacture of wire for precise *resistors*.

MANIAC. A large electronic *digital computer* built by the Los Alamos Scientific Laboratory for its own use.

MAN-MADE NOISE. *Interference* with radio reception caused by sources such as car-ignition systems, diathermy machines, domestic electrical appliances, etc.

MANY-ONE FUNCTION SWITCH. In *computers*, a *function switch* in which a combination of the inputs is excited at one time to produce a corresponding single output.

MARKER. In radio-aids to navigation, a device for providing a signal to designate a small area above it.

MARKER PULSES. In *time division multiplexing*, *pulses* transmitted at regular intervals to synchronize the transmitter and receiver.

MARKING WAVE. In *telegraphic communication*, the emission which takes place during transmission of the active portion of the code.

MASER. Microwave Amplification by Stimulated Emission of Radiation. A general term for a class of amplifiers and oscillators which convert the internal energy of atoms or molecules into *microwave* energy by allowing them to react directly with electromagnetic radiation. If the electromagnetic waves are of the right frequency they will stimulate high energy electrons to release quanta of energy and fall to lower energy states, thereby amplifying the stimulating radiation. This process is not affected by the random motion of electrons, and maser amplifiers generate less *noise* than other types of amplifier. The maser principle was discovered independently by American and Russian physicists in 1951. See entries below and *Laser*.

General symbol

MASER, AMMONIA. The first *maser* to be conceived – by C. H. Townes of Columbia University – used a beam of ammonia molecules which were shot through a small hole into an evacuated cavity. Here they were subjected to electric fields which concentrated the excited molecules and focused them through a hole into another cavity, where they resonated with suitable microwaves and gave up their energy to them. This is the principle of the ammonia *atomic clock*.

MASER, GAS. A *maser* in which the electromagnetic radiation interacts with the molecules of a gas; e.g. the *ammonia maser*.

MASER, OPTICAL. See *Optical maser*.

MASER, SOLID-STATE. A *maser* in which the electromagnetic radiation interacts with a solid, usually crystalline.

MASER RELAXATION. In a *maser*, the processes which tend to restore molecules from an excited energy state to a state of equilibrium.

MASKING (AUDIO MASKING). The amount by which the threshold of audibility of a sound is raised by the presence of another sound; usually expressed in *decibels*.

MASS NUMBER. The total number of *neutrons* and *protons* in the *nucleus*, i.e. the atomic weight.

MASS SPECTROGRAPH. A device for analysing a substance in terms of the ratio of mass to charge of its constituent elementary particles.

One of its main applications is to determine the relative proportions of the different *isotopes* of an element.

MASS SPECTROMETER. A device resembling a *mass spectrograph* but so designed that the particles of a given mass-to-charge ratio are focused on an electrode and detected or measured electrically.

MASSEY FORMULA. An expression for the probability that *secondary emission* of electrons will result when an excited atom approaches the surface of a metal.

MASTER, ORIGINAL. In disc recording, the master record produced by electroforming from the face of a wax or lacquer recording.

MASTER OSCILLATOR. An *oscillator* so arranged as to establish the *carrier frequency* of the output of an amplifier.

MASTER-SLAVE MANIPULATOR. A device for the remote handling of radioactive isotopes.

MASTER-TRIGGER. A *multivibrator* in a *radar* transmitter which determines the *pulse repetition frequency*.

MATADOR. A surface-to-surface *missile* used by the U.S. Air Force.

MATCHED LOAD. A device for terminating a *waveguide* or *coaxial line* so as to absorb the whole of the incident power.

MATCHED TERMINATION. In networks or *transmission lines* a termination which produces no *reflected wave*.

MATCHED WAVEGUIDE. A *waveguide* having no *reflected wave* at any transverse section.

MATCHING. See *Impedance matching, Load matching*.

MATHEMATICAL CHECK. Of a *computer*: a programmed check of a sequence of operations using mathematical properties of the sequence.

MATRICON. A *cathode-ray tube*, invented by A. T. Starr in 1962, which can focus onto its screen 35 beams in a 7 column by 5 rows matrix. By the selective switching of beams a character pattern may be produced on the screen with a brightness of display estimated to be 20 times higher than is possible with conventional cursive systems. See Figure 87.

MATRIX. In *computers*, a *logical* network in the form of a rectangular array of intersections of its input and output leads with elements connected at some of these intersections. It usually functions as an *encoder* or *decoder*.

MATTHIESEN'S RULE. Impurities or lattice imperfections in a crystal increase its *conductivity* over that of a perfect crystal.

MAVAR. Modulating Amplifier using VAriable Reactance: a *parametric amplifier*, usually of the solid-state kind.

MAXWELL. The unit of *magnetic flux* in the c.g.s. *electromagnetic system*.

MAXWELL BRIDGE. A four-arm *ac bridge* in which *inductance* or *capacitance* is measured as the product of two *non-reactive resistors*, and the balance is independent of frequency.

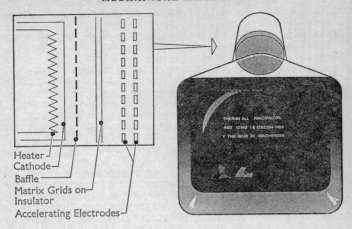

Heater
Cathode
Baffle
Matrix Grids on Insulator
Accelerating Electrodes

Fig. 87. Matricon

MAXWELL'S EQUATIONS. Four classical equations of electromagnetic theory which give the relations between *magnetizing force H, electric field strength E, magnetic induction B, electric displacement D, conduction current density J,* and *electric charge* density ρ, in a medium.

MAXWELL'S RULE. Every part of an electric circuit tends to move in such a direction as to enclose the maximum *magnetic flux*.

MAXWELL-TURNS. A measure of the flux *linkage* in a coil; the product of the *magnetic flux* and the number of turns in the coil.

MAYDAY. International distress call (from the French 'm'aidez' = 'help me').

MC/s. Abbreviation for *megacycles per second*.

MCNALLY TUBE. A *reflex klystron* whose frequency is controllable over a wide range, used as a *local oscillator*.

MEASURING AMPLIFIER. A wide range stable amplifier calibrated in terms of gain.

MECHANICAL ANALOGUE. Of an electrical variable, system, or device, any mechanical variable, system, or device which behaves in an analogous manner to the electrical counterpart; e.g. the mechanical analogue of inductance is a mass or a flywheel. Mechanical analogy can be made mathematically exact.

MECHANICAL IMPEDANCE. Of a mechanical system, the complex ratio of the alternating force applied to the system to the resulting alternating velocity in the direction of the applied force. Compare *Impedance*.

235

MECHANICAL (COUNTING) REGISTER. An electromechanical device for recording or registering counts.

MEDIUM FREQUENCY. Radio frequencies between 300 and 3,000 kilocycles per second. See *Frequency band*.

MEGA-. Prefix denoting one million (10^6).

MEGACYCLES PER SECOND (MC/S). A unit of frequency: one million cycles per second. The 'per second' is frequently dropped.

MEGAHERTZ (MHz). One million *hertz* = one megacycle/second.

MEGA-ELECTRON-VOLT (MEV). One million *electron volts*.

MEGAPHONE, ELECTRICAL. A megaphone which includes a microphone, amplifier, and sound reproducer and is used as a directional loudspeaker.

MEGATRON. A *lighthouse valve*.

MEGAVOLT (MV). A unit of *potential*: one million *volts*.

MEGAWATT (MW). A unit of power: one million *watts*.

MEGGER®. An adjective describing any product of Evershed & Vignoles Ltd; for example, their insulation resistance testing sets.

MEGOHM. One million *ohms*.

MEISSNER EFFECT. An effect observed in a *superconductor* at extremely low temperatures, in which the lines of *magnetic induction* are apparently expelled from the superconductor when it is cooled below the *superconductivity transition temperature* in a magnetic field.

MEKOMETER. An electronic device in which the *Pockels effect* is used to measure distances of the order of 50 metres to an accuracy of 0·05 mm.

MEL. A unit of pitch: a simple tone of frequency 1,000 c/s, 40 decibels above the listener's threshold, produces a pitch of 1,000 mels.

MELTBACK TRANSISTOR. A *junction transistor* in which the junction is formed by allowing the molten doped semiconductor to solidify.

MEMISTOR®. A self-adjusting resistor used for an adaptive memory.

MEMORY. In *computers* and recording instruments, any device into which information can be introduced and later extracted. The mechanism for effecting this is usually an integral part of the computer but this is not essential. See the following: *Bin, Buffer, Circulating memory, Delay line memory, Electrostatic memory, Ferrite bead, Magnetic core, Magnetic drum, Magnetic memory, Magnetic tape, Memistor, Mercury tank, Parallel memory, Permanent memory, Random-access memory, Secondary memory, Serial memory, Static memory, Temporary memory, Williams tube, Working memory, Zero-access memory*.

MEMORY CAPACITY. The amount of information which can be retained in a *memory*, usually expressed as the number of *words* which can be retained. For comparison of different memories this number is expressed in *bits*.

MEMORY LOCATION. A unit *storage* position in the main *internal memory* holding one computer *word*, usually having a specific *address*.

MEMORY REGISTER. A *register* in the *memory* of a computer.

MERCURY (Hg). The only metallic element liquid at ordinary temperatures. Atomic No. 80. Used in arc *rectifiers* and *thyratrons*.

MERCURY (-POOL) ARC. A *cold-cathode arc* using a pool of mercury as the cathode.

MERCURY CELLS. *Electrolytic cells* having mercury cathodes with which alkali metal deposited from the electrolyte forms an amalgam.

MERCURY COMPUTER. A large electronic *digital computer* built by Ferranti Ltd.

MERCURY TANK. A container filled with mercury and incorporating a large number of delay line circuits, used as a method of storing information by circulating it continuously through the tank. See *Circulating memory*.

MERCURY-ARC CONVERTER. A *frequency-arc converter* using a mercury arc power converter.

MERCURY-ARC RECTIFIER. A *rectifier* based on the property that a *mercury arc* enclosed in a tube with a non-emitting anode, allows current to pass in one direction only, i.e. from *cathode* to *anode*.

MERCURY-POOL CATHODE. A gas-tube cathode which consists of a pool of mercury. Electrons are freed by the action of an arc spot on the pool.

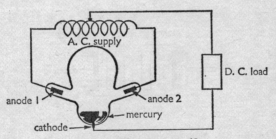

Fig. 88. Mercury-arc rectifier

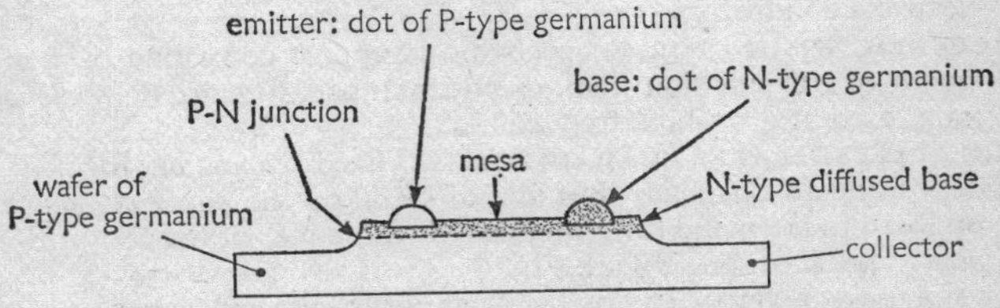

Fig. 89. Diagram of construction of P-N-P mesa transistor

MESA TRANSISTOR. A *transistor* in which a germanium or silicon wafer is selectively etched in steps so that the base and emitter regions appear as raised plateaux above the collector region. See Figure 89.

MESH. A set of branches forming a closed path in a *network*.

MESH CURRENT. A current assumed to exist over the whole cross section of a *mesh*.

MESSAGE. In computer terminology, two or more *words* treated as a unit.

METADYNE. See *Amplidyne*.

METAL RECTIFIER. A rectifier the operation of which depends on the asymmetrical conductance of certain surfaces in contact, at least one of them being metallic.

METAL VALVE (TUBE). A valve (tube) with a metal envelope and the leads passing through glass beads fixed in the housing.

METAL-CERAMIC. See *Cermet*.

METAL-FILM RESISTOR. A *resistor* made by coating a high-temperature insulator, such as mica, ceramic, Pyrex, or quartz, with a metal film by firing or evaporation.

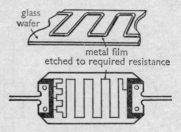

Fig. 90. Film resistor

METALLIC BOND. A bond, characteristic of atoms in a metal, which allows free passage of the *valence electrons* throughout the metal lattice and therefore accounts for the fact that metals are good electrical conductors.

METALLIC INSULATOR. A mechanical support consisting of a short length of transmission line, an odd number of *quarter wavelengths* long, used at ultra-high frequencies.

METALLIZED-PAPER CAPACITOR. A fixed *capacitor* formed by depositing an extremely thin film of metal on the paper *dielectric* as opposed to the metal foils interleaved with paper of the conventional *paper capacitor*. See Figure 91.

METER PROTECTION CIRCUIT. A circuit to prevent heavy overloads of current through the meter, which might damage it. It often takes

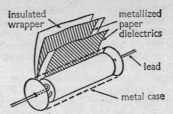

Fig. 91. Construction of metallized-paper capacitor

the form of a *neon lamp* connected in parallel, which breaks down above a predetermined voltage and shunts the current which would otherwise flow through the meter.

METER RESISTANCE. The resistance of a meter as measured at the terminals at a given reference temperature. In the case of a *rectifier instrument* the frequency and wave shape of the applied energy and the indicated value must also be specified.

MEV. Abbreviation for Million Electron Volt.

MF. Abbreviation for *Medium Frequency*.

MHO. The unit of *conductance* or *admittance*. It is the reciprocal of the *ohm*, and is the conductance of a conductor when a potential difference of 1 volt between its ends produces a current of 1 ampere.

MICA. A naturally occurring mineral – one of a family of complex aluminium-potassium silicates – which has a characteristic cleavage into thin transparent sheets. It is an excellent insulator, even at very high temperatures, and is extensively used as a *dielectric* in *capacitors* and as a supporting material for valve electrodes.

MICA CAPACITOR. A *capacitor*, using a *mica dielectric*, which is used where precision, stability, and high *dielectric strength* are more important than small size or low cost. See also *Silver-mica capacitor*

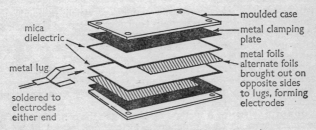

Fig. 92. Simplified construction of mica capacitor

MICR. Abbreviation for Magnetic Ink Character Recognition.

MICRO. A prefix meaning one millionth, sometimes used to mean very small, as in *microwave*.

MICROALLOY DIFFUSED TRANSISTOR. A *microalloy transistor* in which the semiconductor is first subjected to gaseous diffusion to produce a non-uniform base region.

MICROALLOY TRANSISTOR. A *transistor* in which the emitter and collector electrodes are formed by etching small pits in the semiconductor wafer, electroplating, and alloying a thin film of the impurity metal.

MICROAMMETER. A sensitive current meter, usually with a *D'Arsonval* type movement, calibrated in *microamperes*.

MICROAMPERE (μA). One millionth of an *ampere*.

MICROBAR. A unit of pressure in acoustics. It is equal to a pressure of 1 dyne per sq. cm. or nearly one millionth of normal atmospheric pressure.

MICRO-ELECTRONICS. A branch of electronics concerned with the design, production, and application of electronic devices of very small dimensions, in which a high packing density of the component parts is achieved by eliminating individual containers and connecting wires. See also *Molecular electronics*.

MICROFARAD (μF). A millionth of a *farad*; the unit of *capacitance* in the practical and *mksa* system of units.

MICROHENRY (μH). One millionth of a *henry*.

MICROHM. One millionth of an *ohm*.

MICROLOCK. The name given to certain U.S. artificial *satellite* detecting and tracking stations.

MICROMICROAMMETER. An extremely sensitive current meter calibrated directly in micromicroamperes or 10^{-12} amperes.

MICROMICROFARAD. A millionth of a *microfarad* or 10^{-12} farads. Also called a *picofarad*.

MICRO-MINIATURIZATION. See *Micro-electronics*.

MICRON. A unit of length equal to a millionth of a metre or a thousandth of a millimetre.

MICROPHONE. An electro-acoustic *transducer* which responds to sound waves and delivers essentially equivalent electrical waves. The following types of microphone are described in the dictionary: *carbon, cardioid, crystal, ear, electrostatic, glow-discharge, inductor, lip, magnetic, magnetostriction, moving-coil, pressure, ribbon, throat, velocity.*

symbol

MICROPHONICS. The noises resulting from *microphony*.

MICROPHONY. In a *valve*, the production of *noise* as a result of mechanical vibration or shock.

MICROPROGRAMMING. A method of working in the control unit of a *computer*, in which each programme instruction, instead of being used to initiate control signals directly, is first interpreted by a programme which is inherent in the construction of the control unit.

MICRORADIOGRAPH. An optical reproduction of the image produced by, for example, an *electron microscope*.

MICRORADIOMETER. An instrument for the detection of radiant power which uses a *thermopile* connected to the moving coil of a galvanometer by very short leads.

MICROSCOPIC MOBILITY. See *Hall mobility*.

MICROSECOND. One-millionth of a second.

MICROSTRIP. A kind of *microwave transmission line* used in place of conventional *waveguides* for certain applications.

MICROSWITCH. A small, sensitive switch specially designed to make or break with very small applied forces.

MICROVOLT (μv). A millionth of a *volt*.

MICROVOLTMETER. A *voltmeter* calibrated in *microvolts*.

MICROVOLTS PER METRE. A standard unit of radio field strength, measured at any point as the ratio of the voltage at a receiving aerial in *microvolts* to the length of the aerial in metres.

MICROWATT. A millionth of a *watt*.

MICROWATTMETER. An electric *wattmeter* calibrated in *microwatts*.

MICROWAVES. Very short electromagnetic waves whose frequencies extend from about 1,000 megacycles per second to about 300,000 megacycles per second or almost into the *infra-red* region; i.e. from about 30 centimetres to 1 millimetre in wavelength. See *Frequency band, Frequency spectrum*.

MICROWAVE SPECTROSCOPY. The determination and measurement of the *microwave* frequencies which are selectively absorbed by different materials.

MILLER BRIDGE. A *bridge* used to measure the amplification factors of valves.

MILLER EFFECT. The effect of *amplification* on the input *capacitance* of a valve. The effective or Miller input capacitance is always equal to or greater than the sum of the static electrode capacitances because of the effect of *feedback* in the valve.

MILLER INTEGRATOR. An *integrator* which employs a valve to improve the linearity of *pulse-shaping* circuits.

MILLER SWEEP GENERATOR. A *sawtooth sweep generator* employing a *Miller integrator* to improve the linearity of the sweep voltage.

MILLI-. A prefix meaning one-thousandth.

MILLIAMMETER. An *ammeter* calibrated in *milliamperes*.

MILLIAMPERE (mA). One-thousandth of an *ampere*.

MILLIBAR. One-thousandth of a bar; one thousand *microbars*.

MILLIKAN METER. An ionization chamber designed by Millikan, using a *golf-leaf* electroscope.

MILLIMETRE WAVES. *Microwaves* with wavelengths between 10 and 1 millimetres, corresponding to frequencies between 30 and 300 GC/s.

MILLIVOLT (mv). One-thousandth of a volt.

MILLIVOLTMETER. A *voltmeter* calibrated in *millivolts*.

MILLIWATT. One-thousandth of a *watt*.

MILLIWATTMETER. A *wattmeter* calibrated in *milliwatts*.

MILLMAN TRAVELLING WAVE TUBE. A type of *slow-wave tube* which uses equally spaced regions in which the beam and field interact, each separated by a region of field-free electron drift.

MINIMUM DISCERNIBLE SIGNAL (MDS). The input power level of a radio receiver which is just sufficient to produce a discernible signal in the receiver output.

MINIMUM SAMPLING FREQUENCY. In *pulse modulation systems*, the minimum frequency at which the signal may be sampled by the modulating pulse with satisfactory recovery of the signal by the demodulator.

MINITRACK. A system for tracking an earth satellite by radio (U.S.).

MINOR CYCLE. Of a *computer*, the time required to transmit a *machine word*. Compare *Major cycle*.

MINORITY CARRIER. In *semiconductors*, the type of charge *carrier* which makes up less than half of the total number of carriers.

MINORITY EMITTER. Of a *transistor*, an electrode from which a flow of *minority carriers* enters the inter-electrode region.

MIRROR GALVANOMETER. A galvanometer with a small mirror mounted on the moving coil, so that the deflection is optically magnified and measured by a ray of light reflected from the mirror to a scale.

MISCH METAL. A metallic alloy of rare-earth elements used to coat the cathodes of glow-type tubes to reduce the *cathode drop*.

MISMATCH. The condition in which the *impedance* of a *load* does not match the impedance of the source to which it is connected.

MISSILE. See *Guided missile*.

MIXED HIGHS. In *colour television*, the *video* signal channel which carries the information for the fine detail of the picture in the form of an achromatic signal.

MIXER. The part of a *heterodyne* receiver in which the incoming signal and the *local-oscillator* signal are mixed (*modulated*) to produce the *intermediate-frequency* signal.

MIXER VALVE. A valve which performs the *frequency-conversion* function only in a *heterodyne frequency changer* when it is supplied with power from an external oscillator.

MKSA ELECTROMAGNETIC SYSTEM OF UNITS. An absolute system of units based on the metre, kilogram, and second. See *Units*.

MMF. Abbreviation for *magnetomotive force*.

MODE. One of several states of electromagnetic wave *oscillation* which may be sustained in a given *resonant* system; or one of several methods of exciting a resonant system. See *Dominant mode, Magnetron mode, Resonant mode, Transmission mode*.

MODE FILTER. A *waveguide filter* designed to separate waves of the same frequency but with different *transmission modes*.

MODE JUMPING. In *magnetrons*, a sudden change in the *mode* from one pulse to the next.

MODE NUMBER. (1) In *magnetrons*, the number of radians of *phase* shift in one complete circuit of the anode divided by 2π. (2) In *reflex klystrons*, the number of whole cycles during which a meanspeed electron remains in the *drift space*.

MODE SEPARATION. In oscillators, the difference in frequency between resonator *modes* of oscillation.

MODE SHIFT. In a *magnetron*, a change in the *mode* during a pulse.

MODE SKIP. Failure of a *magnetron* to fire on successive pulses.

MODE TRANSDUCER (MODE TRANSFORMER). A device for converting an electromagnetic wave from one *mode* of operation to another.

MODEM. Contraction of Modulator-Demodulator: a device which converts a signal from one kind of equipment into a form suitable for another.

MODER. See *Pulse moder*.

MODIFIER. In *data processing*: a quantity used to alter an instruction in a prescribed way to produce the instruction actually obeyed.

MODULATED AMPLIFIER. An amplifier stage in a transmitter in which the *modulating signal* is introduced to modulate the *carrier*.

MODULATED CARRIER. An r-f *carrier* wave with some of its characteristics varied to correspond with the information transmitted.

MODULATED CONTINUOUS WAVE (MCW). A wave whose *carrier* is modulated by a constant *audio-frequency* tone.

MODULATED QUANTITY. A combination of two or more oscillating quantities which results in the production of new frequency components not present in the original quantities. In *communications*, one of these oscillating quantities is called the *carrier* and the other is called the *signal*.

MODULATED STAGE. The radio-frequency stage to which the *modulator* is coupled, and in which the carrier wave is modulated.

MODULATED WAVE. A combination of two or more waves resulting in the production of frequencies not present in the original waves, the new frequencies being usually the sums and differences of integral multiples of the frequencies in the original waves.

MODULATING ELECTRODE. An *electrode* of a valve to which a potential is applied to control the magnitude of the beam current.

MODULATING SIGNAL. A *signal* which causes a variation of some characteristics of a *carrier*.

MODULATION. The process, or the result of the process, in which some characteristic of one wave is varied in sympathy with some characteristic of another wave. The main types are *amplitude*, *angle*, *frequency*, and *phase modulation*. See also the following: *Carrier, Cathode, Compound, Conductivity, Constant-current, Cross, Double,*

243

Dual, Facsimile, Grid, High-level, Hum, Intensity, Linear, Low-level, Negative, Noise, Percentage, Positive, Pulse, Screen-grid, Series, Single-sideband, Spark-gap, Time, and Velocity modulations, and Intermodulation.

MODULATION CAPABILITY (MAXIMUM PERCENTAGE MODULATION). The maximum percentage modulation possible without objectionable distortion.

MODULATION DEPTH. See Percentage modulation.

MODULATION FACTOR. In an amplitude-modulated wave, the ratio of the maximum departure of the modulation envelope from the carrier level to the carrier amplitude.

MODULATION INDEX. Generally, a measure of the degree of modulation. In particular, for angle modulation, the ratio of the frequency deviation to the frequency of the modulating wave.

MODULATION METER. An instrument for measuring the modulation factor, i.e. the degree of modulation, of a modulated wave train.

MODULATION MONITOR. See Modulation meter.

MODULATION PATTERNS. Patterns displayed on a cathode-ray oscilloscope which are used to determine percentage modulation.

MODULATION TRANSFORMER. An audio-frequency transformer designed to couple the anode of an audio output stage to the grid or anode of a modulated amplifier.

MODULATOR. Any device for effecting the process of modulation. In radar, a device for generating a succession of short pulses which cause a transmitting valve to oscillate at each pulse. See Magnettor, Reactance modulator.

MOIRÉ FRINGES. Interference-fringe effects produced when one optical grating slides over another with its lines inclined at a slight angle to those of the first. The pattern of light and dark bands produced can be detected by a system of photoelectric cells and provide highly accurate methods of measuring small movements.

MOLECULAR ELECTRONICS. The branch of electronics dealing with the production of complex electronic circuits of near-microscopic dimensions by the technique of growing them integrally into the structure of a single crystal. See Micro-electronics.

MOLECULE. The smallest particle into which any substance may be divided and still retain its chemical properties.

MOLYBDENUM (Mo). A metallic element, Atomic No. 42. Similar to tungsten chemically, but softer and more ductile. Used in valve electrodes and filaments.

MONIMAX®. A magnetic alloy of very high permeability composed of 50 per cent iron, 47 per cent nickel, 3 per cent molybdenum.

MONITORING AMPLIFIER. In radio broadcasting and audio-engineering, an amplifier designed to amplify signals with the minimum distortion, used to check programme performance or the level of recording signals.

MONITORING KEY. A key which allows an operator to listen on a telephone circuit without appreciably affecting transmission on the circuit.

MONITRON. An instrument for indicating the level of background radiation in working areas near radiation hazards.

MONOCHROME CHANNEL. In *television*, a channel to carry the *monochrome signal*.

MONOCHROME SIGNAL. (1) In monochrome (black and white) television, a signal wave for controlling the *luminance* values in the picture, i.e. the part carrying the information for *brightness*. (2) In *colour television*, that part of the signal wave which has major control of the luminance values of the picture whether displayed in colour or monochrome, but no control over the chromaticity. (Contrast *Chrominance channel*).

MONOFORMER. In a *cathode ray tube*, a device similar to a *photoformer* with the photocell and external mask replaced by an internal mask controlling the secondary emission and a collector inside the tube.

MONOLITHIC. Descriptive of an *integrated circuit* in which the entire structure is obtained by processing a single chip of crystalline semiconductor.

MONOSCOPE. An *electron-beam tube* in which signals are generated by scanning an electrode of aluminium with a single image printed on it in carbon. Because of the difference in *secondary emission* of aluminium and carbon the image is converted into electric signals identical with those obtained from a camera tube. Single images, e.g. of test patterns, may be televised in this way.

MONOSTABLE. A monostable circuit has only one stable state but can be triggered into a second quasi-stable state by an external pulse.

MORSE TELEGRAPHY. *Telegraphy* in which the signals are formed according to one of the Morse codes.

MOSAIC. In *iconoscopes*, a device for the electrical *storage* of the image to be televised. A typical mosaic consists of a thin sheet of mica on which there are a large number of very small islands of photo-emissive material, each of which is capacitively coupled through the mica to a conducting metallic coating on the other side, called the signal plate.

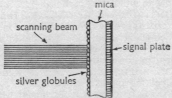

Fig. 93. Diagram of magnified section of iconoscope mosaic

MOSELEY LAW. The frequencies of the characteristic *X-rays* of the elements are directly proportional to the squares of their *atomic numbers*.

MOSFET. Abbreviation for Metal-Oxide Semiconductor Field Effect Transistor.

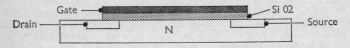

Fig. 94. MOSFET integrated circuit

MOST. Abbreviation for Metal-Oxide Semiconductor Transistor.

MOTIONAL IMPEDANCE. Of a *transducer*, the *impedance* obtained by subtracting the *blocked impedance* from the *loaded impedance*.

MOTOR. A machine which transforms electrical into mechanical energy using the force from a magnetic field set up by an electric current.

MOTOR EFFECT. The repulsive force between adjacent conductors carrying currents in opposite directions.

MOTOR-BOATING. In *low* or *audio-frequency amplifiers*, self-oscillation resulting in a noise like that of a motor-boat engine in the audio output.

MOTOR-CONVERTER. An *induction motor* and a *synchronous converter* mounted on the same shaft, with their rotor windings in series.

MOTOR-FIELD CONTROL. A method of controlling the speed of a motor by changing the magnitude of the field current.

MOTOR-GENERATOR SET. A combination of one or more motors, mechanically coupled to one or more generators.

MOTOR-TORQUE GENERATOR. See *Synchro*.

MOTOR-TORQUE REGULATOR. An electronic device for controlling a motor so as to provide constant torque independent of speed.

MOTZ UNDULATOR. A method studied at Oxford University of producing very short *microwaves* by subjecting a stream of electrons accelerated to *relativistic* velocities to periodic magnetic forces so that they follow an undulating path.

MOVING-COIL INSTRUMENT. See *D'Arsonval galvanometer*.

MOVING-COIL LOUDSPEAKER. A *loudspeaker* in which the moving conductor which actuates the cone is in the form of a moving coil connected to the source of electrical energy, and mounted in the field of a permanent- or electro-magnet. See Fig. 81.

MOVING-COIL MICROPHONE. A *microphone* whose electrical output results from a coil moving in a magnetic field.

MOVING-IRON INSTRUMENT. An instrument whose operation depends on the reaction between current flowing in fixed coils and the magnetic field of soft iron or other magnetic material in the moving system.

MOVING-MAGNET GALVANOMETER. A *galvanometer* with a moving magnet suspended on a fine thread between deflecting coils carrying direct current.

MOVING TARGET INDICATOR (MTI). A device which limits the display of information on a *radar screen* primarily to moving targets.

MSI. Abbreviation for Medium Scale Integration, descriptive of *integrated circuits*.

MU CIRCUIT. In a *feedback amplifier*, the circuit which amplifies the vector sum of the input signal and the fed-back portion of the output signal, in order to generate the output signal.

MULTICHANNEL ANALYSER. See *Pulse height analyser*, *Spectrum analyser*.

MULTI-ELECTRODE VALVE (TUBE). A valve or tube containing more than three electrodes associated with a single electron stream.

MULTIPACTOR. A high vacuum device in which electrons are driven back and forth between two electrodes by a synchrononous radio frequency field.

MULTIPLE-ADDRESS CODE. In *computers*, an instruction, usually a coded representation of the operation to be performed, and including one or more *addresses* of *words* in storage.

MULTIPLE-HOP TRANSMISSION. *Radio transmission* which depends on multiple reflection and refraction of the *sky wave* between earth and *ionosphere*.

MULTIPLE-UNIT VALVE. A valve in which two or more groups of electrodes associated with independent electron streams are contained in one envelope. Examples are *diode-pentode* and *triode-pentode*.

MULTIPLEX. The simultaneous transmission of several functions, such as frequency, amplitude, or waveshape, over one path without any loss of identity of each function.

MULTIPLEX CHANNEL. A *communication* channel for *multiplex* signals.

MULTIPLEX OPERATION (MULTIPLEXING). The operation of a single path for the transmission of more than one set of signals simultaneously. The two main methods are *frequency division multiplexing* and *time division multiplexing*.

MULTIPLEX RADIO TRANSMISSION. *Multiplex operation* of a *radio transmission* system.

MULTIPLIER. (1) An *electron multiplier*. (2) A device with two or more inputs whose output is a representation of the product of the magnitudes represented by the inputs. (3) A device in which the output quantity is some multiple of the input.

MULTIPLIER PHOTOTUBE. A *photomultiplier*.

MULTISTABLE. Having more than one stable state.

MULTITRON. A power amplifying valve developed at the Atomic Energy Research Establishment for efficient pulsed operation at frequencies of the order of 200 Mc/s.

MULTIVIBRATOR. A form of *relaxation oscillator* which includes two stages coupled so that the input of each is derived from the output of the other. A multivibrator is either 'free-running' or 'driven' according to whether its frequency is determined by its own circuit constants or by an external synchronizing voltage. See *Astable circuit, Bistable, Flip-flop, Monostable.*

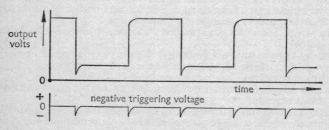

Fig. 95. Typical waveform from a bistable multivibrator with a negative triggering voltage

MUMETAL®. A high *permeability* magnetic alloy containing 78 per cent nickel.

MURRAY LOOP. A loop circuit for testing and locating ground faults on land lines.

MUSA RECEPTION. A special kind of *short-wave* reception which takes advantage of the varying response of an aerial assembly to incident short-waves according to their angle or arrival.

MUSICAL SCALE. A series of notes (audible sensations) arranged from low to high by a specified system of intervals, suitable for musical purposes.

MUTING SWITCH. A switch designed to introduce *noise-suppression* circuits into automatic tuning systems, or in sensitive gramophone reproducers during record changing.

MUTUAL CAPACITANCE. See *Capacitance, mutual.*

MUTUAL CONDUCTANCE. Of a *thermionic valve*: the ratio between the change of anode current due to a small change in control grid voltage and the corresponding change in the control grid voltage itself. Used as a figure of merit for the valve. Symbol g_m. See *Transconductance.*

MUTUAL IMPEDANCE. Between any two pairs of terminals of a network, the ratio of the open-circuit potential difference between either pair of terminals to the current applied at the other pair, all other terminals being open.

MUTUAL INDUCTANCE. The common property of two associated electrical circuits which determines the *electromotive force induced* in one circuit for a given rate of change of current in the other. Symbol: *M*. Practical unit: *henry*.

MUTUAL INDUCTOR. An *inductor* for changing the *mutual inductance* between two circuits.

MV. Abbreviation for *megavolt*.

mv. Abbreviation for *millivolt*.

MW. Abbreviation for *megawatt*.

mw. Abbreviation for *milliwatt*.

MYCALEX®. A moulded insulating material for high-temperature working. It is made of ground mica and ground glass fused together at high temperature and pressure.

MYLAR®. A polyester used as an insulator, as a backing for magnetic tape, and as the dielectric for some kinds of fixed capacitor.

MYO-ELECTRIC CONTROL. A technique for controlling machines with the electric potentials developed during muscle contraction. Its most important application is in controlling artificial limbs and powered assistive devices for the physically handicapped.

MYRIAD. A range of digital computers developed by the Marconi Company Ltd in 1965 which used advanced techniques in the fields of micro-electronics and high-speed silicon logic to produce desk-sized units operating at ten times the speed of earlier types.

N

NAND. A logical operator having the property that if P is a statement, Q is a statement, R is a statement . . . then the nand of P, Q, R . . . is true if, and only if, at least one statement is false, false if all statements are true.

NANO-. Prefix meaning one-thousand-millionth (10^{-9}).

NANOFARAD (NF). 10^{-9} *farads* or 1,000 *picofarads*.

NANOSECOND. One milli-microsecond (10^{-9} seconds).

NATURAL FREQUENCY. (1) Of an aerial, the lowest *resonant frequency* without added *inductance* or *capacitance*. (2) The natural frequencies of a body or system are the frequencies of *free oscillation*.

NATURAL PERIOD. Of a body or system, the period of free oscillation. When the period varies with amplitude, the period for an amplitude approaching zero is taken.

NATURAL RESONANCE. A *resonance* such that the *period* of the driving force is the same as the *natural period* of the system.

NEAR-END CROSS TALK. *Cross talk* propagated in a disturbed communication channel in the direction opposite to the direction of propagation of the current in the disturbing channel.

NEEDLE GAP. A *spark gap* in which the electrodes are needle points.

NEGATIVE. (1) A qualifying term applied to one of two points between which a difference of potential exists, to distinguish the one which corresponds (as far as the tendency to set up a current in an external circuit is concerned) to the zinc plate of a *Daniell cell*. See *Electronegative*. (2) Having an effect which is opposite in direction to that considered positive, e.g. *negative feedback*.

NEGATIVE BIAS. A voltage applied to the *control grid* of a valve to make it negative with respect to the *cathode*.

NEGATIVE COUPLING. Of two coils which are electrically *coupled*, the arrangement which results in the increases in the inducing and *induced* currents and voltages being inversely proportional. Also called negative *mutual inductance*.

NEGATIVE ELECTRICITY. The kind of *electricity* predominating in a piece of resin which has been electrified by rubbing it with wool. It consists of an excess of *electrons*.

NEGATIVE ELECTRODE. (1) The *cathode* of a valve. (2) In a *primary cell*, the conducting material which serves as the *anode* when the cell is discharging and to which the negative terminal is connected.

NEGATIVE ELECTRON. A *negatron* as distinct from a *positron*. See *Electron*.

NEGATIVE FEEDBACK. See *Feedback, negative*.

NEGATIVE GLOW. The luminous glow in a *discharge tube* between *cathode dark-space* and the *Faraday dark-space*. See *Glow discharge*.

NEGATIVE ION. An *ion* with a negative charge, i.e. an *anion*.

NEGATIVE MODULATION. In television systems, the form of *modulation* in which an increase in *brightness* corresponds to a decrease in transmitted power.

NEGATIVE RESISTANCE. In certain devices, when the voltage at a point in the voltage-current characteristic is increased, the current decreases, i.e. the characteristic has a negative slope. Such devices include the *dynatron* oscillator and the *magnetron*.

NEGATRON. A *negative electron*. See *Positron*.

NEON (Ne). One of the inert gaseous elements, Atomic No. 10. Has a characteristic red glow when ionized and is extensively employed in *discharge* tubes (e.g. neon signs). In electronics neon tubes are used as indicators and in voltage regulators.

NEON LAMP. A small glass bulb containing two electrodes and filled with neon gas at low pressure. When voltages exceeding the *ionization potential* are applied to the electrodes, a *discharge* occurs accompanied by a red glow. The voltage across the electrodes then remains constant and the lamp can therefore be used as a *voltage regulator*.

symbol for neon
glow discharge tube

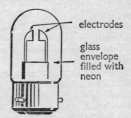

electrodes

glass
envelope
filled with
neon

Fig. 96. Typical neon lamp with bayonet base (actual size)

NEON OSCILLATOR. An *oscillator* based on a *neon lamp* and a capacitor, with sometimes a resistance in circuit.

NEPER. The natural logarithm of the ratio of two currents or voltages.

NERNST BRIDGE. An *ac bridge* for measuring *capacitance,* in which capacitors are used in all four arms.

NERNST EFFECT. In electronic *conductors* and *semiconductors*, a transverse *electric field* is created by the presence of a transverse *magnetic field* if a temperature gradient is maintained across the conductor. See *Ettingshausen effect*.

NERVE CURRENT. A very small natural current flowing through a nerve path.

NESTING STORE. See *Push-down store*.

NET LOSS. Of a circuit, the sum of all the *transmission losses* between the two ends of the circuit, minus the sum of all the transmission *gains*.

NETWORK. A system of interconnected *admittances*. The point where three or more admittances meet is called a junction point or *node* of the network. An admittance connecting any two junction points is called an arm or branch; a *mesh* is a closed path through three or more junction points.

NETWORK, ACTIVE. A *network* containing a source of energy.

NETWORK, PASSIVE. A *network* which does not contain a source of energy.

NETWORK ANALYSER (CALCULATOR). An assembly of electric circuit elements of known value, i.e. *lumped* inductances, resistances, etc., and a means for connecting them together so as to form models of electrical networks. Such an assembly amounts to a simple *analogue computer*, as measurements on the model can provide the corresponding values of electrical quantities in the original network. Network analysers are frequently used to solve problems in power distribution systems.

NETWORK ANALYSIS. The mathematical operation of deriving the electrical properties of a *network* from its configuration, *parameters*, and driving forces.

NETWORK NODE. See *Network*.

NETWORK SYNTHESIS. The operation of designing a *network*, i.e. determining its configuration and the values of its elements, given certain properties of the complete network and specified responses for given driving forces.

NETWORK TRANSFER CONSTANT. A measure of the attenuation and time delay experienced by a signal which passes through a *network*.

NEURISTOR. A semiconductor device which simulates certain aspects of the behaviour of neurons.

NEURO-ELECTRICITY. Electricity generated by the nervous system of an animal or human being.

NEUTRAL. Having no net electric charge.

NEUTRAL CONDUCTOR (LINE). Of a *polyphase* circuit, or a 3-wire *single-phase* circuit, the conductor whose potential is such that the potential differences between it and each of the other conductors are approximately equal in value and phase.

NEUTRALIZATION. A method of counteracting the effect of unwanted *feedback* in a valve or transistor amplifier. Especially of preventing oscillations in an amplifier by introducing into the input a voltage equal in magnitude but opposite in phase to that produced by the *Miller effect* on the anode-grid capacitance. See *Cross, Grid, Hazeltine,* and *Rice neutralization*.

NEUTRALIZING INDICATOR. A device for indicating the degree of *neutralization* in an amplifier.

NEUTRALIZING VOLTAGE. The voltage fed back in the process of *neutralization*.

NEUTRINO. A *neutral* particle with very small mass (less than one hundredth of the mass of an electron) and spin quantum number of $\frac{1}{2}$.

NEUTRODYNE. An *amplifier* in which *neutralization* is effected by voltage fed back from a capacitor.

NEUTRON. A particle in the *nucleus* of an atom, with no electric charge but with a mass equal to that of a *proton*.

NF. Abbreviation for *Noise factor*.

NICHROME®. Trade name of an alloy of nickel and chromium which is widely used for *wire-wound resistors* and heating elements because of its high intrinsic resistance and ability to withstand high operating temperatures.

NICKEL (Ni). A strongly *ferromagnetic* metallic element, Atomic No. 28. It is widely used in electronics, in magnetic alloys, as a conductor for valve pins and transistor leads, for plating, etc.

NICKEL-CADMIUM CELL. A secondary cell which uses a mixture of nickel hydroxide and nickel oxide at the anode and cadmium at the cathode, in an electrolyte of potassium hydroxide.

NICKEL-IRON CELL. A *secondary* cell which uses a mixture of nickel hydroxide and nickel oxide at the anode, and iron oxide at the cathode, in an electrolyte of potassium hydroxide.

NICOL PRISM. A device made from crystals of Iceland spar used to produce plane-polarized light.

NIPKOW DISC. A mechanical scanning device used in early television systems.

NITROGEN (N). Gaseous element forming four fifths of the atmosphere. Atomic No. 7.

NMR. Abbreviation for *Nuclear magnetic resonance*.

NOCTOVISION. A *television* system in which the image to be transmitted is illuminated with infra-red instead of visible light.

NODE. In electricity and electronics, a point of zero current or zero voltage on a conductor. A point in a radio wave where the amplitude is zero. A junction point in a network. See *Current node, Network node, Partial node, Standing wave*.

NODE VOLTAGE. The voltage of some point in a *network* with respect to a *node*.

NODON RECTIFIER. An *electrolytic rectifier* in which the cathode is of aluminium, the anode of lead, and the electrolyte is ammonium phosphate.

NOISE. (1) In electronics, an undesired electrical disturbance within the useful *frequency band*. (2) In acoustics, any extraneous sound tending to interfere with the perception of wanted sounds. *Interference* often, but not always, produces noise. See *Clutter, Cosmic noise, Cross talk, Flicker noise, Galaxy noise, Grass, Howl, Hum, Man-made noise,*

Microphony, Shot noise, Thermal noise, Valve noise and the 'noise' entries following.

NOISE, AMBIENT (ROOM NOISE). Acoustic *noise* in a room or other environment.

NOISE, AMPLITUDE-MODULATION. The *noise* produced by undesired variations of amplitude in a radio-frequency signal.

NOISE, BACKGROUND. Of any system capable of receiving a signal, the total system *noise*, independent of whether or not a signal is present, but excluding the signal as part of the noise.

NOISE, BASIC. At a point in a system: all noise present in a system under test when the system is carrying any wanted signals.

NOISE, CARRIER. The *noise* produced by undesired variations of a radio-frequency signal in the absence of any intended *modulation*.

NOISE, CIRCUIT. In *telephony, noise* produced in the receiver by the telephone system, excluding noise picked up acoustically by the telephone transmitters.

NOISE, CONTACT. *Noise* due to fluctuations in the resistance of a junction between two metals or between a metal and a semi-conductor.

NOISE, COSMIC. See *Cosmic noise.*

NOISE, ELECTRICAL. Unwanted electrical energy, other than *cross talk,* present in a transmission system.

NOISE, FLICKER. See *Flicker noise.*

NOISE, GALAXY. See *Galaxy noise.*

NOISE, GAS. Electrical *noise* due to random motions of molecules of gas in gas-filled or vacuum valves or tubes.

NOISE, IGNITION. See *Ignition interference.*

NOISE, IMPULSE. *Noise* due to transient disturbances which do not overlap. See also *Noise, random.*

NOISE, INDUCED. *Noise voltages* induced in valve electrodes, especially *grids,* by random motions of electrons in the interelectrode space.

NOISE, LINE. *Noise* originating in a *transmission line.*

NOISE, MODULATION. *Noise* associated with a *radio signal,* but not a part of the signal.

NOISE, PARTITION. In valves and tubes, *noise* due to random fluctuations in the distribution of cathode current between the various electrodes.

NOISE, PHOTON. In *photocells, noise* due to fluctuations in the rate of arrival of light *quanta* at the *photocathode.*

NOISE, RANDOM. Noise made up of transient disturbances occurring at random; usually, where the number of such disturbances in unit time is large so that the noise is similar to *thermal noise.*

NOISE, REFERENCE. The magnitude of *circuit noise* which will produce a reading in a *noise meter* equal to that produced by 10^{-12} watt of electric power at 1,000 cycles per second.

NOISE, SCHOTTKY. See *Shot noise.*

NOISE, SET. In a radio receiver, the *random noise* in the receiver, i.e. the total effects of *thermal noise* and *shot noise*.

NOISE, SHOT. See *Shot noise*.

NOISE, SURFACE. In mechanical recording, the *noise* component in the output of a *pick-up* due to irregularities in the contact surface of the groove.

NOISE, THERMAL. See *Thermal noise*.

NOISE, VALVE. See *Valve noise*.

NOISE, WHITE. *Noise* having a wide frequency spectrum (resembling white light in this respect) and including all *audio* frequencies.

NOISE BANDWIDTH. Of an *amplifier,* the width of an ideal *band-pass* characteristic which has the same area and peak value as the *power gain* versus *frequency characteristic* of the amplifier.

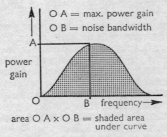

Fig. 97. Noise bandwidth

NOISE-CURRENT GENERATOR. A device which generates current so that the output of current varies with time in a random manner. See *Noise generator*.

NOISE DIODE. A *diode* used as a standard source of electrical *noise*. The noise is produced by random arrival of electrons at the anode when the diode is operated under conditions of *temperature-limited emission*.

NOISE FACTOR (NOISE FIGURE). Of a *transducer* with two *ports* at a specified frequency and with a specified termination, the ratio of actual output noise to that which would remain if the transducer itself were rendered noiseless.

NOISE FIELD INTENSITY, RADIO. A measure of the field intensity at a given point (e.g. a radio receiver) of electromagnetic waves of an interfering character.

NOISE FIGURE, RECEIVER (NF). A figure of merit expressed as the ratio of *noise* in a given receiver to that in a theoretically perfect receiver.

NOISE GENERATOR. An electronic instrument consisting of a stable *noise source* and some means of controlling the amplitude of the *noise voltage,* designed to produce a calibrated *noise* output.

NOISE INTENSITY. The *field strength* of *noise* in a given medium, with reference to a specified *frequency band*. It is not always possible to define some types of noise in this way.

NOISE LEVEL. The value of *noise* integrated over a specified frequency range, with specified frequency *weighting* and integration time. Expressed in decibels relative to a standard level.

NOISE LIMITER. See *Noise suppressor*.

NOISE METER. An instrument for measuring *noise level*, designed to give equal readings for noises which have approximately the same interfering effect.

NOISE MODULATION. *Modulation* with a *noise generator* for testing amplifiers or for modulating the carrier of radio transmitters to produce *jamming*.

NOISE POWER. The power dissipated by a *noise source* in its internal resistance and the resistance of its load.

NOISE POWER, AVAILABLE. The maximum *noise power* available from a given network; equal to the square of the *effective value* of the noise voltage divided by four times the resistance of a load which is matched to the network.

NOISE RATIO. Of a *transducer*, the ratio of the available *noise power* at the output to the noise power at the input.

NOISE SOURCE. The device used in a *noise generator* to generate the random noise signal. See *Noise diode*.

NOISE SUPPRESSOR. A device or circuit designed to remove or limit electrical *noise* in a system.

NOISE TEMPERATURE. At a *port* or pair of terminals at a specified frequency, the temperature of a passive system with an *available noise power* per unit bandwidth equal to that of the actual terminals.

NOISE TEMPERATURE, STANDARD. The standard reference temperature for *noise* measurements is taken as 290 degrees Kelvin (17 degrees Centigrade).

NOISE VOLTAGE. Fluctuations of *electric potential* in a physical system due to spontaneous disturbances in the system. See *Noise*.

NOMINAL LINE WIDTH. In television, the reciprocal of the number of lines per unit length in the direction of the line of progression.

NOMOGRAM. A series of curved or straight lines on a chart with suitable index marks, normally arranged for the solution of equations involving three variables. By setting a straight edge to read two of the variables, the third can be read by the intersection of the straight edge with the third line or curve.

NON-INDUCTIVE CAPACITOR. A *capacitor* constructed so that its incidental *inductance* has been reduced to a minimum.

NON-INDUCTIVE CIRCUIT. A circuit with negligible *inductance*.

NON-INDUCTIVE RESISTOR. A *resistor* designed to have the minimum amount of *inductance*. See *Bifilar winding*.

NON-LINEAR DISTORTION. *Distortion* in a circuit or system in which the ratio of instantaneous voltage to current depends on the values of either voltage or current.

NON-LINEAR NETWORK. A *network* which cannot be specified by linear differential equations with time as the independent variable.

NON-RESONANT LINE. A *transmission line* in which there are no reflected waves and therefore no *standing waves*.

NOR. A logical operator having the property that if P is a statement, Q is a statement, R is a statement . . . then the nor of P, Q, R . . . is true if, and only if, all statements are false; false if, and only if, at least one statement is true.

NOT. A logical operator having the property that if P is a statement then the not of P is true if P is false; false if P is true.

NOT-GATE. An electronic circuit whose output is energized only if its single input is not energized.

NOVAL BASE. A *valve base* for miniature valves, having positions for insertion of nine pins which extend directly through the glass envelope of the valve.

N-P-N TRANSISTOR. See *Transistor, n-p-n*.

NTSC. National Television Standards Committee: the label for the American system of *colour television*.

N-TYPE CONDUCTIVITY. *Conductivity* associated with *conduction electrons* in a *semiconductor*.

N-TYPE SEMICONDUCTOR. See *Semiconductor, n-type*.

NUCLEAR BATTERY. A device for converting *nuclear energy* into electric energy.

NUCLEAR ENERGY. Energy contained in the *nucleus* of an atom.

NUCLEAR MAGNETIC RESONANCE (NMR). The resonance phenomenon observed in the transfer of energy between a radio-frequency alternating magnetic field and a *nucleus* placed in a constant magnetic field strong enough to decouple the nuclear spin from the influence of the atomic electrons.

NUCLEAR REACTOR (ATOMIC PILE). An assembly in which controlled nuclear fission takes place and the energy so released is made available.

NUCLEON. One of a group of particles making up an atomic nucleus, e.g. *proton*, *neutron*.

NUCLEUS. The central region of an atom which carries a *positive charge* and constitutes most of the weight of the atom.

NULL METHOD. An accurate method of electrical measurement used in devices such as *bridges* and *potentiometers,* in which voltages in different circuits are compared and adjusted until they are equal and an instrument detecting the difference between the voltages gives a zero or null reading.

NUMERICAL CONTROL. Control by numbers: descriptive of systems in which *digital computers* are used to control automatic machines, especially machine tools.

NUVISTOR®. A range of reliable small valves with exceptionally rigid construction.

NYQUIST DIAGRAM. Of a *feedback amplifier* or *oscillator,* a graph in rectangular coordinates of the relation between the amplification and the feedback, which may be used as a criterion for *stability* in the amplifier.

O

OBOE. A *radar* navigation and blind bombing system, using technique similar to that of *Loran*.

OCTAL BASE. The standard eight-pin *valve base*.

OCTAVE. In electrical communications, the interval between two frequencies having a ratio of two to one.

OCTODE. A valve containing eight electrodes: anode, cathode, control grid, and five additional electrodes, usually grids.

ODD-EVEN CHECK. In *digital computers,* an automatic check to determine whether the number of 1 or 0 digits in a *word* is odd or even.

ODOGRAPH. An automatic electronic map tracer, which plots the exact course taken by a vehicle in which it is mounted.

OERSTED. The unit of *magnetizing force* in the c.g.s. *electromagnetic system of units*. At any point in a vacuum it is equal to the force in dynes on a unit magnetic pole at the point. Symbol H. Compare *Gauss*.

OFF-LINE EQUIPMENT. In *data processing*: the peripheral equipment or devices which are not in direct communication with the central processing unit of the computer.

OHM. The unit of electrical *resistance*. A constant *potential difference* of one *volt* across a resistance of one ohm will produce a current in it of one *ampere*. See *Mho*.

OHM-CENTIMETRE. The unit of *resistivity*.

OHMIC CONTACT. If the *potential difference* across a contact is proportional to current passing through it, the contact is said to be ohmic.

OHMMETER. An instrument for measuring electrical *resistance,* whose scale is graduated in *ohms* or *megohms*.

OHM'S LAW. The *current* in an electric circuit is directly proportional to the *electromotive force* in the circuit. (This does not apply to all circuits.) In the form $E=IR$, where E is the electromotive force, I is the current, and R the resistance of the circuit, Ohm's Law defines *resistance*.

OIL SWITCH. A switch in which the interruption of the circuit occurs in oil.

OMEGA. An accurate navigation system developed by the U.S. Navy, consisting of 8 low frequency transmitters strategically located round the world and operating on a band between 10 and 14 kilohertz. Pulsed signals from all stations are synchronized by atomic clocks so that their phases differ by less than a microsecond.

ONDOGRAPH. An instrument which draws the waveform of an alternating voltage in a series of steps.

ONDOSCOPE. A *glow-discharge tube* used to detect the presence of high-frequency radiation.

ONE-MANY FUNCTION SWITCH. In *computers,* a *function switch* in which one input only is excited at a time and each input produces a combination of outputs. See *Many-one function switch.*

ON-LINE OPERATION. In computing and data-processing, operation so that, at the end of a given stage, results are available immediately for that stage. Compare *Real-time operation.*

ON-OFF CONTROL. A control system in which the controller has only two choices.

ON-OFF KEYING. A form of *keying* in which the output of a source is alternately transmitted and suppressed to form signals.

OPEN CIRCUIT. A circuit which does not provide a continuous path for the flow of current, or in which continuity is incomplete or interrupted.

OPEN GRID. See *Floating grid.*

OPEN SUB-ROUTINE. A computer *sub-routine* inserted directly into the sequence of linear operations, which must be recopied at each point in the routine where it is needed. Contrast *closed sub-routine,* which can only be entered by a *jump.*

OPEN WIRE. A conductor separately supported above the ground.

OPEN-CIRCUIT IMPEDANCE. Of a line or four-terminal network, the *driving point impedance* when the far end is open.

OPEN-ENDED. Referring to a system or process which can be augmented.

OPEN-LOOP CONTROL SYSTEM. A control system in which no self-correcting action (negative feedback) occurs, as in *closed-loop* control systems.

OPEN-LOOP TEST. Of a complex *servo system,* a test applied to one or more of the feedback loops, with these loops left open.

OPERATING ANGLE. The period, measured in electrical degrees, during which a valve or tube conducts, usually measured from current maximum to *cut-off.*

OPERATING POINT. Of a *valve* or *transistor,* the point on the family of characteristic curves whose coordinates represent the instantaneous magnitudes of electrode potential and current for the particular operating conditions under consideration.

OPERATING RATIO. Of a *computer,* the ratio of the total number of hours of correct operation to the total number of hours of scheduled operation, including maintenance.

OPERATION CODE. In a *computer,* the code contained in an instruction which defines the operation associated with the instruction.

OPERATION NUMBER. In a *computer,* a number used to indicate the relative position of an operation in a programme.

OPERATION TIME. (1) Of a *valve* or *tube,* the time required for the current to reach a stated fraction of its final value, after simultaneous application of all electrode voltages. (2) In *data processing*: the time taken to select, prepare and execute an operation; e.g. addition, subtraction, multiplication, comparison time. It includes *access time*.

OPERATIONAL AMPLIFIER. In *analogue computers*: an amplifier with a high and stable gain, usually *direct-coupled*.

OPPOSITION. Two periodic quantities of the same *period* are said to be in opposition when the *phase* difference between them is one half of a period.

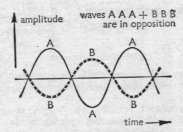

Fig. 98. Opposition

OPTAR. A portable device for guidance of the blind, which uses radar techniques in conjunction with visible light.

OPTICAL AMMETER. An instrument for measuring the current in the filament of an incandescent lamp by comparing its resulting illumination with that produced when a current of known magnitude is used in the same filament.

OPTICAL MASER. A *maser* in which the pumping (i.e. stimulating) frequency is in the visible light or infra-red region. See *Infra-red maser, Laser*.

OPTIMUM BUNCHING. In *velocity-modulated* electron beams, the *bunching* which produces the maximum power at the desired frequency in an output gap.

OPTIMUM PROGRAMMING. In a computer, selection of a *programme* which is generally superior to any other possible programme in at least one respect, usually that of duration.

OPTOPHONE. A *photoelectric* device, used in training the blind, which converts the letters of a printed word into sounds.

OPTO-TRANSISTOR. An optical transistor.

OR-CIRCUIT. In *computer* terminology, synonym for *or-gate*.

OR-GATE. A *gate* circuit which allows two or more *pulse* circuits to be connected together to a common load at its input, and whose output is energized when a pulse is applied to one or more of the inputs.

ORTHICON. A *camera tube* in which a low-velocity electron beam scans a *photo-mosaic* capable of electrical *storage*. See *Image orthicon* and Fig. 99.

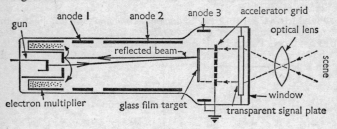

Fig. 99. Image orthicon

ORTHONIK. A very high *permeability* magnet alloy of equal parts of iron and nickel, used where rectangular *magnetization curves* are important.

OSCILLATING CURRENT (VOLTAGE). A current or voltage that alternately increases and decreases in magnitude with respect to time, according to some definite law.

OSCILLATION. Any quantity is in a state of oscillation while the value of that quantity is continually changing so that it passes through maximum and minimum values. See the following: *Damped, Forced, Free, Parasitic, Stable, Steady-state, Sustained,* and *Unstable oscillation.*

OSCILLATOR. A non-rotating device for converting direct-current power into alternating-current power, capable of initiating and sustaining *oscillation* at a frequency determined by the physical constants of the oscillating system. The following types of oscillator are described in this dictionary: *beat-frequency, blocking, coherent Colpitt's, crystal, dynatron, electron-coupled, Hartley, local, magnetostriction, magnetron, master, negative-resistance, pulse, RC, relaxation, squegging, tank circuit transitron, tuned-base, tuned-grid, Van der Pol, velocity-modulation, Wien Bridge.*

OSCILLATORY CIRCUIT. A circuit containing *inductance* and/or *capacitance* and *resistance* connected so that a voltage impulse will produce a current which periodically reverses or oscillates.

OSCILLOGRAM. The record from a recording *oscillograph* or the reading from an *oscilloscope.*

OSCILLOGRAPH. An instrument which records electrical signals on film or sensitized paper, using magnetic or optical means. In the cathode-ray oscillograph the electron beam of a *cathode-ray tube* is used to make a photographic or other record.

OSCILLOSCOPE. An instrument used primarily to provide visible images of one or more electrical quantities which vary rapidly with time or with respect to other electrical quantities. The *cathode-ray oscilloscope* is the usual type and one of its main uses is as indicator for a radar set.

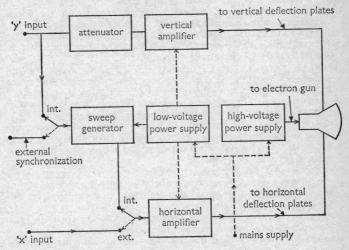

Fig. 100. Block diagram of an oscilloscope

OSCILLOSCOPE, THREE-DIMENSIONAL. An *oscilloscope* which provides a three-dimensional display on a screen consisting of a cube of *electroluminescent* material.

OSMIUM (Os). A metallic element, the heaviest of all the elements. Atomic No. 76.

OUTGASSING. Of a vacuum valve or tube, the removal by heating of some of the air absorbed on the inside surface of the envelope, while pumping to complete the evacuation. Also used in the sense of a slow deterioration of the vacuum owing to the emission of the absorbed gases from the interior of the envelope.

OUTPUT. (1) The power, current, or voltage delivered by a circuit or device. (2) The terminals or other places where the power, voltage, or current may be delivered.

OUTPUT BLOCK. Of a *digital computer,* the part of the internal *storage* reserved for accepting information which may be produced at the output.

OUTPUT GAP. An *interaction gap* which allows usable power to be extracted from an electron stream.

OUTPUT IMPEDANCE. Of a device, the *impedance* presented by the device to the *load*.

OUTPUT METER. A meter coupled to the output circuit of a device to measure the output power.

OUTPUT POWER. See *Power output, average*.

OUTPUT TRANSFORMER. A *transformer* used to couple the anode circuit of a valve or valves to a load, usually a *loudspeaker*.

OUTPUT VARIATION. In television, the change in peak amplitude due to hum, noise, etc., during a period not exceeding one frame in length.

OUTPUT WINDING. See *Power winding*.

OVERALL EFFICIENCY. Of devices such as *power amplifiers* and *induction heaters,* the ratio of the power absorbed by the load to the total power drawn from the mains supply (or internal batteries).

OVERBUNCHING. In *velocity-modulated* electron streams, *bunching* continued beyond the condition of *optimum bunching*.

OVERCOUPLING. The condition in which two resonant circuits are tuned to the same frequency but coupled so closely that two peaks are obtained. This has the effect of giving a broad-band response with uniform impedance.

OVERDAMPING. See *Damping, aperiodic*.

OVERDRIVEN AMPLIFIER. An amplifier whose driving voltage exceeds the value for which the circuit was designed and thus introduces distortion.

OVEREXCITED. Of synchronous motors, a condition in which the back electromotive force is larger than the terminal voltage.

OVERFLOW. In *computers,* the condition arising when the result of an arithmetic operation exceeds the capacity of the number representation; also the carry digit arising from this condition.

OVERLOAD CAPACITY. The current, voltage, or power level beyond which permanent damage occurs to a device or system.

OVERLOAD LEVEL. Of a system, component, etc., the level at which operation ceases to be satisfactory, as a result of overheating, signal distortion, etc.

OVERSCANNING. In a *cathode-ray tube* beam, *scanning* extending beyond the usable dimensions of the screen.

OVERSHOOT. The initial *transient* response to a unidirectional change in input which exceeds the steady-state response. See *Pulse rise time*.

OVERTONE. A physical component of a complex sound having a frequency higher than that of the basic frequency.

OVERVOLTAGE. In a *Geiger counter*, the amount by which the operating voltage exceeds the *Geiger threshold*.

OWEN BRIDGE. An *ac bridge* used to measure *self-inductance* in terms of capacitance and resistance.

OXIDE-COATED CATHODE. See *Coated cathode*.

OXYGEN (O). Very reactive gaseous element forming one fifth of the atmosphere. Atomic No. 8.

P

PACKAGED MAGNETRON. See *Magnetron, packaged*.

PACKING DENSITY. Of a *digital computer*, the amount of information contained in a given dimension of the storage system, e.g. the number of binary digits per inch of magnetic tape.

PAD. A fixed-value *transducer* used to reduce the amplitude of a wave without introducing appreciable distortion.

PADDER. In *ganged tuning,* a small adjustable capacitor connected in series with the main tuning capacitor.

PAIR. In electrical transmission, two similar conductors used to form an electric circuit.

PAIR PRODUCTION. The conversion of a *photon* into an *electron* and a *positron* when the photon meets a strong electric field, such as that in the neighbourhood of a *nucleus* or electron.

PAIRED CABLE. A cable in which all the conductors are arranged in the form of twisted pairs and no two sets of pairs are twisted together.

PAIRING. Faulty *interlaced scanning* in a television picture tube, in which lines of alternate fields tend to pair and coincide, thus halving the vertical resolution.

PAL. Phase Alternation Line: a system of *colour television* developed in Germany and favoured on the Continent. In this system the phase of a quadrature-modulated subcarrier is alternated on successive lines, thus correcting phase errors.

PALLADIUM (Pd). Metallic element of the platinum group. Atomic No. 46.

PAM. Abbreviation for *pulse amplitude modulation*.

PANCAKE COIL. A *coil* in the shape of a pancake, usually with the turns coiled in a flat spiral.

PANORAMIC. In radar, a *radarscope* which displays simultaneously all the signals received at different frequencies.

PANORAMIC RECEIVER. A radio *receiver* whose *tuning* is varied periodically so as to receive a selected band of frequencies. The period of the tuning is determined automatically.

PAPER CAPACITOR. A *capacitor* in which the *dielectric* is made of paper. See Figure 101.

PAR. Precision Approach Radar.

PARABOLIC (PARABOLOID) REFLECTOR. A hollow, concave *reflector* of radio- or microwave shaped so that all rays from the focus will be reflected in a direction parallel to the axis.

PARALLEL. In *computers*, relating to simultaneous transmission of, storage of, or logical operations on the parts of a *word,* using separate facilities for the various parts.

265

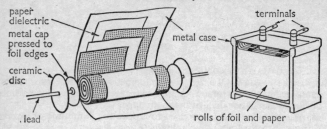

Fig. 101. Construction of tubular and box-type paper capacitors

PARALLEL, IN. Two or more *conductors* are said to be in parallel with one another when the current flowing in the circuit is divided between them; machines, *transformers*, and *cells,* when terminals of the same *polarity* are connected together.

PARALLEL CIRCUIT. A circuit in which two or more elements are connected across the same pair of lines or terminals so that current divides between the components. A circuit consisting of a *resistor*, *capacitor*, and *inductor* in parallel is called a simple parallel circuit.

PARALLEL DIGITAL COMPUTER. A *digital computer* in which the digits are handled in parallel.

PARALLEL FEED. Application of direct voltage to the anode or grid of a valve *in parallel* with the ac circuit, so that the direct and alternating components flow in separate paths. Compare *Series feed.*

PARALLEL MEMORY. In computers, *memory* capacity such that the *access time* for a given unit of information in the memory is nearly the same for all units in the memory.

PARALLEL RESONANCE. *Resonance* in a circuit where the impressed signal is applied across the *inductive* and *capacitive* branches *in parallel.* In most cases the *resonant frequency* is practically the same as it is when the branches are connected for *series resonance.*

PARALLEL TRANSMISSION. A system of *transmission* in which the characters of a word are transmitted (usually simultaneously) over separate lines, in contrast to *serial transmission.*

PARALLEL-PLANE WAVEGUIDE. A pair of parallel conducting planes used for propagating uniform circularity cylindrical waves whose axes are normal to the plane.

PARALLEL T-NETWORK. See *Twin T-network.*

PARALLEL-WIRE LINE. A *transmission line* consisting of two separate parallel wires insulated from each other. See *Microstrip.*

PARALLEL-WIRE RESONATOR. See *Lecher line.*

PARALYSIS. An effect following the application of such a large signal to a valve that its associated circuit capacitances are charged to a point where the circuit cannot fulfil its function.

266

PARAMAGNETIC MATERIAL. A material whose *permeability* is slightly greater than that of a vacuum and which is approximately independent of the *magnetizing force*.

PARAMAGNETIC RESONANCE. When electromagnetic radiation is incident on *paramagnetic material* in a magnetic field a resonance effect is observed as a peak in the absorption of energy from the radiation.

PARAMETER. Generally, a constant with an assigned series of arbitrary values each of which characterizes a member in a system or family of curves, functions, surfaces, etc. In particular, a *network* constant representing one of the resistances, mutual or self inductances, capacitances, etc., in a network. See *Lumped parameter*, *Transistor parameters*.

PARAMETRIC AMPLIFIER. A *microwave amplifier* based on a valve or solid-state device whose *reactance* can be varied periodically by an alternating voltage.

PARAMETRON. A resonant circuit, usually based on magnetic cores, in which either the inductance or the capacitance is made to vary periodically at half the driving frequency. First used in Japan as a digital computer element in which each oscillation of the parametron represents a binary digit.

PARAPHASE AMPLIFIER. An *amplifier* which converts a single input into a *push-pull* circuit.

PARAPHASING CIRCUIT. A circuit to provide *phase-inversion* with single-ended circuits in order to convert a push-pull output without using a *centre-tap transformer*.

PARASITIC OSCILLATION. Undesired or spurious *oscillation* in a circuit, often caused by unwanted *feedback*, resulting in detrimental effects such as instability, overloading, and reduced efficiency.

PARASITIC SUPPRESSOR. A device for the suppression or prevention of *parasitic oscillation*.

PARITY CHECK. In *digital computers*: a *routine* for finding errors in which the total number of 1's or 0's in the *binary code* groups can be checked frequently by inspecting their patterns as they flow through the computer. See *Odd-even check*.

PARTIAL NODE. A *node* in a system of *standing waves* where the wave amplitude is not zero.

PARTITION NOISE. See *Noise, partition*.

PASCHEN'S LAW. The *potential difference* which initiates a *discharge* in a gas is a function of the product of the electrode separation and the gas pressure. See *Hittorf principle*.

PASSIVE COMPONENT. An electronic component which contains no source of power, in contrast to *active* components.

PASSIVE NETWORK. See *Network, passive*, *Transducer, passive*.

PATCH. (1) Connect circuits together temporarily with a *patch cord*. (2) In data processing or digital computing, insert a section of *coding* into a *routine* in order to change the routine.

PATCH BAY. The part of an *analogue computer* which receives the *patch board*.

PATCH BOARD. A board or panel where circuits are terminated in jacks which receive *patch cords*.

PATCH CORD. In *analogue computers* and other electronic equipment, a flexible insulated conductor with terminals at the ends for rapid or temporary connection between the various parts of components.

PAULI EXCLUSION PRINCIPLE. According to this principle no two electrons in an atom may possess the same four *quantum numbers*. Since these numbers determine the behaviour of electrons in atoms, the principle is used to establish the number and arrangement of electrons in the *atomic structure*.

PCM. Abbreviation for *Pulse Code Modulation*.

P-DISPLAY. See *Plan position indicator*.

PEAK CATHODE CURRENT. The maximum instantaneous value of a periodically recurring *cathode* current.

PEAK CLIPPING. See *Clipping circuit*.

PEAK CURRENT. See *Peak value*.

PEAK DISTORTION. The largest total distortion of telegraph signals noted during a period of observation.

PEAK FACTOR. The ratio of the *peak value* of an alternating or pulsating wave to its *root-mean-square* (r.m.s.) value.

PEAK FORWARD VOLTAGE. The maximum instantaneous voltage in the direction in which a device is designed to take current.

PEAK FORWARD ANODE VOLTAGE. The maximum instantaneous *anode* voltage in the direction in which the valve is designed to pass current.

PEAK INVERSE VOLTAGE. The maximum instantaneous voltage in the direction opposite to that in which a device is designed to pass current.

PEAK INVERSE ANODE VOLTAGE. The maximum instantaneous *anode* voltage in the direction opposite to that in which the valve is designed to pass current.

PEAK LIMITER. A device which automatically limits the peak output to a predetermined maximum value. See *Limiter*.

PEAK LOAD. The maximum electrical power load consumed or produced in a stated period of time.

PEAK POWER. See *Peak load*.

PEAK POWER OUTPUT. In a *modulated carrier* system, the output power, averaged over a carrier cycle, at the maximum amplitude which can occur with any combination of transmitted signals.

PEAK PULSE AMPLITUDE. See *Pulse amplitude, peak*.

PEAK SIGNAL LEVEL. In a *facsimile* transmission, the maximum instantaneous signal power or voltage at any point.

PEAK VALUE. Of any quantity, such as *current*, *voltage*, etc., which varies with time, the maximum value which the quantity attains during the time interval under consideration.

PEAK VOLTAGE. See *Peak value*.

PEAKING CIRCUIT. A circuit capable of converting an input wave into a peaked waveform.

PEAKING NETWORK. A coupling network in which an inductance is effectively in series or shunt with the *parasitic* capacitance to increase the amplification at the upper end of the frequency range.

PEAKING STRIP. A device consisting of a *permalloy* coil with a single winding, used to detect the instant at which a rapidly varying magnetic field passes through zero or other *biasing* value.

PEAKING TRANSFORMER. A *transformer* which produces a sharply peaked output voltage waveform whatever the shape of the input waveform.

PEAK-RIDING CLIPPER. A circuit in which the level of voltage *clipping* is automatically determined by the peak amplitude of the *pulse train* applied to the circuit.

PEDESTAL. A flat-topped *pulse* which elevates the base level for another wave. See *Blanking level*.

PEGASUS. A *digital computer* built by Ferranti Ltd.

PELTIER EFFECT. A liberation or absorption of heat which takes place at the junction between two dissimilar metals or *semiconductors* when a current flows across the junction.

PENCIL VALVE (TUBE). A long narrow pencil-shaped valve used for extremely high-frequency applications. See *Lighthouse valve*.

PENETRATION FACTOR (DURCHGRIFF). Of a valve, the ratio of an infinitesimal change in grid voltage to an infinitesimal change in anode voltage for which the anode current is held constant.

PENTAGRID CONVERTER. A *superheterodyne frequency converter* in which both *local oscillation* and *mixing* are effected by a single electron stream in a specially designed five-grid valve.

PENTATRON. A five-electrode valve comprising two *triodes* with a single common cathode.

PENTODE. A five-electrode valve containing an anode, cathode, control grid, screen grid, and suppressor grid.

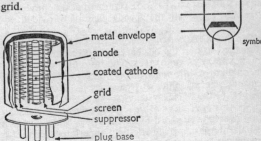

symbol

metal envelope
anode
coated cathode
grid
screen
suppressor
plug base

Fig. 102. Simplified construction of a pentode

269

PENTODE TRANSISTOR. See *Transistor, pentode*.

PERCENTAGE MODULATION. In *amplitude modulation*, the *modulation factor* expressed as a percentage.

PERCEPTRON. An experimental *computer*, currently under development by the American Office of Naval Research, which is intended to have the ability to read written or printed characters and to respond to spoken commands.

PERFECT DIELECTRIC. A *dielectric* in which all the energy used in establishing the field in the dielectric is restored to the electric system when the field is removed.

PERHAPSATRON. An experimental device used at the Los Alamos Laboratory to study *plasma* and controlled *thermonuclear reactions*.

PERIOD. The time required for a single cycle of events which are regularly repeated. In any oscillatory motion the time of a complete *oscillation* and the reciprocal of the *frequency*. Symbol T.

PERIODIC CURRENT. An oscillating current whose values recur at equal intervals of time.

PERIODIC QUANTITY. An oscillating quantity whose values recur for equal increments of the independent variables.

PERIODIC RESONANCE. *Natural resonance*.

PERIODIC TABLE. A table of the chemical *elements* in increasing order of *atomic number*, arranged to show the chemical similarities between groups of elements.

PERIPHERAL EQUIPMENT. In *data processing*: the auxiliary machines, such as *card readers* or high-speed printers, which may be placed under the control of the central computer.

PERMALLOY®. A high *permeability* alloy of nickel and iron.

PERMANENT MAGNET. A piece of *magnetic* material which, having been magnetized, retains a substantial proportion of its magnetization indefinitely.

PERMANENT MEMORY. *Storage* capacity which does not depend on a continuous supply of power, e.g. *magnetic tape, discs*, etc.

PERMANENT-MAGNET LOUDSPEAKER. A moving-conductor *loudspeaker* in which the steady field is provided by a *permanent magnet*.

PERMATRON. A *hot-cathode gas discharge* tube with two electrodes, designed so that the discharge can be controlled magnetically.

PERMEABILITY, ABSOLUTE. The ratio of the *magnetic flux density* in a medium to the *magnetizing force* producing it. Symbol for the absolute permeability of free space μ_0. In the e.m. system of units $\mu_0 = 1$. In the MKS system $\mu_0 = 4\pi \times 10^{-7}$.

PERMEABILITY, FREE SPACE. See *Permeability, absolute*.

PERMEABILITY, INCREMENTAL. The *permeability* of a material to a small alternating magnetic field superimposed on a larger constant magnetic field.

PERMEABILITY, RELATIVE. The ratio of the *magnetic flux density* in a medium to that produced in a vacuum by the same *magnetizing force*. Symbol μ.

PERMEABILITY TUNING. A method of changing the *resonant frequency* of a circuit by varying the position of a core of *ferromagnetic* material in a tuned coil in the circuit.

PERMEAMETER. An instrument for measuring the magnetic characteristics, in particular the *permeability*, of a *ferromagnetic* material.

PERMEANCE. The reciprocal of *reluctance*.

PERMENDUR.® A magnetic alloy of equal parts of cobalt and iron which has an extremely high *permeability* when magnetically saturated.

PERMINVAR.® Magnetic alloys of cobalt, nickel, and iron which exhibit very stable *permeability*.

PERMITTIVITY. Of a *dielectric medium* or space, the ratio of the *electric displacement* to the *electric field strength* at the same point.

PERMITTIVITY, RELATIVE. See *Dielectric constant*.

PERMITTIVITY OF FREE SPACE. The value of *permittivity* in a vacuum.

PERSISTENCE. The length of time during which *phosphorescent* light is emitted from the screen of a *cathode-ray tube*. See *Afterglow*.

PERSISTENCE CHARACTERISTIC. Of a *luminescent* screen, a curve showing the relation between luminance and time after excitation.

PERSISTOR. A device based on the sharp changes in the critical value of current in a metal loop passing from a state of *super-conductivity* to its normal resistive state. In the form of a miniature bi-metallic printed circuit operated at temperatures near absolute zero, it can be used as a storage element or very fast switch for computers.

PERSISTRON. A solid-state electroluminescent and photoconductive display panel which can be used for amplification of light.

PERT. Programme Evaluation and Review Technique: a programme for operational research which solves specific problems related to management.

PERVEANCE. Of a diode or equivalent diode: the quotient of the space-charge limited cathode current and the three-halves power of the anode voltage.

PH. Of an electrolyte, a measure of the hydrogen ion activity in the electrolyte and hence of its acidity or alkalinity.

PHANATRON. A *gas diode*.

PHANTASTRON. A special type of *trigger circuit* which can exist in *astable*, *bistable*, and *monostable* forms, used to generate short *pulses* and short linear *sweep* voltages.

PHANTOM CIRCUIT. A superimposed circuit derived from two suitably arranged pairs of wires, the two wires of each pair being effectively in parallel.

PHASE. (1) In an operation which recurs periodically: *a.* the stage or state to which the operation has proceeded; *b.* the fraction of the *period* which has elapsed measured from some fixed origin. (2) One of the circuits of a *polyphase* system or apparatus; or one of the lines or terminals of the system.

PHASE ANGLE. The angle between two vectors representing two simple periodic quantities which vary sinusoidally and which have the same frequency but differ in *phase*, e.g. the vectors representing an alternating *voltage* and the *current* due to it.

PHASE CONSTANT. The space rate of decrease of the *phase* of a wave.

PHASE CORRECTOR. A network designed to correct for *phase distortion*.

PHASE DELAY. The ratio of the insertion *phase shift*, measured in cycles, to the frequency.

PHASE DEVIATION. In *phase modulation*, the peak difference between the instantaneous *phase angle* of the *modulated wave* and the angle of the *carrier*.

PHASE DIFFERENCE. The difference of *phase* (usually expressed as a time or an angle) between two periodic quantities which vary sinusoidally and have the same frequency.

PHASE DISCRIMINATOR. A device in which *amplitude* variations are derived in response to *phase* variations.

PHASE (FREQUENCY) DISTORTION. The lack of direct proportionality of *phase shift* to frequency, or the effect of such departure on a transmitted signal.

PHASE INVERTER. A stage in an *amplifier* whose chief function is to change the phase of a signal by 180 degrees, usually for feeding one side of a following *push-pull amplifier*.

PHASE LAG. See *Phase difference*.

PHASE METER. An instrument for measuring the difference in *phase* between two alternating quantities of the same frequency.

PHASE MODULATION (PM). *Modulation* in which the *phase angle* of a *sine-wave carrier* is caused to depart from the carrier angle by an amount proportional to the instantaneous magnitude of the modulating wave.

PHASE SHIFT. Any change in the *phase* of a periodic quantity or in the *phase difference* between two or more periodic quantities.

PHASE SPLITTER. A device which produces two or more output waves differing in phase from a single input wave.

PHASE VELOCITY. Of a *sinusoidal wave*, the velocity of travel of an equiphase surface in the direction of the wave normal.

PHASE-SHIFTING NETWORK. A network which shifts the *phase* of one sinusoidal alternating voltage with respect to another.

PHASING. In *television* or *facsimile*, the adjustment of the picture position along the *scanning line*.

PHASITRON. A specially constructed tube in which a rotating electric field is used to *modulate* a high-frequency electron current.

PHON. The unit of *loudness level* of sound. See *Sone*.

PHONOELECTROCARDIOSCOPE. An instrument using a double-beam cathode-ray oscilloscope to show simultaneously the waveforms of two different quantities related to the heart action.

PHONOMETER. An instrument for measuring sound intensity.

PHONON. A *quantum* of thermal energy in a crystal lattice, containing an amount of energy=hv, where h is *Planck's constant* and v is the thermal vibrational frequency.

PHONON MASER. *Maser* action in ruby which is capable of amplifying high-frequency sound waves.

PHOSPHOR. A substance capable of *luminescence*, especially in response to an incident *electron beam*, and consequently used in coating the viewing screen in *cathode-ray tubes*. See *Persistence*.

PHOSPHORESCENCE. Emission of radiation as a result of previous absorption of radiation of shorter wavelengths. The emission may continue for a considerable time after the excitation has ceased, in contrast to *fluorescence*. See *Afterglow, Cathodophosphorescence, Luminescence, Persistence.*

PHOSPHORUS (P). Non-metallic element, Atomic No. 15.

PHOTOACTOR. A device which uses the light from a lamp to operate a *photoconductive cell* in the manner of an electromagnetic relay.

PHOTOCATHODE. A *cathode* from which the electrons are released by the process of photoemission, i.e. by the action of an incident light beam.

PHOTOCELL. Abbreviation for *photoelectric cell*.

PHOTOCONDUCTIVE CELL. A cell consisting of a thin coating of *semiconductor* between two electrodes, on a glass plate, whose electrical *resistance* varies with the incident light.

PHOTOCONDUCTIVE DIODE (PHOTODIODE). A cell in which a single *rectifying contact of semiconductor* materials between electrodes is used to control the flow of current between the electrodes according to the light incident on the cell.

symbol for *pn* light-sensitive diode

PHOTOCONDUCTIVE EFFECT. An increase or decrease in the *conductivity* of a substance, usually a *semiconductor*, with incident radiation. According to the substance, the radiation may extend from the visible spectrum well into the infra-red spectrum. See entries above.

PHOTO-CONVERTER. A device which converts optical patterns into digital electrical signals. The usual form is illustrated in Fig. 103. See *Converter, Digitizer.* See Figure 103.

PHOTOCURRENT. An electric current which varies with illumination.

PHOTODIODE. See *photoconductive diode*.

PHOTOELECTRIC. Descriptive of the effects resulting from the liberation of bound electrons by *photons*, usually of light; these effects include the *photoconductive, photoelectromagnetic, photoemissive,* and *photovoltaic* effects.

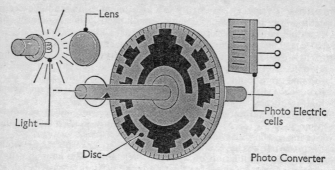

Fig. 103. Photo-converter

PHOTOELECTRIC ABSORPTION. The absorption of *photons* in one of the several *photoelectric* effects.

PHOTOELECTRIC ALARM. An electronic alarm circuit which includes a photoelectric relay.

PHOTOELECTRIC CELL. A cell composed of material, usually a semiconductor such as selenium, germanium, or silicon, which is sensitive to light. Any of the effects listed under *photoelectric* can occur in such a cell, according to the materials used and the design of the cell.

PHOTOELECTRIC CELL, BARRIER LAYER. A *photovoltaic* cell operating on the barrier-layer principle, i.e. current set up by incident light flows from a semiconductor to a metal and is prevented from flowing in the reverse direction by a surface barrier in the semiconductor. The cell therefore acts as an electric generator as long as the illumination continues.

PHOTOELECTRIC CONSTANT. A constant associated with photoemission equal to the ratio h/e, where h is *Planck's constant* and e is the *electronic charge*. See *Physical constants*.

PHOTOELECTRIC EFFECT. A general term covering the effects listed under *photoelectric* and especially the emission of electrons from metal surfaces exposed to light or gamma radiation.

PHOTOELECTRIC EFFECT, ATOMIC. The ejection of bound electrons from an atom by *photons* incident on it, so that the whole energy of a photon is absorbed for each electron ejected.

PHOTOELECTRIC EFFECT, INVERSE. The reverse of the *photoemissive* effect, i.e. the production of radiation from a surface by electrons impinging on the surface.

PHOTOELECTRIC EQUATION. Einstein's equation relating the energy of the incident radiation and the *photoelectric work function* of the substance used to the kinetic energy of an escaped *electron*.

PHOTOELECTRIC GALVANOMETER. A moving-coil galvanometer in which the movements of a mirror mounted on the suspension are detected by the effect of a reflected light beam on a double-cathode *photocell*, so that the sensitivity is enormously increased.

PHOTOELECTRIC SYSTEM. Of protective signalling, an apparatus designed to project a beam of invisible light on to a *photocell* and to produce an alarm condition in the protection circuit when the beam is interrupted. If the light beam is interrupted or *modulated* in the predetermined manner, and the receiving equipment designed to accept only modulated light, the system is called a modulated photo-electric system.

PHOTOELECTRIC THRESHOLD. The *quantum* of energy just sufficient to release an electron in the *photoelectric effect*.

PHOTOELECTRIC WORK FUNCTION. The energy required by a *photon* to eject an *electron* from a metal.

PHOTOELECTROMAGNETIC EFFECT. If a *semiconductor* is placed in a magnetic field parallel to its surface, an electric current may be generated by illuminating the semiconductor surface strongly.

PHOTOEMISSIVE CELL. A *photoelectric cell* which employs a *cathode*.

PHOTOEMISSIVE EFFECT. The *emission* of electrons from a substance by subjecting it to *electromagnetic* radiation such as light, X-rays, etc. The wavelength of the radiation must be less than a critical value which depends on the material, and for some metals it is in the visible spectrum.

PHOTOFORMER. A *cathode-ray tube,* used in *analogue computers,* which generates a voltage function with an opaque mask and a *feedback loop,* so that the output voltage is constrained to vary as some desired function of the input.

PHOTOGLOW TUBE. A gas *phototube* which employs a *glow discharge* to increase the sensitivity.

PHOTO-ISLAND GRID. See *Mosaic.*

PHOTOMULTIPLIER (TUBE). An *electron multiplier* tube in which the electrons initiating the cascade are due to *photoemissive effect.*

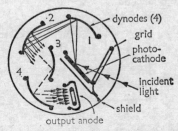

Fig. 104. Simplified action of photomultiplier

PHOTON. A fundamental particle or *quantum* of electromagnetic radiation.

PHOTOPOSITIVE. Descriptive of a *photoelectric* material whose *conductivity* increases with increasing incident radiation.

PHOTOSENSITIVE. Having a *photoelectric* reaction when subjected to radiation.

PHOTOTRANSISTORS. *Semiconductor* devices which are sensitive to light.

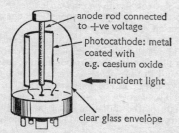

symbol for *pn* photo transist.

PHOTOTUBE. An electron tube in which one of the electrodes is irradiated (usually with visible light) so as to cause it to emit electrons. See *Photomultiplier, Gas phototube, Vacuum phototube.*

anode rod connected to +ve voltage

photocathode: metal coated with e.g. caesium oxide

← incident light

clear glass envelope

Fig. 105. Photocathode in a typical vacuum phototube

PHOTOVARISTOR. A *varistor,* made of materials such as cadmium sulphide, whose current-voltage relation is made to depend on its illumination.

PHOTOVOLTAIC CELL, BACK WALL. A cell which combines the *photovoltaic effect* with the *barrier-layer* principle. See Fig. 106.

PHOTOVOLTAIC CELL, FRONT WALL. A cell which reverses the arrangement of the back wall photovoltaic cell and has the barrier layer at the front junction.

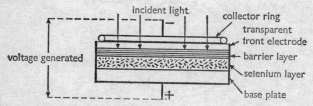

incident light

collector ring
transparent front electrode
barrier layer
selenium layer
base plate

voltage generated

Fig. 106. Construction of selenium photovoltaic cell

PHOTOVOLTAIC EFFECT. An electromotive force may be generated by allowing the junction between dissimilar materials, e.g. a metal and a *semiconductor,* to be exposed to radiation.

PHYSICAL CONSTANTS. A table of physical constants used in electronics is provided in Fig. 107.

Charge on electron (e)	=	$1 \cdot 602 \times 10^{-19}$ coulombs
Mass of electrons (at rest) (mo)	=	$9 \cdot 11 \times 10^{-28}$ grammes
Ratio of electronic charge to mass (e/mo)	=	$1 \cdot 759 \times 10^{18}$ coulombs/gramme
Radius of electron (at rest)	=	$2 \cdot 818 \times 10^{-13}$ centimetres
Energy of 1 electron-volt	=	$1 \cdot 602 \times 10^{-12}$ ergs
Wavelength of electrons of v volts	=	$\dfrac{12 \cdot 24}{v.} \times 10^{-8}$ centimetres
Velocity of light (c)	=	$2 \cdot 99776 \times 10^{10}$ centimetres/second
Planck's constant (h)	=	$6 \cdot 6 \times 10^{-27}$ erg-second
Boltzmann constant (k)	=	$1 \cdot 38 \times 10^{-16}$ erg/degree
Gas constant (R)	=	$8 \cdot 314 \times 10^{7}$ ergs/gm. molecule
Avogadro's number	=	$6 \cdot 028 \times 10^{23}$ per gm. molecule
Joule's constant (mechanical) equivalent of heat	=	$4 \cdot 18$ Joules/calorie
Photoelectric constant (h)	=	$4 \cdot 135 \times 10^{-7}$ erg-sec/coulomb
Magnetic moment of electron	=	$(0 \cdot 92838 + 0 \cdot 00006) \times 10^{-20}$ erg gauss^{-1}
Faraday constant	=	$9652 \cdot 2 + 0 \cdot 2$ c.m.u./gm. molecule

Fig. 107. Physical constants

PHYSIOLOGICAL MONITOR. Generally, any instrument for measuring and recording physiological processes. In particular, an electronic instrument used in an operating theatre to monitor the condition of a patient under anaesthesia.

PI NETWORK. A *network* of three *impedance* branches connected in series with each other to form a closed circuit, so that the three junction points form an input terminal, an output terminal, and a common input and output terminal, respectively.

PI-MODE. Of a *magnetron*: the mode of operation in which the phase difference between adjacent cavities is π radians.

PICCOLO. A code name for a radar *jammer* which uses a series of high-power magnetrons.

PICK-UP. A *transducer* which converts intelligence, recorded or otherwise, into electric signals. Especially a gramophone pick-up.

PICK-UP, CAPACITOR. A gramophone *pick-up* whose operation depends on variation of its *capacitance.*

PICK-UP, CRYSTAL. A gramophone *pick-up* whose operation depends on the properties of a *piezoelectric crystal.*

PICK-UP, ELECTRONIC. A gramophone *pick-up* in which the output voltage is due to the motion of an electrode in a valve.

PICK-UP, GRAMOPHONE. An electromechanical *transducer* actuated by modulations in the groove of the recording medium, which transforms this mechanical input into an electrical output.

PICK-UP, INDUCTANCE. A *telemetering pick-up* which depends for its operation on the variation of its *inductance*.

PICK-UP, LIGHT BEAM. A gramophone *pick-up* in which a beam of light is a coupling element for the *transducer*.

PICK-UP, MAGNETIC. A gramophone *pick-up* in which the electric output is due to the relative motion of a magnetic field and a coil or conductor located in the field.

PICK-UP, VARIABLE RELUCTANCE. A gramophone *pick-up* which depends on the variation in the *reluctance* of a magnetic circuit.

PICK-UP, VARIABLE RESISTANCE. A *telemetering* or gramophone *pick-up* whose operation depends on changes in its electrical resistance.

PICK-UP, VIBRATION. A *transducer* which converts mechanical vibrations into a corresponding oscillating electric current.

PICK-UP CARTRIDGE. The removable part of a *pick-up* containing the electromechanical translating elements and the reproducing stylus.

PICO. Prefix meaning a million millionth or 10^{-12}.

PICOFARAD. A *micro-microfarad*, 10^{-12} *farad*.

PICTURE ELEMENT. In *television*, any segment of a *scanning line* whose dimension along the line is exactly equal to the line width.

PICTURE FREQUENCIES. In *facsimile transmission*, frequencies which result solely from *scanning* subject copy.

PICTURE INVERSION. In *facsimile transmission*, a process which causes reversal of the black and white shades of the recorded copy.

PICTURE SIGNAL. In *television* or *facsimile transmission*, the signal which results from the *scanning* process.

PICTURE TRANSMISSION. The electric transmission of a picture having a gradation of shade values. See *Facsimile, Television*.

PICTURE TUBE. A *cathode-ray tube* used in *television*.

PIEZOELECTRIC CRYSTAL. A crystalline dielectric which exhibits the *piezoelectric effect*. Examples are Rochelle salt and quartz.

PIEZOELECTRIC EFFECT. Certain crystals develop an electric charge or potential difference across some of the crystal faces when the crystal is subjected to mechanical strain; and conversely they produce mechanical forces when a voltage is applied in a suitable manner. See *Piezoelectric (crystal) unit*.

PIEZOELECTRIC (CRYSTAL) UNIT. An assembly of *piezoelectric crystal* elements, suitably mounted and housed, and adjusted to resonate to a desired frequency, with means for connecting it to an electric circuit. Such a device is used for frequency control and measurement, for *crystal filters*, or for interconversion of electric and elastic waves.

PIEZOELECTRICITY. Electricity resulting from the *piezoelectric effect*.

PILOT LAMP. A small lamp which indicates the condition of an associated circuit or control.

PILOT SPARK. In a *gas discharge tube,* a small preliminary discharge to a secondary anode which provides an ionized path for the main discharge to the primary anode.

PILOT-WIRE REGULATOR. In transmission circuits, an automatic device to control adjustable gains and losses associated with the circuit and to compensate for transmission changes due to temperature variations.

P-I-N DIODE. A *semiconductor* diode in which the junction consists of a layer of almost *intrinsic semiconductor,* formed between *P-type* and *N-type* regions. It behaves as a normal P-N junction at low frequencies but as a controllable variable resistance at ultra-high frequencies.

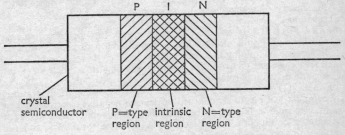

Fig. 108. The junction in a P-I-N diode

PINCH EFFECT. An effect observed in a molten conductor carrying a heavy current: the conductor becomes constricted and may even be momentarily ruptured. This is due to the attractive forces between parallel currents in the same direction and is of great importance in *thermonuclear* reactions.

PINCH-OFF. The equivalent of collector cut-off in a *field-effect transistor.*

PINCUSHION DISTORTION. In *cathode-ray tubes:* the type of distortion which makes a rectangle appear pincushion-shaped.

Fig. 109. Pincushion distortion

PIRANI GAUGE. A hot-wire vacuum gauge for measuring very low gas pressures as a function of the resistance of the hot filament.

PLAN POSITION INDICATOR (PPI). In *radar*, a presentation in which the signal appears as a bright spot, with the range indicated by its distance from the centre of the screen and the bearing by the radial angle.

PLANAR DIODE. A *diode* whose electrodes are planar and parallel.

PLANAR TECHNIQUE. The formation of *p-type* and/or *n-type* regions in a *semiconductor* crystal by diffusing impurity atoms into the crystal through holes in an oxide mask on the surface.

PLANAR TRANSISTOR. A *transistor*, manufactured by selective etching of thin wafers of suitably treated *semiconductor*, whose electrodes form essentially parallel planes.

PLANCK'S CONSTANT. A universal constant, the factor which relates the energy of a *photon* to its frequency. Symbol: *h*. See *Physical constants*.

PLANE-POLARIZED WAVE. A *transverse* wave in which the displacements at all points along a line in the direction of propagation lie in a plane containing this line.

PLANETARY ELECTRON. An *electron* moving in a shell or orbit around an atomic nucleus.

PLASMA. (1) A completely *ionized* gas at extremely high temperatures composed of positively charged nuclei and negative electrons. (2) In an electric *arc*, the region between the cathode and anode in which gaseous conduction takes place. (3) A synonym for *positive column*.

PLASMATRON. A *gas discharge tube* in which the conducting path is provided by a *plasma* previously generated in the tube.

PLATE. (1) One of the conductive electrodes in a *capacitor*. (2) One of the electrodes in a storage battery. (3) The American name for the *anode* of a valve.

PLATEAU. (1) In *Geiger counters*, the portion of the characteristic curve of *counting rate* versus voltage which is substantially independent of the applied voltage. (2) In a *cold-cathode gas discharge*, the part of the voltage-current characteristic in which the tube voltage drop is constant for wide changes of current.

PLATED CIRCUIT. A *printed circuit* produced by electrodeposition of a conductive pattern on an insulating base.

PLATED MAGNETIC WIRE. A magnetic wire with a core of non-magnetic material and a plated surface of *ferromagnetic* material.

PLATINOTRON. A *microwave* valve similar to a *magnetron*.

PLATINUM (Pt). A noble metallic element, Atomic No. 78. Used for electrical contacts which must withstand high temperatures, and for electrodes which must withstand chemical attack.

PLAYBACK. The operation of an input medium containing recorded instructions, particularly on magnetic tape or punched tape, to generate coded signals for use as the input to a computer or control system. See *Tape reader*.

PLUG-BOARD. See *Patch board*.

PLUG-IN. A 'plug-in' device is one whose connections may be completed through pins, plugs, jacks, sockets, or any other kind of quick connectors. The term is widely used now to describe standardized sub-assemblies, especially in *computers,* which can be inserted and removed rapidly in the main equipment.

PLUMBICON®. A modern *camera tube* developed by the Philips Co. which has a photoconductive target of lead monoxide in place of the antimony trisulphide target used with the *Vidicon.* (See Fig. 110.) It has higher sensitivity, faster response, and lower dark current than the Vidicon.

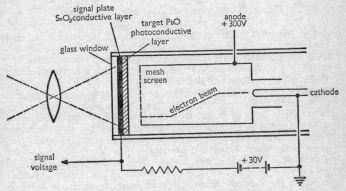

Fig. 110. Diagram of Plumbicon television camera tube

PLUMBING. Colloquial term applied to the metal parts and connections of *waveguide* and *coaxial* equipment.

P-N BOUNDARY. In *transistors,* a surface in the transition region between *p-type* and *n-type semiconductors* at which the *donor* and *acceptor* concentrations are equal.

P-N JUNCTION. See *p-n boundary* and Fig. 72.

P-N-P TRANSISTOR. See *Transistor, p-n-p.*

POCKELS EFFECT. The *electro-optical effect* in the case when the transparent dielectric is a *piezoelectric crystal.*

POINT CONTACT (SEMICONDUCTOR). Pressure contact between a *semiconductor* and a metallic point. See *Cat's whisker* and following entry.

POINT-CONTACT TRANSISTOR. A transistor having a *base electrode* and two or more *point-contact* electrodes. See Figure 111.

POINT SOURCE. A source of radiation at a great distance from the observer; or an ideal source of infinitesimal size.

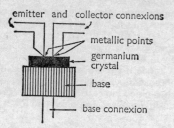

Fig. 111. Construction of point-contact transistor

POISON. Material which reduces the ability of a *cathode* to emit electrons.

POISSON'S EQUATION. A fundamental equation of *electrostatics*.

POLARIS. A surface-to-surface *guided missile* with a range of about 1,500 miles, developed by the U.S. Navy.

POLARITY. In electricity, a property of points or *poles* in an electric or magnetic field or circuit with respect to the flow of electrons or of magnetic lines of force; it is characterized by the existence of opposite charges, positive and negative, or of opposing magnetic poles. *Negative* particles flow towards points of *positive* polarity and away from points of negative polarity.

POLARIZATION. See *Dielectric polarization, Magnetic polarization.*

POLARIZED ELECTROLYTIC CAPACITOR. An *electrolytic capacitor* in which the dielectric film is formed near one electrode only, so that impedance to current flow is not the same in both directions.

POLARIZED PLUG. A plug constructed so that it can be inserted into a socket in only one position.

POLAROGRAPH. An instrument for rapid chemical analysis of solutions, using specially designed electrodes. One electrode works at constant potential with respect to the solution being analysed, and the other – the reference electrode – produces a potential which is a measure of the concentration polarization of the ions in solution.

POLE. (1) A *magnetic pole*. (2) One electrode of an electric *cell*.

POLE PIECE. A piece of *ferromagnetic* material forming one end of a magnet, shaped to control the distribution of *magnetic flux* in its vicinity.

POLYPHASE. A qualifying term applied to a system or apparatus, to denote one in which there are two or more alternating voltages, usually (but not necessarily) displaced in *phase* relative to one another by equal portions of a *period*. In a *two-phase* system the displacement is one quarter of a period, in a *three-phase* system it is one third of a period, and so on. Compare *Single-phase.*

POOL RECTIFIER. A *rectifier* with a pool cathode.

PORT. Of an electronic system, circuit, or device, a place where energy may be supplied or withdrawn, or where variables of the system may be measured or observed.

PORTABLE BATTERY. A storage battery designed for convenient transportation.

POSITIVE. A qualifying term applied to one of two points between which a *potential* exists, to distinguish that one which corresponds, as far as the tendency to set up a current in an external circuit is concerned, to the copper plate of a *Daniell cell*. See *Electropositive*.

POSITIVE COLUMN. The luminous glow between the *Faraday dark-space* and the *anode*, in a *cold-cathode glow discharge tube*.

POSITIVE ELECTRICITY. The kind of electricity appearing in a piece of glass which has been rubbed with silk. It is now known that the charge is due to a deficiency of *electrons* which have *negative charge*. See *Negative electricity*.

POSITIVE ELECTRODE. (1) The *anode* of a valve; or any *electrode* at a higher potential than a reference electrode. (2) In a *primary cell*, the conducting material which acts as a *cathode* while the cell is discharging and to which the positive terminal is connected.

POSITIVE ELECTRON. See *Positron*.

POSITIVE ION. An *ion* which has a *positive* charge. See *Cation*.

POSITIVE (PICTURE) MODULATION. In an *amplitude-modulated television* system, that form of *modulation* in which an increase in brightness corresponds to an increase in transmitted power.

POSITIVE RAYS. See *Canal rays*.

POSITIVE-GRID OSCILLATOR VALVE. A *triode*, oscillating in such a way that the *quiescent* voltage on the grid is more positive than that on either of the other electrodes.

POSITRON. A nuclear particle with the same mass as an *electron* but with a *positive* charge equal in magnitude to the electronic *negative* charge. Unlike negative electrons (negatrons), positrons are short-lived and combine with electrons to produce annihilation radiation. See *Pair production*.

POST-DEFLECTION ACCELERATION (PDA). In a cathode-ray tube: acceleration of the electron beam after it emerges from the deflecting system, by a high voltage electrode which is usually applied in the form of a graphite track around the inside surface of the tube. See Fig. 69.

POT. Colloquial for *potentiometer*.

POTASSIUM (K). Very reactive metallic element. Atomic No. 19.

POTENTIAL. The work done when a unit electric charge or unit magnetic pole is brought to the point, whose potential is being measured, from 'infinity', i.e. from some location which is isolated from the system under consideration. See the following types of potential: *Contact, Counter starting, Critical, Deformation, De-*

283

ionization, Diffusion, Electric, Extinction, Floating, Ionization, Radiation, Sticking, Stopping, Striking, Zero.

POTENTIAL BARRIER. Of a *semiconductor junction*: the potential difference across the junction in an unbiased state, due to the diffusion of charge carriers from opposite sides of the junction.

POTENTIAL DIFFERENCE. The algebraic difference between the individual *potentials* of two points. Measured by the work done in transferring unit charge from one point to the other. Symbol: V. Practical unit: *volt*.

POTENTIAL GRADIENT. At a point, the *potential difference* per unit length, measured in the direction in which it is a maximum. Practical unit: volt per unit length.

POTENTIAL TRANSFORMER. See *Voltage transformer*.

POTENTIOMETER. An instrument for measuring electrical quantities by balancing an unknown *potential difference* (or *e.m.f.*) against a known potential difference.

POTENTIOMETER, COORDINATE. A *potentiometer* in which two alternating voltages in *quadrature* are used to develop the balancing potential. An example is the *Gall potentiometer*.

POTENTIOMETER, COSINE. A *voltage divider* in which an applied direct voltage gives an output proportional to the cosine of the angular displacement of a shaft. Compare *Potentiometer, sine*.

POTENTIOMETER, GALL. See *Gall potentiometer*.

POTENTIOMETER, MAGNETIC. See *Magnetic potentiometer*.

POTENTIOMETER, SINE. A *voltage divider* in which an applied direct voltage gives an output proportional to the sine of the angular displacement of a shaft.

POULSEN ARC CONVERTER. A high-frequency *arc oscillator* which can generate frequencies up to 100 kc/s.

POWDERED-IRON CORE. A *magnetic core* used for high frequency inductors and transformers where low loss is required, and composed of finely divided iron particles in a plastic or ceramic binding material.

POWER. The rate of doing work or transforming energy. Electrical unit: *watt*. See *Units*.

POWER AMPLIFIER. An *amplifier* designed to increase the power of a signal applied to its input, especially where the output is not applied to the grid of another valve but to an aerial, loudspeaker, relay system, etc.

POWER DETECTION. A form of *detection* in which the power output of the detecting device supplies a substantial amount of power directly to the load, such as a loudspeaker or recorder.

POWER EFFICIENCY. Of an electro-acoustic *transducer* (e.g. a *loudspeaker*), the ratio of the electric power at the electric terminals of the transducer to the acoustic power available to the transducer.

POWER FACTOR. The ratio of the actual power of an *alternating* or *pulsating current*, as measured by a *wattmeter*, to the apparent power indicated by ammeter and voltmeter readings. It is also the cosine of the *phase angle* between a *sinusoidal* voltage and the resulting current, the ratio of the *resistance* to *impedance*, and so a measure of the loss in an *insulator*, *inductor*, or *capacitor*.

POWER FACTOR, DIELECTRIC. The cosine of the *dielectric phase angle*, or the sine of the *dielectric loss angle*.

POWER FREQUENCY. The frequency at which domestic and industrial electricity supply is distributed and used, i.e. 50 cycles per second in the British grid system.

POWER GAIN. See *Amplification, power*.

POWER LEVEL. At any point in a *transmission system*, the ratio of the power at that point to an arbitrary reference level, usually expressed in *decibels referred to one milliwatt*.

POWER LOSS. Of a *transducer*, the ratio of the power absorbed by the input to the power delivered to a specified *load impedance*.

POWER OUTPUT, AVERAGE. Of an *amplitude-modulated transmitter*, the radio-frequency power delivered to the transmitter output terminals, averaged over one modulation cycle.

POWER OUTPUT, MAXIMUM UNDISTORTED. Of a *transducer*, the maximum power delivered under specified conditions with a total *harmonic* not exceeding a specified percentage.

POWER OUTPUT, PEAK. See *Peak power output*.

POWER PACK. A unit for converting power from an ac or dc supply into ac or dc power at voltages suitable for operating electronic devices.

POWER RELAY. A *relay* which functions at a predetermined value of power.

POWER SUPPLY. Any means for supplying electric power to a circuit, e.g. generators, batteries, rectifiers, converters, etc.

POWER SUPPLY, FILAMENT (HEATER). The means for supplying power to heat the *filament* of a valve. See *Filament transformer*.

POWER SUPPLY, FULL-WAVE. A *full-wave rectifier* and associated filter which supply power for the high-voltage circuit of electronic equipment.

POWER SUPPLY, VIBRATOR. A power supply in which the varying current required for the input of a *step-up transformer* is provided by a *vibrator*.

POWER SUPPLY, VOLTAGE-REGULATED. A power supply consisting of a *rectifier*, *filter*, and *stabilizer* for high direct voltage to electronic circuits.

POWER TRANSFORMER. In electronic equipment, the main transformer, usually provided with a number of output windings for high tension supply and for filament supply.

POWER TRANSISTOR. A *transistor* designed to handle relatively high power, up to 100 watts, or to give a relatively high undistorted *power gain*. Such transistors usually require cooling. See Figure 112.

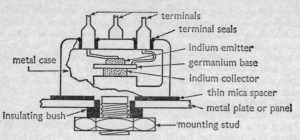

Fig. 112. Construction and mounting of a typical germanium power transistor

POWER VALVE. A valve designed to handle a larger power than the ordinary voltage-amplifier valve.

POWER WINDING. Of a *saturable reactor*, the winding which handles the power to be controlled; it is often also the output winding.

POWER-FACTOR METER. A direct-reading instrument for measuring *power factor*.

PPI. Abbreviation for *plan position indicator*.

PPM. Abbreviation for *pulse position modulation*.

PRACTICAL UNITS. Units which have been adopted for practical use because the centimetre-gramme-second (c.g.s.) units are often inconveniently large or small. See *Units*.

PREAMPLIFIER. An *amplifier* used in front of the main amplifier. In *radar* the preamplifier is separated from the rest of the receiver and located to provide the shortest possible path from aerial to input circuit to achieve the best possible *signal-to-noise* ratio. Preamplifiers are also housed in probes or pick-up devices which are connected to the main amplifier by a cable, so that the initial input is obtained as close as possible to the source.

PRECISION APPROACH RADAR. See *Radar, precision approach*.

PRE-CONDUCTING CURRENT. In a *gas discharge tube*, the current of very low value which flows during the period of the discharge which is not self-sustaining.

PRE-EMPHASIS NETWORK. A network inserted in a system in order to emphasize one range of frequencies with respect to another.

PREFERRED VALUES. Manufacturers and users of electrical and electronic components such as fixed *resistors* and *capacitors* find that there are advantages in standardizing component values and adopt preferred values so that each value differs from the preceding one by a constant multiple.

PRESELECTOR. A tuned r-f amplifier used before the frequency converter in a *superheterodyne* receiver to increase the overall sensitivity and selectivity.

PRESET GUIDANCE. In *guided missiles*, the form of guidance in which the control mechanism is set, before launching, for a predetermined path and there is no provision for subsequent adjustment.

PRESSDUCTOR®. A transducer for measuring weight in which the stress in a supporting member is measured magnetically.

PRESSURE MICROPHONE. A *microphone* whose electrical output corresponds to the instantaneous pressure of the impressed sound waves.

PRESSURE-TYPE CAPACITOR. A high-voltage *capacitor*, in which the *dielectric* is an inert gas under pressure.

PRESTORE. In a *computer*: to store a quantity in a location before it is required in a routine.

PRE-TR CELL. In *radar* receivers, a gas-filled radio-frequency switching valve (or tube) which protects the *TR cell* from excessive power and the receiver from frequencies other than the fundamental.

PRF. Abbreviation for *pulse repetition* (or recurrence) *frequency*.

PRIMARY BATTERY. A *battery* made up of *primary cells*.

PRIMARY CELL. A *cell* in which the electrochemical action producing the current is not normally reversible. Such a cell cannot be recharged by an electric current and is frequently used as a *standard cell*.

PRIMARY ELECTRON. An *electron* released from an atom by internal forces and not, as with *secondary electrons*, by external radiation.

PRIMARY EMISSION. *Electron emission* due directly to the temperature or irradiation of a surface, or the application of an electric field to the surface. Compare *secondary emission*.

PRIMARY RADIATION. *Radiation* direct from the source.

PRIMARY STANDARD. The standard for a given unit which defines this unit for the whole world.

PRIMARY VOLTAGE. (1) The *voltage* across the input winding of a *transformer*. (2) The voltage of a *primary cell*.

PRIMARY WINDING. The input winding of a *transformer*.

PRINCIPAL AXIS. Of a crystal, the longest axis.

PRINTED CIRCUIT. An electric circuit or circuit element in which the wiring connections and certain fixed components are 'printed' on a circuit board. The usual process starts with a board of insulating material which is coated with metal, such as copper, and the portions of the metal which represent wiring and components are supplied with a protective coating by a photographic method. The unit is then immersed in acid which removes the unprotected metal, leaving the circuit elements. See Figure 113.

PRINTOSCOPE®. An electrostatic character writing tube similar to the *Charactron*.

PRINT-THROUGH. In *magnetic-tape* recording, a form of distortion due to a strongly magnetized layer of tape affecting adjacent layers.

PROBE. (1) A test lead which contains at its end or along its length a passive or active network for measuring or monitoring. (2) A resonant

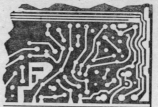

Fig. 113. Typical printed circuit (under-chassis wiring)

conductor inserted in a *waveguide* or *cavity resonator* for injecting or withdrawing energy.

PROGRAMME (or PROGRAM). In a *digital computer*, a set of instructions in the proper sequence for the collation of a set of data or the solution of a problem. See *Coded programme.*

PROGRAMME CONTROL. A general term for the type of control in which the target or objective is automatically held or changed with time according to a prescribed programme.

PROGRAMME SIGNAL. In *audio* systems, the complex electric wave corresponding to speech, music, and associated sounds intended for audible reproduction.

PROGRAMME TAPE. In *computing* and *data-processing*, a tape containing the sequence of instructions.

PROGRAMMED CHECK. In *computers*, a series of tests inserted into the programme of the problem, and accomplished by the appropriate use of the machine's instructions.

PROGRAMMER. One who prepares the sequence of instructions for a *computing* or *data-processing* problem, but does not necessarily perform any *coding*.

PROGRESSIVE SCANNING. In *television*, a rectilinear *scanning* process in which the distance from centre to centre of successively scanned lines is equal to the nominal line width.

PROPAGATION. In electrical practice, the travel of waves through or along a medium.

PROPAGATION LOSS. The transmission loss suffered by radiant energy.

PROPORTIONAL CONTROL. A control system in which the correcting effort is proportional to the magnitude of the difference between the actual value of the controlled quantity and the desired value.

PROPORTIONAL COUNTER. A gas-filled radiation *counter* in which the magnitude of the pulse generated per count is proportional to the energy of the particle or photon being counted.

PROTECTIVE SIGNALLING. The generation, transmission, and reception of signals involved in the detection and prevention of loss or damage due to fire, burglary, and other destructive conditions.

PROTECTOR TUBE. A *cold-cathode glow discharge* tube in which a low-voltage breakdown between electrodes is used to protect circuits against voltage overloads.

PROTON. A positively charged elementary particle which is a constituent of every atomic *nucleus*. Its charge is equal in magnitude and sign to that of the *positron* but its mass is about 1836 times the mass of the *electron*. The *atomic number* gives the number of protons in the nucleus, as well as the number of electrons surrounding it.

PROXIMITY EFFECT. The redistribution of current in a conductor brought about by the presence of another conductor.

PSEUDO-CODE. An arbitrary *code* which must be translated into *computer code*.

PSOPHOMETRIC VOLTAGE. The voltage at 800 Hz at a point in a telephone system, which, if it replaced the disturbing voltage, would produce the same degree of interference with a telephone conversation as the disturbing voltage.

P.T.F.E. Polytetrafluorethylene. See *Teflon*.

PTM. Abbreviation for *pulse time modulation*.

P-TYPE CONDUCTIVITY. *Conductivity* associated with *holes* in a *semiconductor*. See *Acceptor impurity*.

P-TYPE SEMICONDUCTOR. See *Semiconductor, p-type*.

PUBLIC ADDRESS (PA) SYSTEM. A system designed to amplify sounds for an assembly of people.

PULSATING CURRENT. A *current* which undergoes regularly recurring variations of magnitude. The term is usually confined to a unidirectional current.

PULSATING QUANTITY. A *periodic* quantity which can be considered as the sum of an alternating component and a continuous component; the average value cannot be zero.

PULSE. A variation of a quantity whose value is normally constant. The essential characteristics of a pulse are: a rise, a finite duration, and a decay.

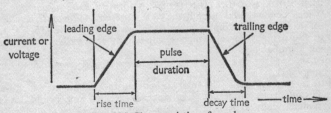

Fig. 114. Characteristics of a pulse

PULSE, LINE-SYNCHRONIZING. In *television,* the *pulse* transmitted at the end of each line to synchronize the start of the *scan* for the next line.

PULSE, SAWTOOTH. A *pulse* whose waveshape has the profile of a sawtooth. A particular case of *triangular pulses*.

PULSE, SELECTOR. A *pulse* used to identify and select one event in a series of events.

PULSE, SYNCHRONIZING. In any *pulse-modulated* radar or radio system, a *pulse* transmitted to synchronize the transmitter and receiver at the start of each train of pulses or at definite frequent intervals.

PULSE, TRAPEZOIDAL. A *pulse* with a trapezoidal waveshape.

PULSE, TRIANGULAR. A *pulse* with a triangular waveshape, especially an isosceles triangle. See *Pulse, sawtooth*.

PULSE AMPLITUDE. A general term indicating the magnitude of a *pulse*, measured with respect to the normally constant value and ignoring spikes and ripples, unless otherwise stated.

PULSE AMPLITUDE, AVERAGE. The average of the instantaneous amplitude taken over the *pulse* duration.

PULSE AMPLITUDE, PEAK. The maximum absolute peak value of the *pulse* excluding unwanted portions such as *spikes*. The amplitude chosen should be illustrated pictorially.

PULSE AMPLITUDE, RMS. The square root of the average square of the instantaneous amplitude, over the *pulse* duration.

PULSE AMPLITUDE MODULATION (PAM). *Modulation* in which the modulating wave is caused to *amplitude-modulate* a *pulse carrier*.

PULSE BAND WIDTH. The smallest frequency interval outside which the amplitude of the *pulse frequency spectrum* is less than a specified fraction of the amplitude at a specified frequency.

PULSE CARRIER. A *carrier* consisting of a series of *pulses*.

PULSE CHARACTERISTICS. A *pulse* may be treated as being composed of a dc component, a *fundamental frequency*, and an infinite number of *harmonics*.

PULSE CODE. A *pulse train modulated* so as to present information, e.g. the *Morse* code.

PULSE CODE MODULATION (PCM). A form of *pulse modulation* in which a code is used; usually the code represents the *quantized* values of instantaneous samples of the signal wave. See Figure 115.

PULSE CODER. A circuit which sets up a number of *pulses* arranged in an identifiable pattern.

PULSE CREST FACTOR. The ratio of the pulse *peak amplitude* to *rms amplitude*.

PULSE DECAY TIME. The interval between the instants at which the instantaneous amplitude last reaches specified upper and lower limits: usually 90 per cent and 10 per cent of the *peak pulse amplitude*.

PULSE DELAY. Of a *transducer*, the interval of time between a specified point on the input *pulse* and a specified point on the output pulse.

PULSE DEMODER. A circuit adjusted to respond only to *pulse* signals which have certain spacings between pulses. Contrast *Pulse moder*.

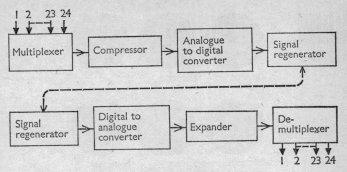

Fig. 115. Simplified P C M system

PULSE DETECTOR. A *detector* which provides separation and detection of a train of modulated *pulses*.

PULSE DISCRIMINATOR. A device which responds only to a *pulse* having a particular characteristic such as *amplitude*, *period*, etc.

PULSE DROOP. See *Pulse flatness deviation*.

PULSE DURATION. The time between the first and last instants at which the instantaneous amplitude reaches a stated fraction of the *peak pulse amplitude*.

PULSE DURATION MODULATION. See *Pulse width modulation*.

PULSE DUTY FACTOR. The ratio of the average *pulse duration* to the average *pulse spacing*.

PULSE FALL TIME. See *Pulse time constant of fall*.

PULSE FLATNESS DEVIATION. The difference between the maximum and minimum amplitudes of a *pulse*, divided by the maximum amplitude, all taken between the first and last *knees* of the pulse.

PULSE FREQUENCY DIVIDER. A circuit for dividing the number of incoming pulses so that a relatively slow recorder can respond and store the information transmitted.

PULSE FREQUENCY MODULATION (PFM). A form of *pulse time modulation* in which the *pulse repetition frequency* of the *carrier* is varied in accordance with the amplitude and frequency of the modulating signal.

PULSE FREQUENCY SPECTRUM. The frequency distribution of the *sinusoidal* components of the *pulse* in relative *amplitude* and relative *phase*.

PULSE GENERATOR. An electronic circuit or device which generates *pulses* of current or voltage. See *Pulse oscillator*, *Impulse generator*, *Surge generator*.

PULSE HEIGHT. The voltage *amplitude* of a *pulse*.

291

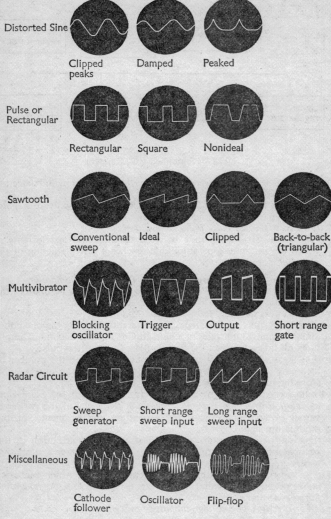

Fig. 116. Typical AC and pulse waveforms

PULSE HEIGHT ANALYSER. A circuit for sorting *pulses* into selected ranges of *amplitude* and recording the numbers in each range.

PULSE HEIGHT DISCRIMINATOR. A circuit designed to select and pass voltage *pulses* of a specified minimum *amplitude*.

PULSE INTERROGATION. The triggering of a *transponder* by a *pulse* or *pulse mode*.

PULSE INTERVAL. The *pulse spacing*.

PULSE JITTER. Minor variations of *pulse spacing* in a *pulse train*.

PULSE LEADING-EDGE. The portion of a *pulse* which first increases in *amplitude*.

PULSE LENGTH. The *pulse duration* usually measured in microseconds or the equivalent distance in yards or miles, represented by the pulse signal on a *radar* screen.

PULSE MODE. A series of *pulses* arranged in a particular pattern for selecting and isolating a communication channel.

PULSE MODE MULTIPLEX. A device for selecting two or more channels on the same *carrier* frequency by means of *pulse modes*.

PULSE MODER. A device for producing a *pulse mode*. Contrast *Pulse demoder*.

PULSE MODULATION. *Modulation* of a *carrier* by *pulses*, or modulation of a *pulse carrier*. See Figure 117.

PULSE OPERATION. Any method of operation in which the energy is delivered in pulses.

PULSE OSCILLATOR. An *oscillator* operating at recurrent intervals by self-generated or externally applied *pulses*.

PULSE POSITION MODULATION (PPM). *Pulse time modulation* in which each instantaneous sample of a modulating wave is caused to *modulate* the position in time of a *pulse*.

PULSE POWER, PEAK. The power at the maximum of a *pulse* of power, excluding *spikes*.

PULSE RADIO. A system of *radio communication* which employs a modulated *pulse carrier*.

PULSE REGENERATION. The process of restoring a series of *pulses* to their original timing, form, and magnitude.

PULSE REPETITION (RECURRENCE) FREQUENCY. The rate at which *pulses* are transmitted by a *pulse-modulated* system, usually given in cycles per second.

PULSE REPETITION (RECURRENCE) PERIOD. The reciprocal of the *pulse repetition frequency*.

PULSE RISE TIME. The interval between the instants at which the instantaneous amplitude of a *pulse* first reaches specified lower and upper limits (usually 10 per cent and 90 per cent) of the *peak pulse amplitude*.

PULSE SEPARATION. See *Pulse spacing*.

PULSE SHAPER. Any *transducer* used for changing one or more of the characteristics of a *pulse*. The term includes *pulse regenerator*.

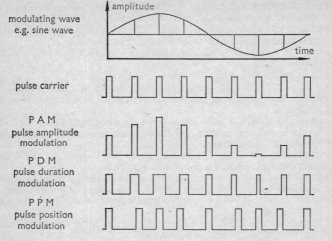

Fig. 117. Pulse modulation – three kinds of pulse modulation of a pulse carrier by a sine wave

PULSE SPACING. The interval between the corresponding pulse times (e.g. *pulse rise time*) of two *pulses*.

PULSE SPIKE. An unwanted *pulse* of relatively short duration superimposed on the main pulse.

PULSE TILT. See *Pulse flatness deviation*.

PULSE TIME CONSTANT OF FALL (FALL TIME). The time required for the *pulse* to fall from 70·7 per cent to 26 per cent of its maximum amplitude (excluding spike). Compare *Pulse decay time*.

PULSE TIME MODULATION (PTM). *Modulation* in which the values of instantaneous samples of the modulating wave are caused to modulate the time of occurrence of some characteristic of a *pulse*. *Pulse position modulation* and *pulse width modulation* are particular cases of ptm.

PULSE TIME MULTIPLEX. Any system of *multiplex* operation which combines *time-division multiplex* and *pulse modulation*.

PULSE TRAILING EDGE. The major portion of the decay of a *pulse*.

PULSE TRAIN. A group of *pulses* of similar characteristics.

PULSE TRANSFORMER. A *transformer* specially designed to operate over the wide range of frequencies involved in *pulse* transmission.

PULSE VALLEY. The part of a *pulse* between two specified maxima.

PULSE WIDTH MODULATION. *Pulse time modulation* in which the value of each instantaneous sample of the modulating wave is caused to vary the width or duration of the *pulse*.

PULSE-FORMING LINE. An *artificial line,* consisting of *lumped inductances, capacitances,* etc., in tandem, used to form high-voltage short-duration pulses in *radar modulators*.

PULSER. A generator used to produce high-voltage, short-duration pulses.

PUMPED VALVE (TUBE). A valve or tube which is continuously evacuated during operation.

PUNCHED TAPE. An input medium comprising a paper or plastic tape in which coded instructions in the form of holes can be punched to store and transmit information. Standard punched tape is 1 inch in width and can accommodate eight tracks.

PUNCH-THROUGH VOLTAGE. Of a *transistor*: the value of the collector-base voltage above which the emitter-base voltage increases almost linearly with increasing collector-base voltage.

PUSH-DOWN STORE (NESTING STORE). In *data processing*: a store which works as though it comprised a number of registers arranged in a column, with only the register at the top of the column connected to the rest of the system.

PUSH-PULL AMPLIFIER. An *amplifier* in which two identical signal branches operate in *phase* opposition and input and output connections are each balanced to ground.

PUSH-PULL CIRCUIT. A circuit containing two like elements which operate in 180 degrees *phase* relation to produce additive output components of the desired wave and cancellation of certain unwanted products. See *Push-pull amplifier*.

PUSH-PUSH AMPLIFIER. An *amplifier* which employs two similar valves or transistors, with their grids connected in phase-opposition and their anodes connected in parallel to a common load. It is used to emphasize even-order harmonics.

PYROELECTRIC EFFECT. Certain crystals develop electric charges when they are heated or cooled unequally.

Q

Q. A very widely used symbol in electronics for the quality factor or 'goodness' of a circuit, circuit element, material, component, or device. It is a measure of the relation between the stored energy and the rate of dissipation of energy in those structures and is often used as a *parameter* for *inductors, capacitors, tuned circuits, cavity resonators,* and *dielectrics.*

Q, CAPACITOR. See *Q, reactor.*

Q, DAMPING FACTOR. Of a *series resonant* circuit, a quality factor expressed as the rate of decay of oscillations due to loss of energy in the series resistance. See *Damping.*

Q, DIELECTRIC. Of a *dielectric* material, the ratio of the *displacement current* density to the *conduction current* density.

Q, INDUCTOR. See *Q, reactor.*

Q, LOADED. Of an *impedance*, the value of the *Q* of the impedance when connected or coupled under working conditions.

Q, NON-LOADED. The value of *Q* of an *impedance* without external connection or coupling.

Q, PARALLEL CIRCUIT. The quality factor of a *parallel* circuit which expresses the energy stored as a fraction of the energy wasted.

Q, PHASE ANGLE. The *Q* of a *series resonant* circuit expressed in terms of the *phase angle* between the current and voltage in the circuit.

Q, POWER FACTOR. The *Q* of a *series resonant* circuit expressed in terms of the *power factor* of the circuit.

Q, REACTOR. At any frequency, the ratio of the *reactance* to the *effective series resistance* of the *reactor* at that frequency.

Q, SELECTIVITY. A measure of the *selectivity* of a *series resonant* circuit, based on the power dissipated at two selected frequencies. These usually correspond to the half-power points on the *resonance curve* compared to the power at resonance.

Q-METER. An instrument which measures the *Q* of a circuit or component. It usually includes a means for making the measured component a part of a *resonant* circuit, a variable *oscillator* for injecting a signal of suitable frequency into this circuit, and a *valve voltmeter* for measuring the voltage multiplication at resonance.

QAVC. Abbreviation for *quiet automatic volume control.*

Q-POINT. See *Quiescent point.*

QUAD. Of cables, a method of construction in which four separately insulated conductors are twisted together.

QUADRANT ELECTROMETER. A type of *electrometer* consisting of a flat cylindrical metal box in the form of four individually isolated quadrants, inside which a thin metal vane is suspended horizontally by a fine wire. Electric *charges* can be measured by the deflection of the vane when a charge is applied between it and the quadrants.

QUADRATURE. The *phase* relation between two *periodic* quantities of the same period when the phase difference between them is one quarter of a period (usually 90 degrees).

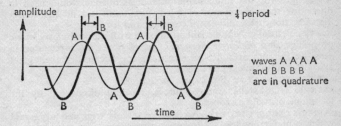

Fig. 118. Quadrature

QUADRATURE COMPONENT. The *reactive* component of a current or voltage due to *inductive* or *capacitive reactance* in a circuit.

QUADRUPOLE AMPLIFIER. A low-noise *parametric amplifier* using an electron beam tube in which a fast beam is subjected to the influence of a 4-pole field.

QUALITY CONTROL. In large-scale production, e.g. of electronic equipment, a system of manufacture and inspection based on scientific sampling and checking at every stage in the manufacturing process, which ensures that the finished product complies with all specifications.

QUANTICON. A photoconductive *camera tube* made in Czechoslovakia.

QUANTIZATION. In communications, a process by which the range of values of a wave is divided into a finite number of smaller sub-ranges each of which is represented by an assigned or 'quantized' value within the sub-range.

QUANTIZATION DISTORTION (NOISE). In communications, the inherent distortion involved in the process of *quantization*.

QUANTUM. If an observable quantity is always a multiple of a definite unit, then that unit is called the quantum of the quantity. See *Planck's constant, Quantum theory*.

QUANTUM EFFICIENCY, PHOTOTUBE. See *Quantum yield*.

QUANTUM NUMBER. The number of *quantums* of a quantity. It normally takes the form of an integer or an odd half-integer.

QUANTUM THEORY. According to Planck's quantum theory, energy exists in individual packets – quanta – of radiation and each packet contains an amount of energy equal to hv, where h is *Planck's constant* and v is the frequency of the radiation.

QUANTUM YIELD. (1) Generally the number of reactions of a specific type induced by *photons,* per photon absorbed. (2) Of a *phototube,* the average number of electrons emitted photoelectrically from a *photocathode* per incident photon of a given wavelength.

QUARTER-WAVE LINE. A section of *transmission line* one quarter of a wavelength (of the fundamental design frequency) long, used in filters for the suppression of even *harmonics* and in *impedance-matching* networks, and *aerial* coupling and feed.

QUARTZ. Natural crystalline silicon dioxide with marked *piezoelectric* properties, and *dielectric strength.* Quartz fibres are often used in sensitive instruments, such as electrometers, because of their great elasticity and physical and chemical stability.

QUARTZ OSCILLATOR. See *Crystal oscillator.*

QUASARS. Quasi-stellar radio sources, recently discovered by the technique of radio astronomy.

QUASI-BISTABLE CIRCUIT. An *astable* circuit triggered at a rate which is high compared to its natural frequency.

QUENCH FREQUENCY. The number of times per second a circuit is made to start and stop *oscillation.*

QUENCHED-SPARK CONVERTER. A generator of power source which uses the oscillatory discharge of a *capacitor* through an *inductor* and a *spark-gap* as a source of radio-frequency power.

QUENCHING. The process of terminating a discharge in a *Geiger counter* by inhibiting *re-ignition.*

QUENCHING CIRCUIT. A circuit which reduces or reverses the voltage applied to a *counter tube* so as to inhibit multiple discharge from a single *ionizing event.*

QUIESCENT CURRENT. The current flowing to or from an electrode under specified normal conditions, in the absence of a signal applied to the input electrode.

QUIESCENT PERIOD. The period between *pulse* transmissions.

QUIESCENT POINT. The condition, or region on an operating characteristic, where a valve or other circuit element is not performing its active function in the circuit.

QUIESCENT PUSH-PULL AMPLIFIER. A *push-pull amplifier* in which the control grids are so biased that very little anode current flows when there is no signal.

QUIESCENT-CARRIER TELEPHONY. *Carrier telephony* in which the carrier is suppressed when no *modulating* signals are being transmitted.

QUIET AUTOMATIC VOLUME CONTROL. *Automatic volume control* designed to operate only for signal strengths exceeding a certain value, so that *noise* (or other weak signal) is suppressed.

QUIETING SENSITIVITY. See *Receiver quieting sensitivity.*

R

RADAC. RApid Digital Automatic Computing system for fast, accurate analysis of complex data, such as that required for fire control against rockets and missiles.

RADAN. A RAdar Doppler Automatic Navigation system which is independent of ground-based equipment.

RADAR. RAdio Detection And Ranging, originally designed to detect and find the range of moving objects by transmitting a beam of radio-frequency energy in the general direction of the objects and measuring the time taken for the reflected part of the energy (the echo) to return to the source of transmission. This technique has now been enormously improved and extended to include the location, measurement, and recording of the position, velocity, shape, size, and nature of stationary and moving objects; and for the navigation and guidance of ships, aircraft, moving vehicles, missiles, and artificial satellites.

RADAR, AIRBORNE. A *radar* set carried by an aircraft to provide information about the relative position of fixed identification points or of other aircraft.

RADAR, AIRPORT SURVEILLANCE (ASR). A *radar system* used to detect aircraft within a certain radius of an airfield and to present continuously to the operator information about distance and azimuth, but not elevation, of these aircraft. See also *Radar, precision approach.*

RADAR, CONTINUOUS WAVE. A *radar* system in which a transmitter sends out a continuous flow of energy at radio-frequency to the target, which returns a small fraction to the transmitter. Compare *Radar, pulse.*

RADAR, DOPPLER. See *Doppler radar.*

RADAR, EARLY WARNING (EWR). A *radar* system set up near the periphery of a defended area to warn of aircraft or other objects approaching the area.

RADAR, FIRE-CONTROL. A radar used for directing gunfire against targets which it observes.

RADAR, FREQUENCY MODULATED. See *Frequency modulated radar.*

RADAR, GROUND SURVEILLANCE. A *radar* set operated at a fixed point for observation and control of the position of aircraft or other vehicles in the vicinity.

RADAR, PRECISION APPROACH (PAR). A *radar* system used as an aid to the control of air traffic which presents to the controller accurate information about the location of incoming aircraft.

RADAR, PRIMARY. *Radar* in which the signals are broadcast and reflected from a target. Compare *Radar, secondary.*

RADAR, PULSE. *Radar* in which individual sharp bursts of energy are sent out from a transmitter and individual reflected echoes are detected by a receiver in the intervals between transmitted pulses.

RADAR, SEARCH. A *radar* primarily intended to indicate targets as soon as possible after they enter the surveillance area.

RADAR, SECONDARY. A *radar* in which the received *pulses* have been transmitted by a *responder* triggered by a *primary radar*.

RADAR, V-BEAM. A *volumetric radar system* for the determination of distance, height, and bearing, using two fan-shaped beams.

RADAR, VOLUMETRIC. A *radar system* capable of producing three-dimensional position data on a number of targets.

RADAR ALTIMETER. An *altimeter* which indicates the true altitude of an aircraft by measuring the time for a radio-frequency *pulse* to travel from the aircraft to the ground and back.

RADAR BAND. See *Frequency band* and Fig. 57.

RADAR BEACON. See *Beacon, radar*.

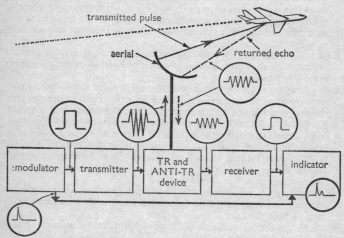

Fig. 119. Radar system: functional diagram of a pulse-modulated radar system showing pulse shapes at different points

RADAR BOMBSIGHT. An airborne *radar* set used to find the target, solve the bombing problem, and drop the bombs.

RADAR CAMERA. A special camera used for photographing the images on a radarscope.

RADAR CONTROL. Control of an aircraft, guided missile, gun battery, or any other system by *radar*.

RADAR ECHO. The returned *radar* signal after reflection from the target.

RADAR EQUATION. The equation for the signal power received from a reflected *radar* signal.

RADAR FENCE. A network of radar warning stations round a protected area.

RADAR HOMING. (1) Homing on the source of a radar beam. (2) Guiding a missile to a target by locking a *radar* in the missile on to that target.

RADAR ILLUMINATION. Illumination of a target by a high-powered radar transmitter which is independent of any missile seeking the target.

RADAR INDICATOR. A *cathode-ray tube* used to provide a visual presentation of the echoes returned from a target.

RADAR PERFORMANCE FIGURE. The ratio of the *pulse* power of the *radar* transmitter to the power of the minimum signal which can be detected by the receiver.

RADAR PHOTOGRAPH. A photograph of the images on a *radarscope*.

RADAR PLOT. A plan showing the positions of aircraft or ships made from data obtained by *radar*.

RADAR POWER, PEAK. The maximum power of a *radar* pulse, usually, a much greater value than the average power.

RADAR PULSE. A *pulse* of r-f energy used in *radar*.

RADAR RANGE. The maximum distance at which a *radar* set is normally effective in detecting objects; more specifically the distance at which it can detect a specified object at least half the time.

RADAR RECEIVER. A device for receiving and amplifying *radio* and *radar* waves and passing them to an indicator.

RADAR RELAY. An equipment for relaying the *radar video* and *synchronizing* signals to a remote location.

RADAR RESOLUTION. The ability of a *radar* to differentiate targets solely by distance measurements, generally expressed as the minimum radial distance separating targets at which they can be separately distinguished.

RADAR SCAN. The motion of a radar beam in space while searching for a target; also the path described and the process of directing the beam. See Figure 120.

RADAR SCREEN. A *cathode-ray* screen in a *radar* set.

RADAR SET. See *Radar system*.

RADAR STORM DETECTION. Storms accompanied by precipitation may be detected by the reflection of *radar* signals from liquid or frozen water drops in the storm area.

RADAR SYSTEM. A system of devices which make use of the *radar* principle. The essential components are: a generator of *microwave* signals such as a *magnetron oscillator*; a *modulator* which *pulse-modulates* the magnetron so that bursts of r-f energy are fed to the radar aerial; the aerial which transmits the pulses and receives reflected pulses; a *local oscillator* which mixes with the received

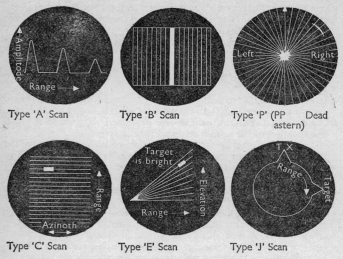

Fig. 120. Some types of radar scan

signals in a *mixer*; a *receiver* to detect and amplify the *heterodyne* output from the mixer; a *cathode-ray tube* indicator which displays the received signals in the correct time sequence; and a *duplexer* to prevent the transmitted signals going directly to the mixer. See Fig. 119.

RADAR TRANSMITTER. The portion of a *radar system* which transmits the pulses.

RADARSCOPE. The *cathode-ray oscilloscope* or screen in a *radar* set.

RADECHON. A *storage tube* in which a fine mesh grid is the controlling element. See Fig. 121.

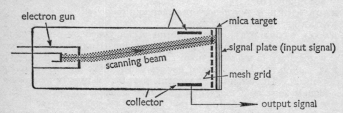

Fig. 121. Principle of radechon storage tube

RADIAC. RAdio-activity Detection, Identification, and Computation.

RADIAC COMPUTER. A computer for processing data from a *radiac*.

RADIAL BEAM TUBE. A vacuum tube in which a radial beam of electrons is rotated past anodes arranged circumferentially by a rotating magnetic field. Used mainly as a high-speed switch or *commutation switch*.

RADIATION. Energy travelling in the form of *electromagnetic waves*, e.g. radio, radar, infra-red, and light waves. By extension, corpuscular emission of, for example, alpha, beta, or neutron radiation. See *Frequency spectrum of electromagnetic waves*.

RADIATION, ANNIHILATION. Electromagnetic radiation produced by the union and consequent annihilation of e.g. an *electron* and a *positron*.

RADIATION, CERENKOV. The visible light produced when charged particles, e.g. a beam of electrons, travel through a transparent medium with velocities exceeding that of light in the medium.

RADIATION, CHARACTERISTIC. Radiation which follows removal of an electron from an atom and which results from an electron of higher energy level dropping to a lower energy level to fill the vacancy.

RADIATION, SECONDARY. Particles or *photons* produced by the interaction between matter and the radiation regarded as primary, e.g. *Compton recoil electrons*, *delta rays*.

RADIATION COUNTER. An instrument for detecting or measuring radiation by a counting process.

RADIATION EFFICIENCY. Of an aerial, the ratio of the power radiated to the total power supplied to the aerial at a given frequency.

RADIATION LOSS. That part of the *transmission loss,* due to radiation, of radio-frequency power from a transmission system.

RADIATION POTENTIAL. The potential difference in volts, or the corresponding energy in electron volts, necessary to cause an atom to emit radiation of a frequency characteristic of the material.

RADIATION PYROMETER (THERMOMETER). An instrument which makes use of the heating effect of radiant power impinging on a detector such as a thermopile, bolometer, or thermocouple, to measure the temperature of the radiant source. For visible radiation a *photoelectric* device may be used.

RADIATION RESISTANCE. Of an aerial, the power radiated by the aerial divided by the square of the effective aerial current.

RADIATION THERAPY. Treatment of disease with any type of radiation.

RADIO. The use of *electromagnetic waves* to transmit or receive electric signals or impulses without connecting wires; also, the *transmission* and *reception* of such signals or impulses. In this broad sense 'radio' includes *radar* and *television*; the chief distinctions are to be found in the methods of transmission and reception, and the frequencies normally used.

RADIO ALTIMETER. A *radar altimeter*.

RADIO ASTRONOMY. The study of astronomical bodies and events by the radio waves which they emit.

RADIO BEACON. See *Beacon, radio*.

RADIO BEAM. A radio wave in which most of the energy is confined within a relatively small angle in at least one plane.

RADIO BOMB. A bomb provided with a radio-controlled time or proximity fuse.

RADIO BROADCASTING. *Radio transmission* intended for general reception. Allocation of frequency channels for broadcasting, which are normally about 10 kilocycles apart, is decided by international agreement, but standards of transmission are usually dealt with on a national basis.

RADIO CHANNEL. Any channel suitable for transmission of radio signals.

RADIO CIRCUIT. A radio system for carrying out one communication at a time in either direction, or in both directions, between two points.

RADIO COMMUNICATION. Communication by radio waves which are not guided by wires or waveguides between transmitter and receiver. This definition includes communication by *radiofacsimile, radiotelegraphy, radiotelephone,* and *television*.

RADIO COMPASS. A radio receiver which is mounted on a ship together with a directional aerial, and displays its heading with respect to a radio transmitter.

RADIO CONTROL. The control of any device or mechanism by radio waves.

RADIO FREQUENCY. A frequency useful for *radio transmission*. The present practical limits are roughly 10 kilocycles per second to 100,000 megacycles per second.

RADIO HORIZON. In propagation of *radio waves* over the earth, the line which bounds that part of the earth's surface reached by *direct rays*.

RADIO INTERFERENCE. Any undesired disturbance in radio reception, whether it be in the transmitter, transmission medium, or receiver. Also the cause of that disturbance. See *Cross talk, Interference, Man-made noise*.

RADIO LINK. Any part of a system of *telecommunication* in which the intelligence is transmitted by *radio* as distinct from other parts which depend on wires or cables.

RADIO PILL. A capsule containing a miniature radio transmitter designed to respond to conditions in the human digestive system. It may be swallowed and recovered after passing through the alimentary tract.

RADIO PROXIMITY FUSE. A radio device contained in a missile to detonate it within predetermined limits of distance from a target by means of electromagnetic interaction with the target.

RADIO RANGE. A radio facility which provides radial position lines for the lateral guidance of aircraft, by the use of special transmitted characteristics recognizable as bearing information.

RADIO RECEIVER. A device for converting *radio waves* into perceptible signals.

RADIO RELAY SYSTEM. A point-to-point radio transmission system in which signals are received and re-transmitted by one or more intermediate radio stations.

RADIO SET. A *radio receiver*; sometimes a combined transmitter and receiver.

RADIO SILENCE, INTERNATIONAL. A three-minute period set aside at 15 and 45 minutes past the hour for listening on the distress call frequency of 500 kc/s.

RADIO SPECTRUM. See *Frequency spectrum of electromagnetic waves*.

RADIO STATION INTERFERENCE. Selective *interference* caused by the radio waves from a station or stations other than that from which reception is desired.

RADIO SYSTEM. The various electronic circuits and devices, including aerial systems, required to convey intelligence from one place to another by radio transmission.

RADIO TRANSMISSION. The *transmission* of signals by *electromagnetic waves* other than light and heat waves.

RADIO TRANSMITTER. A device for transmitting *radio-frequency* power.

RADIO WAVE. An *electromagnetic wave* used in *radio*.

RADIOACTIVE ISOTOPE. An *isotope* which decays spontaneously at a definite rate and with the emission of radiation.

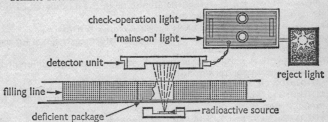

Fig. 122. Radioactive isotope used in a package monitor

RADIOACTIVE STANDARD. A sample of radioactive material, with a precisely known rate of decay, used as a radiation source for calibrating radiation-measuring equipment.

RADIOACTIVE TRACER. A small quantity of *radioactive isotope* in a form suitable for following biological, chemical, or other processes.

RADIOACTIVE VALVE (TUBE). A valve or tube which contains a radioactive material, usually to create an initial *ionization*.

RADIOACTIVITY. Spontaneous disintegration of atomic nuclei accompanied by emission of corpuscular or electromagnetic radiation. To be considered as radioactive, a process must have a measurable lifetime.

RADIO-DOPPLER. The direct determination of the radial component of the relative velocity of an object by an observed change in the frequency of radio waves due to such velocity. See *Doppler effect*, *Doppler radar*.

RADIOFACSIMILE. *Facsimile* by *radio*.

RADIO-FREQUENCY ALTERNATOR. A rotating generator for producing *radio-frequency* power.

RADIO-FREQUENCY AMPLIFIER. An *amplifier*, usually tuned, for *radio-frequency* waves.

RADIO-FREQUENCY CHOKE. An *inductor* used to impede the flow of *radio-frequency* currents, without any appreciable effect on direct or low-frequency currents.

RADIO-FREQUENCY CONVERTER. A power source for production of electrical power at frequencies about 10 kc/s.

RADIO-FREQUENCY GENERATOR. In industrial and *dielectric heating*, an equipment comprising a powerful valve oscillator, amplifier, and control equipment together with a suitable power supply.

RADIO-FREQUENCY HARMONIC. In *radio* or *television*, a *harmonic* of the fundamental frequency of the *carrier* wave.

RADIO-FREQUENCY HEATING. See *Dielectric heating* and *Induction heating*.

RADIO-FREQUENCY PULSE. A *radio-frequency carrier*, *amplitude-modulated* by a *pulse*, so that the amplitude of the modulated carrier is zero before and after the pulse.

RADIO-FREQUENCY RESISTANCE. See *Effective resistance*.

RADIO-FREQUENCY TRANSFORMER. A transformer designed for use with *radio-frequency* currents.

RADIOGRAPH. A photographic image formed as a result of differential absorption of penetrating radiation.

RADIOLOCATION. The original term for *radar*.

RADIOLOGY. The science and applications of radiation applied to *radiotherapy*.

RADIOLUMINESCENCE. Emission of light due to radiations from radioactive substances.

RADIOPHARE. See *Beacon, radio*.

RADIOSCOPE. An *electroscope* used to measure the quantity of a radioactive material.

RADIOSONDE. An automatic radio transmitter carried on a balloon, parachute, kite, or aircraft for the purpose of transmitting meteorological data.

RADIOTELEGRAPHY. *Radio communication* employing coded signals such as International Morse.

RADIOTELEPHONE. A combined *radio transmitter* and *receiver* used for voice communication, sometimes in conjunction with a telephone wire system.

RADIOTHERAPY. *Radiation therapy.*

RADIUM (a). Radioactive metallic element, Atomic No. 88.

RADOME. A protective covering for a radar aerial system which is transparent to the frequency in use.

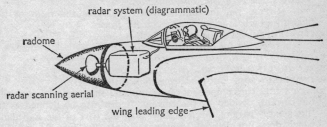

Fig. 123. Radome

RADON (Rn). Inert gaseous element, Atomic No. 86.

RAMAC®. Trade name for a magnetic disc *memory* unit associated with *digital computers* built by the International Business Machines Corporation.

RANDOM ERRORS. Errors which can be predicted only on a statistical basis.

RANDOM EVENTS. Events whose occurrence does not affect in any way whatever any future event.

RANDOM-ACCESS MEMORY. In computer technique, a *memory* designed so that the location or *address* of a given unit of stored information is independent of the information stored.

RANGE TRACKING. The procedure of adjusting the *gate* of a *radar set* so that it opens at the correct instant to accept the signal from a target whose range is changing.

RANGE-HEIGHT INDICATOR. A type of cathode-ray display which indicates the height of a target as seen by a *radar* employing vertical scanning.

RASCAL. A *digital computer* designed by the R.A.F. research establishment at Farnborough.

RASTER. In *television,* a predetermined pattern of *scanning* lines which provides substantially uniform coverage of the area to be televised.

RASTER BURN. In *camera tubes,* a change in the characteristics of that area of the target which has been scanned, resulting in a spurious signal.

RAT RACE. See *Hybrid ring junction.*

RATE CONTROL. Control of the rate of change of the independent variable in an automatic control system.

RATED CURRENT. The designated limit of the current which an electrical device will carry continuously without exceeding the limit of observable temperature rise.

RATING. Of a machine, device, or equipment, a designated limit of operating characteristics, such as *load, voltage, frequency*, based on specified conditions.

RATIO DETECTOR. The *detector* normally used in *frequency modulation* receivers. See *Frequency discriminator*.

RATIO METER. An instrument which measures the ratio of two quantities by balancing electromagnetic forces.

RATIONALIZED SYSTEM OF UNITS. See *Units*.

R-C. Abbreviation for *resistance-capacitance*.

R-C AMPLIFIER. *Resistance-capacitance-coupled amplifier*.

R-C COUPLING. *Resistance-capacitance coupling*.

R-C OSCILLATOR. *Resistance-capacitance oscillator*.

REACTANCE. The component of the applied voltage in *quadrature* with the current, divided by the current. It can also be defined as the 'imaginary' part of the *impedance*, i.e. the part of the impedance which does not dissipate power. Symbol: X. Thus if a complex impedance is given by $Z = R + jX$, where R is the *resistance*, and j a mathematical operator, then X is the reactance.

REACTANCE CHART. A chart in which logarithmic scales are arranged in such a way that it is possible to read directly the *reactance* of a given *inductor* or *capacitor* at any frequency and conversely the inductance or capacitance of a given reactance.

REACTANCE MODULATOR. A device used for *modulation*, whose *reactance* may be varied in accordance with the instantaneous amplitude of the modulating wave applied to it.

REACTANCE VALVE (TUBE). A valve which presents a pure *reactance* to the circuit in which it is connected: i.e. the current through it is 90 degrees out of *phase* with the voltage across it. Reactance valves are used as *modulators* in some systems of *frequency modulation*.

REACTATRON. A low-noise microwave amplifier using a semi-conductor diode.

REACTIVATION. The process of applying an abnormally high voltage to the *thoriated-tungsten filament* of a valve in order to bring a fresh layer of thorium atoms to the surface and so improve the emission.

REACTIVE. Having inductive or capacitive *reactance*.

REACTOR. A device for introducing *reactance* into an electrical circuit. *Capacitors* and *inductors* are particular examples of reactors.

REACTOR, NUCLEAR. See *Nuclear reactor*.

REACTOR, SATURABLE. See *Saturable reactor*.

READ. (1) In *charge-storage tubes*, generate an output corresponding to the stored charge pattern. (2) In *computers*, acquire information usually from some form of *storage*. See *Write*.

READ-AROUND NUMBER. In *charge-storage tubes*, the number of times a particular spot or location may be consulted before *spill* causes a loss of information in surrounding spots.

READING RATE. In *computing* and *data-processing*, the number of *words*, characters, fields, cards, etc., sensed by an input sensing device per unit of time.

READING SPEED. In a *memory*, the rate of reading successive storage elements.

READ-OUT. (1) A display – usually of digits in line – of processed data. (2) In *computers*: to read from the internal store and transfer to an external store.

READ-OUT PULSE. In a *computer memory*, a pulse applied to a number of binary memory cells to find which cell contains a *bit* of information.

REAL-TIME OPERATION. The operation of a computer – especially an *analogue computer* – during the actual time that the related physical processes take place so that the results can be used to guide the physical process. Compare *On-line operation*.

REBECCA-EUREKA. A British homing and navigation *radar* system which was one of the first to use the *radar beacon*.

RECEIVER. (1) The part of a *communication* system in which electric waves or currents are converted into perceptible signals. (2) Any electromechanical *transducer* which converts electrical energy into sound waves. The following types of receiver are described in this dictionary: *ac/dc, amplitude-modulation, beacon, communication, crystal-set, direct-current, facsimile, frequency-modulation, lin-log, panoramic, radar, radio, radio set, regenerative, superheterodyne, superregenerative, television, thermophone, tuned radio-frequency.*

RECEIVER BANDWIDTH. The band of frequency for which the receiver performance for a selected characteristic, e.g. sensitivity, falls within the specified limits.

RECEIVER GATING. The application of operating voltages to one or more stages of a receiver, only during the part of a cycle of operation when reception is required.

RECEIVER MAXIMUM SENSITIVITY. The smallest input signal which will produce a specified output power from the receiver.

RECEIVER NOISE FIGURE. See *Noise figure, receiver*.

RECEIVER PRIMARIES. In a *colour-television receiver*, the colours produced by the receiver, usually red, green, and blue, which produce other colours when mixed in the right proportions.

RECEIVER QUIETING SENSITIVITY. Of an *FM receiver*, the minimum input signal required to give a specified output *signal-to-noise* ratio under specified conditions.

RECIPROCITY THEOREM. In an electrical *network*, if an electromotive force E at one point in a network produces a current I at a second point, then the same voltage E applied at the second point will produce the same current I at the first point.

RECOIL ELECTRON. An electron set in motion by a collision.

RECOMBINATION. The return of an ionized atom or molecule to its electrically neutral state, by the gain of an electron in the case of positive ions and by the loss of an electron for negative ions.

RECOMBINATION RATE. The rate of decrease with time of either positive or negative *ions* in an *ionized* gas.

RECOMBINATION RATE, SURFACE. The rate at which *free electrons* and *holes* recombine within the volume of a *semiconductor*.

RECOMBINATION VELOCITY. On a *semiconductor* surface, the ratio of the normal component of the *electron* (or *hole*) current density at the surface to the excess electron (or hole) charge density at the surface.

RECORDING. (1) The process of registering a receiver signal on a record sheet. (2) The process of making graphical, magnetic, mechanical, or photographic records of currents and voltages in various media.

RECORDING, BLACK. In an *amplitude-modulation* system, the form of recording in which the maximum received power corresponds to the maximum density of the recording medium. In a *frequency-modulated* system, the lowest received frequency corresponds to the maximum density of the recording medium. Compare *Recording, white*.

RECORDING, CONSTANT-AMPLITUDE. Mechanical recording in which, for a fixed signal amplitude, the resulting recorded amplitude is independent of frequency.

RECORDING, CONSTANT-VELOCITY. Mechanical recording in which, for a fixed signal amplitude, the resulting recorded amplitude is inversely proportional to the frequency.

RECORDING, DIRECT. The type of recording in which a visible record is produced in response to the recording signals, without the need for any further processing.

RECORDING, ELECTROCHEMICAL. Recording by means of a chemical reaction brought about by the passage of signal-controlled current through the sensitized portion of the record sheet.

RECORDING, ELECTROMECHANICAL. Recording by means of a signal-actuated mechanical device.

RECORDING, ELECTROSTATIC. Recording by means of a signal-controlled *electrostatic field*.

RECORDING, ELECTROTHERMAL. Recording produced principally by signal-controlled thermal action.

RECORDING, FACSIMILE. In *facsimile* transmission, the process of converting the electrical signal into an image of the recording medium by any of the recording methods listed above.

RECORDING, LATERAL. A *mechanical recording* in which the *modulation* of the recording groove is perpendicular to the motion of the recording medium and parallel to the surface of the medium. See *Recording, vertical*.

RECORDING, MAGNETIC. See *Magnetic recorder, Magnetic recording medium*.

RECORDING, MECHANICAL. See *Recording, electromechanical.*

RECORDING, PHOTOGRAPHIC (PHOTOSENSITIVE). Recording by the exposure of a photosensitive surface to a signal-controlled light beam or spot.

RECORDING, VERTICAL. A *mechanical recording* in which the *modulation* of the groove is perpendicular to the surface of the recording medium. Also called *hill-and-dale recording.*

RECORDING, WHITE. In an *amplitude-modulation system,* that form of recording in which the maximum received power corresponds to the minimum density of the medium. In *frequency modulation,* the lowest received frequency corresponds to the minimum density of the recording medium.

RECORDING CHANNEL. One of a number of independent recorders in a recording system, or independent tracks on a recording medium.

RECORDING HEAD. See *Magnetic head.*

RECORDING INSTRUMENT. An instrument for the storage of information concerning the relation between variables.

RECORDING SPOT. In *facsimile* transmission, the image area formed at the recording medium by the *facsimile recorder.*

RECORDING STYLUS. A tool which inscribes the groove in a recording medium.

RECOVERY TIME. (1) In a *counter,* the minimum time interval between two distinct events which will allow both to be counted. (2) In a radar *duplexer*: the minimum time after a transmitted radar *pulse* before an echo can be detected. (3) In a *gas tube,* the time required for the control electrode to regain control after interruption of anode current.

RECTANGULAR HYSTERESIS LOOP. A *hysteresis loop* in which the saturated parts of the curve are nearly vertical and the unsaturated parts nearly horizontal. This is an important characteristic of *ferrites.* See *Switching time, ferromagnetic.*

RECTANGULAR PULSE. A *pulse* whose waveshape is rectangular in profile. See *Square wave.*

RECTIFICATION. The conversion of *alternating current* into *unidirectional* current by a *rectifier.* See *Full-wave rectification, Half-wave rectification.*

RECTIFICATION EFFICIENCY. The ratio of the direct-current power output to the alternation-current power input of a *rectifier.*

RECTIFIER. A device for converting an *alternating* or *oscillating current* into a unidirectional current, by the inversion or suppression of alternate half-waves. See the following types of rectifier: *Barrier-layer, Bridge, Caesium-vapour, Copper-oxide, Crystal, Diode, Electrolytic, Full-wave, Half-wave, Ignitron, Mercury-arc, Selenium, Silicon, Tantalum, Voltage doubler, Voltage multiplier,* and also the entries below.

General symbol for rectifier equipment

RECTIFIER, CAESIUM-VAPOUR. A *rectifier* with

a single anode using caesium vapour as the conducting gas.

RECTIFIER, CONTACT. A *rectifier* consisting of two different solids in contact, in which rectification is due to greater *conductivity* across the contact in one direction than in the others.

RECTIFIER, DRY-DISC. A *rectifier* made of a series of discs of metal, metal oxides, or other compounds. The discs are drilled through their centres and locked in contact along a stud or rod.

RECTIFIER, JUNCTION. A *rectifier* employing a *semiconductor diode*, e.g. *copper-oxide* or *selenium rectifier*.

RECTIFIER, LINEAR. A *rectifier* with a straight-line conduction characteristic in one direction and zero conduction in the other.

RECTIFIER, METAL(LIC). A combination of metals or metallic compounds which has the property of unidirectional conduction and may be used for *rectification*. See *Rectifier, dry-disc*, *Selenium rectifier*, *Silicon rectifier*.

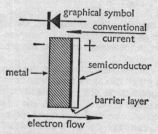

Fig. 124. Elements of a barrier-layer metal rectifier

RECTIFIER, SINGLE-PHASE. A *rectifier* which operates from a *single-phase* ac supply voltage.

RECTIFIER, THREE-PHASE. A *rectifier* which operates from a *three-phase* ac supply voltage.

RECTIFIER, VALVE (VACUUM TUBE). A *rectifier* in which *rectification* is accomplished by the unidirectional flow of electrons from a heated electrode in an evacuated envelope.

RECTIFIER CATHODE. The electrode of a gas *rectifier tube* into which current flows from the *arc*. It is usually a pool of mercury and constitutes the *positive* direct-current terminal.

RECTIFIER FILTER. A filter designed to reduce the voltage fluctuations from a rectifier. It is usually inserted between the rectifier and its load resistance.

RECTIFIER INSTRUMENT. An instrument for measuring *alternating current* and voltage, combining an instrument sensitive to *direct current*, e.g. a moving coil meter, and a *rectifier*.

312

RECTIFIER LEAKAGE CURRENT. Alternating current which passes through a *rectifier* without being rectified.

RECTIFIER RIPPLE FACTOR. The ratio (in per cent) of the *RMS value* of the ac component of the rectifier load voltage to the algebraic average of the total voltage across the load.

RECTIFIER TRANSFORMER. A *transformer* whose *secondary winding* supplies energy to the main anodes of a *rectifier*.

RECTIFIER UNIT. An equipment comprising a *rectifier* with its essential auxiliaries and the rectifier transformer equipment.

RECTILINEAR SCANNING. In *television*, the process of *scanning* an area in a predetermined sequence of narrow, straight, parallel strips.

RED-TAPE OPERATION. In a computer, internal operations in the computer which do not contribute directly to the result.

REDOX SYSTEM. A chemical system of oxidation and reduction which makes use of the potential generated at an electrode of noble metal, e.g. platinum, in a suitable electrolyte, as the working basis of the *solion*.

REDUNDANCY. In the transmission of information, the fraction of the gross *information content* which can be eliminated without loss of essential information.

REDUNDANT CHECK. A check using extra digits to help detect mistakes in the operation of *computers*. See *Forbidden-combination check*, *Summation check*.

REED RELAY. A switch comprising (typically) two magnetic tongues with electroplated gold contacts, sealed into a glass tube filled with protective gases. It is used in increasing numbers in electronic telephone exchanges. See Fig. 125.

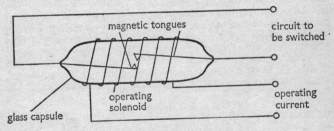

Fig. 125. Principle of dry reed relay

REFERENCE BLACK LEVEL. In *television,* the level at the point of observation corresponding to the specified maximum excursion of the picture signal in the black direction. See *Black level*.

REFERENCE NOISE. See *Noise, reference*.

REFERENCE TUBE. See *Voltage-reference tube*.

REFERENCE WHITE LEVEL. Substitute 'white' for 'black' in *reference black level*.

REFLECTED POWER. Power flowing from the load back to the generator.

REFLECTED-BEAM KINESCOPE. A *cathode-ray tube* in which the electron beam is reflected back from the transparent glass front to the inside rear curved surface to provide a deflection angle of 180°.

REFLECTION COLOUR TUBE. A *colour picture tube* which produces an image by the technique of electron reflection in the screen region.

REFLECTION ERROR. In radio navigation, the error due to the presence of wave energy reaching the receiver by undesired reflections.

REFLECTION FACTOR. Of an electric circuit, the ratio of a current delivered to a load whose *impedance* is not matched to the source, to the current that would be delivered to a load of matched impedance.

REFLECTOMETER. An instrument which measures the ratio of the reflected electromagnetic energy to the incident energy at a reflecting surface or discontinuity.

REFLEX BUNCHING. *Bunching* occurring in an electron stream which has been made to reverse its direction in the *drift space*. See *Reflex klystron*.

REFLEX CIRCUIT. A circuit through which a signal passes for amplification both before and after a change in its frequency.

REFLEX KLYSTRON. A *klystron* which employs a reflector electrode (repeller) in place of a second cavity, to re-direct *velocity-modulated* electrons through the resonant cavity which produced the modulation. By varying the position of the repeller the frequency of oscillation can be controlled. See Figure 126.

REGENERATION. (1) In an *amplifier*, the process by which a part of the power in the output circuit is made to react on the input circuit so as to reinforce the initial power, and thus increase the final amplification. See *Feedback, positive*. (2) In *charge-storage tubes*, replacing the charge to compensate for loss of charge by decay and reading. (3) In computers, the process of restoring a *storage* device, which may have deteriorated, to its latest undeteriorated state. See *Rewrite*.

REGENERATIVE AMPLIFIER. See *Regeneration*.

REGENERATIVE DIVIDER. A *frequency divider* which employs *modulation, amplification*, and selective *feedback* to produce the output.

REGENERATIVE RECEIVER. An *amplitude-modulation* radio *receiver* which employs *regeneration* to increase its sensitivity. See *Super-regenerative receiver*.

REGISTER. (1) An electromechanical recording device which marks a paper tape in accordance with electrical impulses received from transmitting circuits. The marks may be made by inking or by cutting holes. (2) In *computers*, a device capable of retaining information,

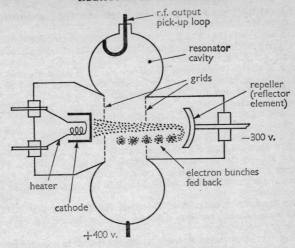

Fig. 126. Diagram of reflex klystron

often that contained in a small subset (e.g. one *word*) of the aggregate information. See *Delay-time register*, *Mechanical register*, *Shift register*, *Storage*.

REGISTER LENGTH. The number of characters which a *register* can store.

REGULATION. In *glow discharge* or *gas tubes*, the difference between the maximum and minimum tube voltage drops for a specified range of anode current.

RE-IGNITION VOLTAGE. The voltage required to re-establish the discharge in a gas discharge tube if applied during a period of *de-ionization*.

REJECTION BAND. In a *transducer*, a band of frequencies which are attenuated or eliminated.

RELATIVE ADDRESS. In a *digital computer*, a label used to identify a *word* in a *routine* or *sub-routine* with respect to its position in that routine or sub-routine. See *Absolute address*.

RELATIVE CODING. In *digital computers*, *coding* in which *relative addresses* are used.

RELATIVISTIC PARTICLE. A particle whose velocity is so large, i.e. exceeding one tenth of the velocity of light, that its mass in motion is significantly greater than its rest mass. This description often applies to an *electron*.

RELAXATION INVERTER. An *inverter* which uses a *relaxation oscillator* to convert dc power to ac power.

RELAXATION OSCILLATOR. A device which generates non-sinusoidal waves by gradually charging and then quickly discharging a *capacitor* or an *inductor* through a *resistor*.

RELAY. A device by means of which one circuit is indirectly controlled by a change in the same or in another circuit. An electrical relay normally includes an *electromagnet* and an *armature* to open and close contacts. In a different sense it is the process by which a signal, especially an electronic signal, is passed on. See *Radio relay system*.

symbol

RELAY, ALLSTROM. A relay which employs a light beam and a photocell to obtain a high degree of sensitivity.

RELAY, CONTROL. A *relay* which functions so as to initiate or permit the next desired operation in a control circuit or scheme.

RELAY, CURRENT. A *relay* which functions at a predetermined value of current.

RELAY, DIFFERENTIAL. A *relay* which functions by reason of the difference between two quantities of the same nature, e.g. current or voltage.

RELAY, DIRECTIONAL. A *relay* which functions in accordance with the direction of current, power, phase-rotation, voltage, etc.

RELAY, ELECTROMAGNETIC. An electromagnetically operated switch, normally composed of one or more coils which control one or more armatures, each of which actuates electric contacts. See *Solenoid* and Fig. 147.

RELAY, FREQUENCY. A *relay* which operates with a selected change of its supply frequency.

RELAY, FREQUENCY-SELECTIVE. A *relay* which operates only when excited with a supply of a predetermined frequency.

RELAY, LIGHT. A circuit in which a phototube is used to trigger a mercury-vapour tube which in turn operates the relay.

RELAY, LOCKING. A *relay* which renders some other relay or device inoperative under predetermined conditions.

RELAY, POLARIZED. A *relay* in which the movement of the armature depends on the direction of the current in the armature control circuit. See *Relay, directional*.

RELAY, RADAR. See *Radar relay*.

RELAY, RESET. A *relay* whose contacts must be reset electrically to their original positions, following an operation.

RELAY, SELF-LOCKING. A relay which, having been operated by another relay or device, remains operated when the other relay or device no longer provides the excitation for it.

RELAY, SLOW-OPERATING. A *relay* which has an intentional delay between energizing and operation.

RELAY, TIME-DELAY. See *Time-delay relay*.

RELAY, VOLTAGE. A *relay* which functions at a predetermined value of voltage.

RELUCTANCE. The ratio of *magnetic force* to *magnetic flux* in a magnetic circuit. The reciprocal of *permeance*.

RELUCTIVITY. The reciprocal of *permeability*.

REMANENCE. The *magnetic induction* which remains in a magnetic circuit after removal of an applied *magnetomotive force*.

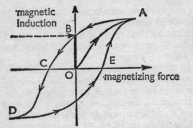

one complete cycle of magnetization O A B C D E starting from unmagnetized condition at O

O B is the value of remanence

Fig. 127. Remanence

REMITRON. A *gas tube* used for counting in computers.

REMOTE CONTROL. A system of control in which the control is exercised from a distance, especially by electrical or electronic apparatus.

REMOTE CUT-OFF VALVE. See *Variable-mu valve*.

REPEATABILITY. The measure of a meter's ability to provide repeated readings with the application of a given energy.

REPEATER. An assembly of apparatus or a device for receiving either one-way or two-way signals and delivering corresponding signals which are amplified, or reshaped, or both.

REPEATER, CARRIER. A *repeater* used in *carrier* transmission.

REPEATER, FOUR-WIRE. A *telephone repeater* used in a *four-wire circuit*.

REPEATER, PULSE. A device used for receiving *pulses* from one circuit and transmitting corresponding pulses to another circuit.

REPEATER, REGENERATIVE. A *repeater* which performs *pulse regeneration*.

REPEATER, TERMINAL. A *repeater* used at the end of a trunk or line.

REPEATER, TWO-WIRE. A *telephone repeater* used in a two-wire circuit.

REPEATING COIL. An *audio-frequency transformer* used to connect two sections of telephone line.

REPELLER. See *Reflex klystron*.

RESET. (1) Restore a *storage* device to a prescribed state. (2) Place a binary cell in the initial or 'zero' state.

RESET CONTROL CIRCUIT. In *magnetic amplifiers*, the *saturable reactor* winding, signal circuit, and any other elements used in resetting the core flux to a given level.

RESET RATE. The number of corrections per minute made by a control system.

RESIDUAL CURRENT. The current flowing through a *thermionic diode* when there is zero anode voltage, due to the velocity of electrons emitted by the *cathode*.

RESIDUAL MAGNETISM. The magnetism remaining in a substance after the *magnetizing force* has been removed.

RESIDUAL RESISTANCE. Inherent *resistance* of a *conductor* which does not change with variations in temperature, ascribed to irregularities in the molecular structure.

RESISTANCE. The tendency of all materials (except *superconductors*) to resist the flow of an electric *current* and to convert electrical energy into heat. Its magnitude depends on many factors which include: the nature of the material; its physical state, dimensions, temperature, and thermal properties; the frequency of the current and often its magnitude; the presence of impurities; the condition of the surface and sometimes the illumination. The rate of conversion of energy (power) when a current passes through a resistance is equal to the product of the square of the current and the resistance. The term is commonly used in connection with circuits in which the resistance is independent of the current. See *Ohm's law*. The practical unit is the *ohm*. Symbol: *R*. See also *Conductance, Impedance, Resistivity*.

RESISTANCE, FOUR-TERMINAL. See *Four-terminal resistor*.

RESISTANCE, HIGH FREQUENCY. See *Skin effect*.

RESISTANCE BOX. A laboratory instrument for providing an accurately calibrated range of resistances in a compact form. It usually consists of a number of accurately made *wire-wound resistors* with provision for connecting them cumulatively in series so that the resistance presented by the terminals of the box can be adjusted in integral steps. See *Decade box*.

RESISTANCE BRAZING. An electric brazing process in which the heat is obtained by resistance to the flow of a current.

RESISTANCE COUPLING. See *Resistive coupling*.

RESISTANCE FURNACE. A furnace in which the heat is generated by the internal ohmic resistance of the furnace.

RESISTANCE LAMP. An electric lamp used to prevent the current in a circuit from exceeding a desired limit.

RESISTANCE STANDARD. See *Standard resistor*.

RESISTANCE TEMPERATURE COEFFICIENT. See *Temperature coefficient of resistance*.

RESISTANCE THERMOMETER. An electrical thermometer which utilizes the change in *resistance* of a conductor with temperature to measure the temperature of the environment in which it is placed.

RESISTANCE WELDING. Welding in which the metals to be joined are melted locally at the welds by passing large electric currents through them for short periods, usually with electronic control.

RESISTANCE WIRE. Wire made from a metal of high *resistivity* and low *temperature coefficient* of *resistance*, e.g. nichrome and *constantan*.

RESISTANCE-CAPACITANCE-COUPLED AMPLIFIER. An amplifier in which the *coupling* between stages is effected by *resistance* and *capacitance*.

RESISTANCE-CAPACITANCE COUPLING (R-C COUPLING). *Coupling* between two or more circuits, usually *amplifiers*, by a combination of *resistive* and *capacitive* elements.

RESISTANCE-CAPACITANCE (R-C) FILTER. A filter made up entirely of series resistance and shunt capacitance elements.

RESISTANCE-CAPACITANCE OSCILLATOR (R-C OSCILLATOR). Any *oscillator* in which the frequency of oscillation is determined by *resistance* and *capacitance*.

RESISTIVE COMPONENT. Of an *impedance*, the part which dissipates power as heat.

RESISTIVE COUPLING. The association of two or more circuits mainly by means of *resistance* common to the circuits.

RESISTIVITY (SPECIFIC RESISTANCE). The reciprocal of the *conductivity* of any material. The *volume resistivity* of a substance is the *resistance* of a piece of unit cross section and unit length at $0°$ C. *Surface resistivity* is the *resistance* of unit length and unit width of a surface. Symbol: ρ. Unit: ohm-centimetre.

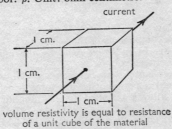

volume resistivity is equal to resistance
of a unit cube of the material

Fig. 128. Resistivity

RESISTOR. A device used primarily for its *resistance*. See the following types of resistor: *Bias, Bifilar winding, Boro-carbon, Carbon, Carbon-film, Composition, Dropping, Fixed, Flexible, Four-terminal, Grid,* symbol *Instrument shunt, Metal-film, Non-inductive, Galvano-meter shunt, Standard, Variable, Wire-wound.*

RESISTOR COLOUR CODE. See *Colour code.*

RESNATRON. A water-cooled *tetrode* valve designed to generate large amounts of *ultra-high frequency* energy. It contains two *resonant cavities*: one between the cathode and control grid, and the other between an accelerating grid and anode.

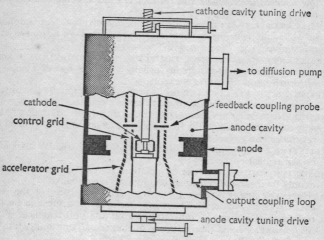

Fig. 129. Construction of resnatron

RESOLVER. A device whose input is a vector quantity and whose outputs are the components of the vector.

RESOLVING TIME. In *radiation counters*, the minimum achievable pulse spacing between counts.

RESONANCE. Resonance exists between a system which is oscillating or vibrating and a periodic driving force which maintains those oscillations or vibrations, when a small amplitude of the periodic force produces a relatively large amplitude of oscillation or vibration. See *Natural resonance, Parallel resonance, Series resonance.*

RESONANCE BRIDGE. An *ac bridge* in which the balance is a function of *resonance* of a circuit.

RESONANCE CURVE. A graph showing the response of a *tuned circuit* to various frequencies in and near the resonant frequency.

RESONANT CAVITY. See *Cavity resonator.*

RESONANT CIRCUIT. A circuit containing both *inductance* and *capacitance* and therefore capable of *resonance.*

RESONANT FREQUENCY. For a given system, the frequency at which the system is in *resonance.*

RESONANT GAP. In *TR cells,* the small region in the interior part of the resonant structure where the electric field is concentrated.

RESONANT LINE. A *transmission line* in which the *inductance* and *capacitance* are so distributed that it is *resonant* at the operating frequency.

RESONANT MODE. A *mode* of resonance of a linear device, with a characteristic field pattern, which, when not coupled to other modes, can be represented by a single tuned circuit.

RESONANT WINDOW. In *switching tubes,* a resonant iris sealed into the envelope of the tube.

RESONATOR. A system or device in which some physical quantity can be made to oscillate by the oscillations in another system, e.g. *cavity resonator.*

RESPONDER (BEACON). That part of a *transponder* which automatically transmits a reply to the *interrogator-responser.*

RESPONSER. A receiver intended to receive and display the signals from a *transponder.*

REST MASS. Of an *electron,* its mass when at rest.

RESTORE. In computers, return a variable *address* or *word* or *cycle index* to its initial value.

RETARDING-FIELD OSCILLATOR, An oscillator based on the electron transit time of a *positive-grid oscillator valve.*

RETENTION TIME. Of a *charge-storage tube,* the maximum time between *writing* into a storage element and *reading* an acceptable output from that element.

RETENTIVITY. Of a *magnetic* material, the property measured by the residual induction corresponding to the *saturation induction* for the material.

RETRACE LINE. The line traced on the screen of a *cathode-ray tube* by the *flyback.*

RETURN INTERVAL. In a *cathode-ray tube,* the interval corresponding to the direction of the *sweep* which is not used for delineation.

RETURN TRACE. The path of the *scanning* spot during the *return interval.*

Fig. 130. Direction of return trace of the spot in a television raster

REVERSE CURRENT. Current flowing on application of a *reverse voltage.*

REVERSE EMISSION. Flow of electrons in the reverse direction – anode to cathode – in a valve, during a part of the cycle when the anode is negative with respect to the cathode.

REVERSE RECOVERY TIME. Of a *semiconductor diode*, the time for the *reverse current* or *voltage* to reach a specified value after instantaneous switching from a steady forward current to a reverse bias in a given circuit.

REVERSE VOLTAGE. In the case of two opposing voltages, the voltage of that *polarity* which produces the smaller current.

REVERSIBLE CELL. An electric *cell* in which the value of the applied external voltage determines the rate and direction of its chemical reaction.

REWRITE. The process of restoring to a *storage* device, whose information may be destroyed by *reading*, the information present before the reading was made.

REX®. An electronic telephone exchange system developed by A.E.I. Ltd, based on the use of *reed relays*.

R-F. Abbreviations for *radio frequency*.

RHEOSTAT. An adjustable *resistor* whose resistance may be changed while it is in circuit.

RHEOTRON. A type of *betatron*.

RHODIUM (Rh). Metallic element similar to platinum. Atomic No. 45.

RHUMBATRON. Name formerly used for a hollow-cavity resonator used to produce high-frequency oscillations.

RIBBON MICROPHONE. A *microphone* in which the moving conductor is in the form of a ribbon directly driven by sound waves.

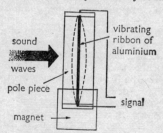

Fig. 131. Principle of ribbon microphone

RICE NEUTRALIZATION. A method of *grid neutralization* used in single-stage *power amplifiers* to neutralize the feedback due to capacitance between grid and anode of the valve.

RICHARDSON–DUSHMAN EQUATION. The equation relating current density with temperature, surface state, and *work function* in *thermionic emission*.

RIEKE DIAGRAM. A polar coordinate load diagram for microwave oscillators, especially *magnetrons* and *klystrons*.

RING CIRCUIT. See *Hybrid ring junction.*

RING COUNTER. A re-entrant multistable circuit in which any number of stages are arranged in a circle so that a unique condition exists in one stage and each *pulse* at the input causes this condition to transfer one unit round the circle.

RING SCALER. A *scaler* consisting of a series of electron tubes in a circle operating on the principle of a *ring counter.*

RINGING CIRCUIT. A *resonant* circuit used to demonstrate the performance of a triode valve as a switch.

RING-SEALED VALVE (TUBE). See *Disc-sealed valve.*

RIPPLE CURRENT. The alternating-current component of a direct current when it is small compared with the direct current as, for example, in the output of a generator.

RIPPLE FILTER. A *low-pass filter* designed to reduce the *ripple current*, while passing freely the direct current from a *rectifier*, generator, etc.

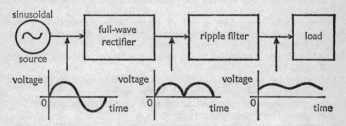

Fig. 132. Waveforms showing effect of ripple filter in a full-wave rectifier and filter circuit

RIPPLE FREQUENCY. The frequency of the *ripple current* in *rectifiers*, generators, and similar devices. For a *full-wave rectifier* it is usually twice the supply frequency.

RIPPLE VOLTAGE, PER CENT. The ratio, expressed in per cent, of the *effective value* of the ripple voltage to the average value of the total voltage.

RISE TIME. See *Pulse rise time.*

RISING SUN MAGNETRON. See *Magnetron, rising sun.*

R.M.S. See *Root-mean-square value.*

ROENTGEN. A unit of exposure dose of *X-ray* or *gamma radiation.*

ROOT-MEAN-SQUARE (R.M.S.) VALUE. Of *amperes*, *volts*, or other recurring variable quantities, the square root of the mean (average) value of the squares of the instantaneous values taken over a complete cycle.

ROPE. In radar, long strips of metal foil used as confusion reflectors. See *Window*.

ROTARY CONVERTER. A mechanically coupled ac electric motor and dc generator used to convert an alternating into a direct current.

ROTARY ENCODER. A coded disc for converting analogue data in the form of the angular position of a shaft into digital data in coded form. See *Digitizer* and *Photo-converter*.

ROTARY SPARK GAP. A device associated with a *pulse-forming line* to produce high voltage pulses of very short duration for *pulse modulation* of *microwave oscillators*.

ROTATING JOINT. A coupling between two *waveguide* structures, designed to allow transmission of electromagnetic energy while one of the structures is rotating.

ROTATING-ANODE TUBE. An *X-ray tube* in which the anode rotates so that the cathode rays are always beamed on a fresh area of the anode surface.

ROTATOR. In *waveguides*: a means of rotating the plane of polarization. In rectangular waveguides this is accomplished by twisting the guide itself.

ROUTINE. In *computers*, a set of instructions arranged in proper sequence to cause the computer to perform a desired operation, e.g. the solution of a mathematical problem.

RUBIDIUM (Rb). Metallic, highly reactive, element of the alkali group, Atomic No. 37.

RUMBLE. In gramophone recording, low-frequency vibration of mechanical origin superimposed on the record.

RUN. In *computers*, the performance of one or more *routines* or a *programme* without human intervention.

R-Y SIGNAL. In *colour television*: the signal which when combined with the luminance colour signal (Y) produces the red primary signal (R).

S

SAMOS. Satellite And Missile Observation System.

SAMPLE INTELLIGENCE. A part of a signal used as evidence of the quality of the whole.

SAMPLING ACTION. In a *feedback control system*, action in which the difference between the *set point* and the value of the controlled variable is measured, and the correction made, only at intermittent intervals.

SAMPLING CIRCUIT. A circuit whose output is a series of distinct values, representative of the input values at a series of different instants.

SAMPLING GATE. In radio aids to navigation, a device which extracts information from the input signal only when activated by a *selector pulse*.

SANATRON. A variable time-delay circuit in which two pentodes and two diodes are used to produce very short gate waveforms whose time durations vary linearly with a reference voltage.

SANDWICH SEAL. A vacuum-sealing device, consisting of a thin lamina of copper between two glass cylinders, designed to withstand the effects of thermal expansion in devices such as disc-seal triodes and other low power valves for use at high frequencies.

SATURABLE REACTOR. An electric *reactor* used in control circuits, whose action depends upon the state of *magnetic flux* saturation of its *core*. It is used where accurate and continuous control is necessary, but where the waveform of the current is not important.

symbol

SATURATION INDUCTION. The maximum intrinsic *magnetic induction* possible in a material.

SATURATION RESISTANCE. Of a *transistor*: the resistance between collector and emitter terminals, under specified conditions of base current and collector current, when the collector current is limited by the external circuit.

SATURATION SIGNAL. In *radar*, a signal whose amplitude is greater than the dynamic range of the receiving system.

SATURATION VOLTAGE. Of a *transistor*: the voltage between collector and emitter terminals under specified conditions of base current and collector current, when the collector current is limited by the external circuit.

SAWTOOTH OSCILLATOR. See *Relaxation oscillator*.

SAWTOOTH PULSE. See *Pulse, sawtooth*.

SAWTOOTH WAVE. A periodic wave whose amplitude varies approximately linearly with time between two limiting values, the interval

325

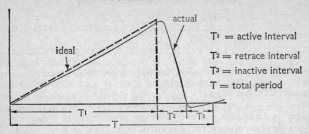

Fig. 133. Approximate sawtooth wave generated by an electronic circuit compared with an ideal wave

required for one direction of progress being longer than that for the other.

SCALE FACTOR. In *analogue computers*, the factor which relates the magnitude of a variable to its representation in the machine. In *digital computers*, the arbitrary factor sometimes associated with the numbers in the computer to adjust the position of the radix point so that significant digits occupy specified columns. See *Machine units*.

SCALE-OF-TEN CIRCUIT. A *decade scaler*.

SCALE-OF-TWO CIRCUIT. A *binary scaler*. A *flip-flop* or *trigger circuit*.

SCALE-OF-TWO COUNTER. A counter based on a *binary scaler*.

SCALER. A device which produces an output *pulse* whenever a pre-scribed number of input pulses have been received. See *Decade scaler*, *Ring scaler*.

SCALING CIRCUIT. See *Scaler*.

SCALING FACTOR. In a *scaler*, the number of input pulses per output *pulse*.

SCALLOPING. In a *travelling-wave tube*, axial variation of the focusing field which causes a corresponding variation in the diameter of the *electron beam*.

SCAN. (1) In *television*, the action of an *electron beam* which sweeps in sequence across each point, line, or field of the *mosaic* in a *camera tube*, or the screen in *picture tubes*. (2) In *radar*, sweep an airspace or region with a succession of directed beams from a radar aerial system.

SCANNER. See *Flying-spot scanner*.

SCANNING. In *television* or *facsimile* the process of analysing or synthesizing successively the light values of the elements making up a picture area, according to a predetermined method. See *Coarse*, *High-velocity*, *Interlaced*, *Low-velocity*, *Progressive*, *Rectilinear*, *Spiral* and *Television scanning*.

SCANNING LINE. In television, a single continuous narrow strip determined by the process of *scanning*. The number of scanning lines is numerically equal to the ratio of *line frequency* to *frame frequency*.

SCANNING LINEARITY. In television, the uniformity of *scanning* speed during the *trace interval*.

SCANNING SPOT. In *television*, the part of the scanned area which is being explored at any instant in the *scanning* process. In *facsimile* transmission, the area on the subject copy viewed instantaneously by the pick-up system of the scanner.

SCATTERING LOSS. That part of the electromagnetic transmission power loss due to scattering in the medium or to roughness in the reflecting surface.

SCENIOSCOPE. An *image iconoscope.*

SCHEMATIC (DIAGRAM). See *Circuit diagram.*

SCHERING BRIDGE. An *ac bridge* for the measurement of *capacitance* and *dissipation factor.*

SCHMIDT OPTICAL SYSTEM. An optical system used in certain astronomical telescopes which has been applied to reflective projection television systems.

SCHMITT LIMITER. A *bistable pulse* generator in which a constant amplitude output pulse exists only so long as the input voltage exceeds a predetermined value.

SCHOTTKY EFFECT. The continued slight increase in valve current over the saturation current value as the *anode* potential is raised, due to the increased *thermionic emission* resulting from an electric field at the surface of the *cathode.*

SCHOTTKY EMISSION. See *Schottky effect.*

SCHOTTKY NOISE. See *Shot noise.*

SCINTILLATION. (1) A flash of light produced in a *phosphor* by an *ionizing* event. (2) On a *radar display*, a rapid apparent displacement of the target from its mean position. (3) In *radio* propagation, a small random fluctuation of the received field about its mean value, analogous to the twinkling of light from a star.

SCINTILLATION COUNTER. The combination of *phosphor, photomultiplier, amplifier,* and *counting circuits* used for counting *scintillations.*

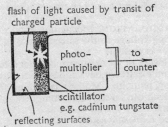

Fig. 134. Principle of scintillation counter

SCINTILLATION CRYSTALS. Materials, e.g. anthracene, used to produce the *scintillations* in *scintillation counters*.

SCINTILLATION SPECTROMETER. A *scintillation counter* adapted to the study of energy distribution.

SCOPHONY SYSTEM. A type of projection *television* system in which the normal *fluorescent screen* is replaced by light from a carbon arc or lamp, which is modulated by a *Kerr cell* in response to the *video signal* and projected on to a viewing screen. Compare *Skiatron*.

SCOTOPHOR. A material, such as potassium chloride, used instead of a *phosphor* on *cathode-ray tube* screens where daylight viewing and long persistence are required. It darkens under electron bombardment to give a black on white picture which can be erased by heating the screen.

SCR. Abbreviation for *Silicon controlled rectifier*.

SCRAMBLING CIRCUIT. A transmitting circuit used to render signals unintelligible unless they are received by the corresponding unscrambling circuit.

SCREEN. (1) Of a *cathode-ray tube*: the surface of the tube on which the visible pattern is produced. (2) Material so disposed in relation to a field as to reduce its penetration into an assigned region.

SCREEN GRID. Of a valve, a grid between the *anode* and a *control grid*, usually maintained at a fixed positive potential, for the purpose of reducing the electrostatic influence of the anode in the space between screen grid and *cathode*.

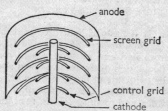

Fig. 135. Section of one type of screen grid

SCREEN-GRID MODULATION. *Modulation* produced by application of the modulating voltage to the screen grid of a multigrid valve in which the *carrier* is present.

SEAL-OFF. The closure of an exhaust tube connected to the envelope of a high-vacuum electronic device, designed so that the device can be sealed during pumping. Modern technique favours the use of copper tubes which can be sealed by compression between a pair of cylindrical rollers.

SEARCH COIL. See *Exploring coil*.

SEARCH RADAR. A *radar* intended primarily to cover a large region of space and to display targets as soon as possible after they enter the region.

SECAM. A system of *colour television* developed intensively in France as an alternative to the *NTSC* and *PAL* systems. It transmits only one colouring signal on any particular scanning line and thus requires the use of a delay line and commutation switch in the receiver to provide both the necessary colour-difference signals in the receiver. A variation of *SECAM*, unofficially christened *SEQUAM*, has now been produced in Russia, and may be accepted in Europe as a compromise system.

SECAR. Secondary Radar. See *Radar, secondary*.

SECOND ANODE. Of a *cathode-ray tube* employing *electrostatic deflection*, the electrode used to accelerate the electron beam before it is deflected and after it has been focused by the first *anode*.

SECOND DETECTOR. The *detector* which separates the intelligence signal from the i-f signal in a *superheterodyne receiver*.

SECONDARY BATTERY. A battery made up of *storage cells*.

SECONDARY ELECTRON. An *electron* emitted from an atom, molecule, or surface as a result of a collision with a charged particle or *photon*.

SECONDARY EMISSION. The ejection of *electrons* from solids or liquids as a result of the impact of charged particles. See *Field-enhanced secondary emission, Photomultiplier*.

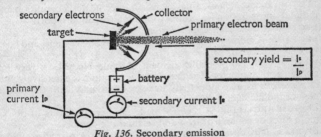

Fig. 136. Secondary emission

SECONDARY EMISSION MULTIPLIER. See *Electron multiplier*.

SECONDARY EMISSION RATIO. The average number of *electrons* emitted from a surface per incident *primary electron*.

SECONDARY MEMORY. Of a *computer, storage* capacity outside the computer, but controlled by and directly linked to it.

SECONDARY WINDING. Of a *transformer*, the winding which receives energy by *electromagnetic induction* from the *primary* winding.

SECTOR DISPLAY. A range-amplitude display used with a radar set which receives signals from a continuously rotating aerial.

SECTOR SCAN(NING). A *radar scan* through a limited angle in azimuth or depth.

SEEBECK EFFECT. See *Thermoelectric effect*.

SEED. A small single crystal of a semiconductor from which is grown the large single crystal for the manufacture of *semiconductor devices*.

SEGMENT. In computer terminology, part of a *routine* consisting of a number of *sub-routines*, each of which can be completely stored in the *internal memory*, and each containing the necessary instructions to jump to other segments.

SELECTANCE. A measure of the falling off in response of a *resonant* device with departure from resonance.

SELECTION CHECK. In *computers*, an automatic check to verify that the correct *register*, or other device, is selected in the performance of an instruction.

SELECTIVE FADING. Of a *radio* signal, *fading* which affects the received frequencies unequally within a specified band.

SELECTIVE INTERFERENCE. *Radio interference* whose energy is concentrated in a narrow band of frequencies.

SELECTIVE NETWORK. A *transducer* which causes an *insertion loss* or *phase shift*, or both, in a given circuit, varying with frequency in some desired way.

SELECTIVITY. Of a receiver, the characteristic which determines the extent to which the receiver can differentiate between the desired signal and disturbances of other frequencies. See *Q*.

SELECTOR. In a *computer*, a device which interrogates a condition and initiates an operation according to the interrogation report.

SELECTOR PULSE. See *Pulse, selector*.

SELECTRON. A *storage tube* which can store 256 binary digits and has a very small *access time*.

SELENIUM (Se). Non-metallic element, Atomic No. 34. A very important semiconductor. The grey form of selenium is extensively used in *photocells* as its conductivity increases markedly with the brightness of the incident light.

SELENIUM RECTIFIER. A *rectifier* made of alternate layers of one or more discs or squares of iron and selenium in contact.

SELF-BIAS. *Bias* developed as a result of the flow of valve current

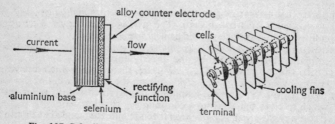

Fig. 137. Selenium rectifier – construction of single cell and stack

330

through a *resistor* in the *cathode* circuit, or through a grid resistor in the grid circuit, and not by the use of a battery. See *Cathode bias*, *Grid bias*.

SELF-CAPACITANCE. The *capacitance* inherent in a conductor resistor, inductor, or other element; usually expressed as an equivalent parallel-connected capacitance.

SELF-IMPEDANCE. At any pair of terminals of a *network*, the ratio of an applied potential difference to the resultant current at these terminals, all other terminals being open. See *Impedance*.

SELF-INDUCTANCE. See *Inductance*.

SELSYN. A *synchro* generator or motor, designed for such applications as electronic control of machine tools and synchronizing gunfire with a radar scanner.

SEMICONDUCTOR. A solid or liquid electronic *conductor* whose *resistivity* lies between that of metals and that of insulators and in which the concentration of electrical *charge carriers* increases with increasing temperature over a certain range. For ordinary temperature ranges, semiconductors, in contrast to metals, have a negative *temperature coefficient* of *resistance*. The electrical properties of semiconductors are extremely sensitive to the presence of impurities and to previous treatment. Commonly used semiconductors include germanium, silicon, selenium, silicon carbide, cuprous oxide, lead telluride, etc. See entries following.

symbol

SEMICONDUCTOR, COMPENSATED. A *semiconductor* in which one type of impurity (e.g. *acceptor*) partially cancels the opposite type of impurity (e.g. *donor*).

SEMICONDUCTOR, DEFECT. One in which there is a slight deficiency of one of the normal constituents and a slight excess of the other. This is one of the three types of semiconductor, the other two being *extrinsic* and *intrinsic*.

SEMICONDUCTOR, DOPED. See *Doping*.

SEMICONDUCTOR, EXTRINSIC. See *Extrinsic semiconductor*.

SEMICONDUCTOR, INTRINSIC. See *Intrinsic semiconductor*.

SEMICONDUCTOR, N-TYPE. An *extrinsic semiconductor* in which the *conduction electron* density exceeds the *hole density*.

SEMICONDUCTOR, P-TYPE. An *extrinsic semiconductor* in which the *hole* density exceeds the *conduction electron* density.

SEMICONDUCTOR ACCEPTOR IMPURITY. See *Acceptor impurity*.

SEMICONDUCTOR DEVICE. A device whose essential characteristics are due to the flow of charge carriers within a *semiconductor*.

SEMICONDUCTOR DIODE. A two-electrode *semiconductor device* which has an asymmetrical voltage-current characteristic. There are two types: *point-contact* and *junction*. See *Crystal rectifier*. See Figure 138.

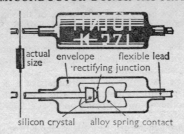

Fig. 138. Semiconductor diode (silicon junction type)

SEMICONDUCTOR DONOR IMPURITY. See *Donor*.

SEMICONDUCTOR GENERATION RATE. The time rate of generation of electron-hole pairs in a *semiconductor*.

SEMICONDUCTOR JUNCTION. A *junction* between *semiconductors*, usually between *n-type* and *p-type*.

SEMICONDUCTOR JUNCTION, ALLOY. A *junction* formed by alloying one or more impurities to a *semiconductor* crystal.

SEMICONDUCTOR JUNCTION, COLLECTOR. A *junction* normally biased in the high-resistance direction, through which the current can be controlled by the introduction of *minority carriers*.

SEMICONDUCTOR JUNCTION, DIFFUSED. A *junction* formed by the diffusion of an *impurity* with a *semiconductor* crystal.

SEMICONDUCTOR JUNCTION, DOPED. A *junction* produced by addition of an impurity to the melt during growth of a *semiconductor* crystal.

SEMICONDUCTOR JUNCTION, EMITTER. A *junction* normally biased in the low-resistance direction to inject *minority carriers* into an inter-electrode region.

SEMICONDUCTOR JUNCTION, FUSED. A *junction* formed by re-crystallization on a base crystal from a melt containing the *semiconductor* and one or more components.

SEMICONDUCTOR JUNCTION, GROWN. A *junction* produced during growth of a crystal from a melt.

SEMICONDUCTOR JUNCTION, N-N. A *junction* between two regions having different properties in a *n-type semiconductor*.

SEMICONDUCTOR JUNCTION, N-P. A *junction* between *n-type* and *p-type* regions of a *semiconductor*.

SEMICONDUCTOR JUNCTION, N-P-N. A double junction formed by introducing a thin slice of *p-semiconductor* between two blocks of *n-type* material of the same semiconductor. See *Transistor, n-p-n*.

SEMICONDUCTOR JUNCTION, P-N-P. Substitute 'n' for 'p', and 'p' for 'n' in the above definition. See *Transistor, p-n-p*.

SEMICONDUCTOR JUNCTION, P-P. Substitute 'p' for 'n' in *semiconductor, n-n*.

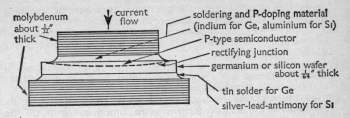

Fig. 139. Typical construction of semiconductor rectifying cell using p–p junction

SEMICONDUCTOR JUNCTION, RATE-GROWN. A *grown junction* produced by varying the rate of growth of the crystal.

SEMICONDUCTOR TRAP. An irregularity in the lattice of a crystalline *semiconductor* which produces a negative *ion* space (a trap for free electrons) or positive ion space (a *hole* trap).

SEMITRANSPARENT CATHODE. In *camera tubes* or *phototubes*, a *photocathode* in which radiant flux incident on one side produces *photoelectric emission* from both sides.

SENSITIVITY. (1) Of a *radio receiver* or similar device, the minimum input signal required to produce a specified output signal having a specified *signal-to-noise* ratio. (2) Of a meter movement, the current, voltage, etc., required for full-scale deflection.

SENTINEL. In computing and data-processing, a symbol marking the beginning or end of some element of information.

SEQUAM. See *SECAM*.

SEQUENCE-CONTROLLED CALCULATOR. See *Automatic sequence-controlled* calculator.

SEQUENCER. A machine widely used in *data-processing* systems which puts items of information into a particular order.

SEQUENTIAL COLOUR TRANSMISSION. See *Colour television, Dot-, Field-,* and *Line-sequential colour television.*

SEQUENTIAL CONTROL. Of a *computer*, the manner of operation in which the instructions are set up in sequence and fed to the computer during the solution of a problem.

SEQUENTIAL TRANSMISSION. A method of transmitting pictures in which the picture elements are selected in regular time intervals and delivered in correct sequence to the communication channel. This is the current technique for *television* transmission.

SERIAL. Of a *computer*, the general technique of transmission or storage of a *word*, or logical operations on a word, in time sequence, using the same facilities successively. Contrast *Parallel digital computer.*

SERIAL MEMORY. A *memory* in which the *bits* or *words* are located and appear one after the other, the *access time* consequently depending on the particular location. Compare *Parallel memory*.

SERIAL TRANSFER. In *computers* and *data-processing*, the *transfer* of information in which the characters move one after the other along a single path.

SERIAL TRANSMISSION. A system of transmitting information in which the characters of a word are sent in sequence over a single line as contrasted with *parallel transmission*.

SERIES CIRCUIT. An electrical circuit whose component parts are connected end-to-end to form a single continuous path for the current. An *inductor, capacitor*, and *resistor* in series constitute a 'simple' series circuit. Compare *Parallel circuit*.

SERIES FEED. Application of direct voltage to the *anode* or *grid* of a valve, through the same *impedance* in which the signal current flows.

SERIES MODULATION. *Modulation* in which the anode circuit of a modulating valve and a *modulating amplifier* valve are in *series* with the same anode voltage supply.

SERIES RESONANCE. The *steady-state* condition existing in a circuit made up of *inductance* and *capacitance* in *series*, when the current in the circuit is in *phase* with the voltage across it.

SERIES STABILIZATION. A method of *stabilization* using amplifier *feedback* in which the amplifier circuit and the feedback circuit are in series at each end of the amplifier. Compare *Shunt stabilization*.

SERIES-PARALLEL NETWORK. Any *network* containing only *resistors, inductors*, and *capacitors*, which is constructed by connecting branches successively in *series* and/or in *parallel*.

SERVICE AREA. Of a *radio broadcast transmitter*, the area which the broadcast is designed to serve for a given effective radiated power and a given aerial height above average ground level.

SERVICE BAND. A band of frequencies accommodating several channels allocated to a given class of radio communications.

SERVICE ROUTINE. In *digital computers*, a *routine* designed to assist the operator in the running of a problem.

SERVO. Short for *servomechanism, servomotor*, etc.

SERVO AMPLIFIER. An electronic *amplifier* forming part of a *servo system*.

SERVO CONTROL. Control of anything with a *servomechanism*.

SERVO LINK. A mechanical power amplifier, e.g. a relay and motor-driven *actuator*, which enables low-power signals to operate control mechanisms requiring relatively large power.

SERVO SYSTEM. A dynamic, closed-circuit, automatic control system, designed so that the output element or quantity follows the input to the system as closely as is required. It is distinguished from other automatic control systems in having at least one *feedback loop* which provides an input signal proportional to the difference between actual output and desired output.

SERVO-AMPLIDYNE SYSTEM. A system in which an *amplidyne* is used to amplify mechanical power in conjunction with a control amplifier.

SERVOMECHANISM. A *feedback control system* in which one or more of the signals in the system represent mechanical motion.

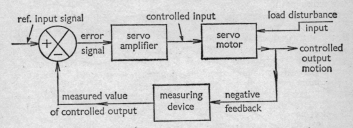

Fig. 140. Block diagram showing elements of a servomechanism

SERVOMECHANISM, DC/AC. A *servo system* in which a direct current *error signal* is converted into alternating current in order to amplify it.

SERVOMOTOR. A motor which supplies power for moving a *servomechanism*.

SET. In *computers*, place a *storage* device in a prescribed state.

SET POINT. In *automatic control systems*, the position to which the control-point setting mechanism, e.g. a scale reading, is set.

SHADING. (1) In *camera tubes*, generation of a non-uniform level of background illumination in the reproduced image, not present in the original image. (2) In *television*, compensation for the spurious signal generated in a *camera tube* during *trace intervals*. (3) A method of controlling the directivity pattern of a *transducer* by controlling the distribution of phase and amplitude of the transducer action over the active surface.

SHADOW EFFECT. In propagation of electromagnetic waves, the effect, usually a loss of signal strength, caused by the topography of the region between transmitting and receiving points.

SHADOW MASK. In *colour picture tubes*, an electrode system in the form of a perforated electrically conductive sheet which uses masking to effect selection of colours. See also Figure 26 and Fig. 141.

SHAPED-BEAM TUBE. A *cathode-ray tube* in which the cross-section of the electron beam is made to correspond to the shape of various characters and each character is formed all at once on the screen of the tube.

SHEATH. In a *gas discharge tube*, a region in which a *space charge* is produced by an accumulation of electrons or ions. See *Anode sheath, Cathode dark-space*.

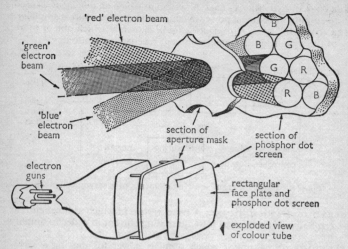

Fig. 141. Shadow mask in a colour television tube (see also Fig. 26)

SHEPHERD TUBE (VALVE). An all-metal *microwave* tube employing *velocity modulation,* used as a *local oscillator* in a microwave receiver.

SHF. *Super High Frequency.*

SHIELD. Any screen, mesh, housing, or other structure designed to reduce the effects of *electric* or *magnetic fields* on objects placed behind or inside the shield. See *Electrostatic screen, Magnetic shielding.*

SHIELD GRID. In *gas tubes,* a grid which shields the control grid from electrostatic fields, thermal radiation, and materials produced by thermionic emission.

SHIELDED (SCREENED) PAIR. A two-wire *transmission line* enclosed in a metal sheath.

SHIFT. In *computer* terminology, the displacement of an ordered set of characters one or more columns to the right or left. If the characters are the digits of a number in a *fixed-point* digital computer, this is equivalent to multiplication by a power of the radix.

SHIFT PULSE. A drive pulse used to bring about a *shift* in a digital computer.

SHIFT REGISTER. In *computers,* a device to store information in the form of pulses, or delay it as required.

SHM. Abbreviation for *simple harmonic motion.*

SHOCK EXCITATION. See *Impulse excitation.*

SHORAN. A *radar* short-range navigation system for precise determination of position in aerial navigation and missile guidance. The general technique is similar to that of *Loran*.

SHORT. Colloquial for *short circuit*.

SHORT CIRCUIT. A connection, whether accidental or intentional, between two points of different potential in a circuit, by a path of relatively low resistance.

SHORT CIRCUIT IMPEDANCE. Of a line or *four-terminal network*, the *driving-point impedance* when the far end is short-circuited.

SHORT WAVE. A radio frequency wave shorter than those used for the standard medium frequency broadcast band. In current use the term is used for frequencies between about 1·5 megacycles/second and 30 megacycles/second. See *Frequency band*.

SHORT-WAVE CONVERTER. A *frequency converter* which is used to convert short-wave signals into signals acceptable to a standard broadcast receiver. It employs a *local oscillator* and *mixer* to give a *beat* signal in the required frequency band.

SHOT NOISE. *Noise* due to the random emission of electrons from a cathode in a *thermionic valve*.

SHUNT. In general, any device connected *in parallel* with any other device. In a special sense, a *galvanometer shunt*.

SHUNT FEED. See *Parallel feed*.

SHUNT STABILIZATION. A method of *stabilization* using amplifier *feedback* in which the amplifier circuit and feedback circuit are in parallel. Compare *Series stabilization*.

SI UNIT SYSTEM. The metric system of International units to which the United Kingdom will eventually convert. Tables of the basic units, some of the derived units, and some multiples and sub-multiples will be found in Figure 142.

SIDE FREQUENCY. One of the frequencies of a *sideband*.

SIDEBAND (S). In a *modulated carrier*, the frequency band on either side of the carrier frequency which contains the frequencies of the waves produced by the process of *modulation*; or the wave components lying within such a band.

SIDEBAND, VESTIGIAL. See *Vestigial sideband*.

SIDEBAND ATTENUATION. A form of attenuation in which the transmitted relative amplitude of the component(s) of a *modulated* signal, excluding the *carrier*, is smaller than that produced by the process of *modulation*.

SIDETONE. The transmission and reproduction of sounds through a local path from the transmitter to the receiver of the same telephone station.

SIGMA CIRCUIT. An electronic circuit used in nuclear reactor control systems for the rapid shut-down of the reactor.

SIGN DIGIT. A character used to designate the algebraic sign of a number in a *digital computer*.

BASIC SI UNITS

Physical quantity	Name of unit	Unit symbol
Length	Metre	m
Mass	Kilogramme	kg
Time	Second	S
Electric current	Ampere	A
Thermodynamic temperature	Degree Kelvin	°K
Luminous intensity	Candela	cd

SOME DERIVED UNITS

Force	Newton	$N = kg\,m/s^2$
Work, energy, quality of heat	Joule	$J = Nm$
Power	Watt	$W = J/S$
Electric charge	Coulomb	$C = As$
Electrical potential	Volt	$V = W/A$
Electric capacitance	Farad	$F = As/V$
Electric resistance	Ohm	$\Omega = V/A$
Frequency	Hertz	$Hz = S^{-1}$
Magnetic flux	Weber	$Wb = Vs$
Magnetic flux density	Tesla	$T = Wb/m^2$
Inductance	Henry	$H = Vs/A$
Luminous flux	Lumen	$lm = cd\,sr$
Illumination	Lux	$lx = Gn/m^2$

MULTIPLES AND SUB-MULTIPLES OF UNITS

Factor by which the unit is multiplied	Prefix	Symbol
$1\ 000\ 000\ 000\ 000 = 10^{12}$	tera	T
$1\ 000\ 000\ 000 = 10^{9}$	giga	G
$1\ 000\ 000 = 10^{6}$	mega	M
$1\ 000 = 10^{3}$	kilo	k
$100 = 10^{2}$	hecto	h
$10 = 10^{1}$	deca	da
$0 \cdot 1 = 10^{-1}$	deci	d
$0 \cdot 01 = 10^{-2}$	cento	c
$0 \cdot 001 = 10^{-3}$	milli	m
$0 \cdot 000\ 001 = 10^{-6}$	micro	μ
$0 \cdot 000\ 000\ 001 = 10^{-9}$	nano	n
$0 \cdot 000\ 000\ 000\ 001 = 10^{-12}$	pico	p
$0 \cdot 000\ 000\ 000\ 000\ 001 = 10^{-15}$	femto	f
$0 \cdot 000\ 000\ 000\ 000\ 000\ 001 = 10^{-18}$	atto	a

Fig. 142. SI unit system

SIGNAL. An electrical quantity which conveys information by varying in some manner.

SIGNAL ELECTRODE. In *camera tubes*, an electrode from which the signal output is obtained.

SIGNAL ELEMENT. The part of a signal which occupies the shortest interval of the signal code.

SIGNAL ELONGATION. The elongation of the envelope of a signal due to delay in the arrival of some of the components.

SIGNAL FREQUENCY SHIFT. In a *facsimile* system using *frequency shift*, the numerical difference between the frequencies corresponding to white signal and black signal at any point in the system.

SIGNAL GENERATOR. An electronic instrument which generates an electric signal whose characteristics may be adjusted and controlled, so that it can be used for the supply of a standard voltage of known amplitude, frequency, and waveform. The essential parts are a variable *oscillator* and an *attenuator*, and most signal generators also have provision for internal or external *modulation*. See *Pulse generator, Square-wave generator*.

SIGNAL LEVEL. At any point in a *transmission system*, the difference between the measure of the signal at that point and the measure of an arbitrary signal chosen as reference.

SIGNAL OUTPUT CURRENT. Of *camera tubes* or *phototubes*, the absolute value of the difference between output current and *dark current*.

SIGNAL WAVE. A *wave* whose characteristics permit some intelligence or message to be conveyed.

SIGNAL WINDING. Of a *saturable reactor*, a control winding to which the signal wave is applied.

SIGNAL-TO-NOISE RATIO. (1) In *radio*, the ratio of the magnitude of the signal to that of the *noise*, usually expressed in *decibels*. (2) In *television,* the ratio in decibels of the maximum peak-to-peak voltage of the *video* television signal, including *synchronizing pulse*, to the *r.m.s.* voltage of the noise. (3) In *camera tubes*, the ratio of peak-to-peak signal output current to r.m.s. noise in the output current.

SILICA GEL. Chemically activated silicon dioxide used to absorb moisture in containers of electronic equipment.

SILICON (Si). An abundant non-metallic element, Atomic No. 14. It is an *intrinsic semiconductor* and is one of the most important materials in the manufacture of transistors.

SILICON CONTROLLED RECTIFIER. The solid-state equivalent of a *thyratron*: a 3-junction semiconductor device which normally represents an open circuit but which switches rapidly to the conducting state of a single-junction *silicon rectifier* when an appropriate signal is applied to the gate terminal.

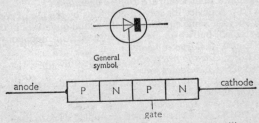

Fig. 143. Basic structure and graphical symbol of a p-n-p-n silicon controlled rectifier

SILICON DIODE. A *crystal diode* using silicon as the *semiconductor*.

SILICON RECTIFIER. A metallic *rectifier* consisting of silicon in contact with a good conductor.

SILICON RESISTOR. A resistor of special silicon material with a positive temperature coefficient of resistance which does not change appreciably with temperature. It is used as a temperature sensing element.

SILICON SOLAR CELL. An element of a *solar battery*, consisting of alternate layers of p and n silicon which form the p-n junctions at which radiant energy is converted into electricity.

SILICONES. Complex compounds of silicon used in electronics as moisture-repellent insulating mediums.

SILVER (Ag). Noble metallic element, Atomic No. 47. Better than all other metals as a conductor of electricity and heat.

SILVER-MICA CAPACITOR. A high-precision stable *fixed capacitor* made by silvering thin sheets of mica.

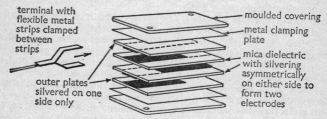

terminal with flexible metal strips clamped between strips

moulded covering

metal clamping plate

mica dielectric with silvering asymmetrically on either side to form two electrodes

outer plates silvered on one side only

Fig. 144. Construction of silver-mica capacitor

SIMPLE HARMONIC MOTION (SHM). Periodic oscillatory motion in a straight line in which the restoring force is proportional to the displacement.

SIMPLEX OPERATION. In general communication, a method of operation in which communication takes place between two stations in one direction at a time. Compare *Duplex operation*.

SIMULATOR. A device for solving problems by the use of components which obey the same equations as the actual quantities in the system being studied, but which are cheaper, easier, or more convenient to use. See also *Analogue computer*.

SINE POT. See *Potentiometer, sine*.

SINE WAVE. See *Sinusoidal wave*.

SINGING. In a *transmission* system, an undesired self-sustained *oscillation* in the system.

SINGING ARC. See *Duddle arc*.

SINGING POINT. Of a circuit which is coupled back to itself, the point at which the gain is just sufficient to make the circuit break into *oscillation*.

SINGLE-ADDRESS CODE. In *computers*, an *instruction code* which contains only one *address*.

SINGLE-ENDED AMPLIFIER. An *amplifier* in which the signal input and output have one side connected to earth.

SINGLE-ENDED VALVE (TUBE). A *valve* (tube) in which all electrode connections are made to base pins.

SINGLE-PARITY CHECK. In *computers*, a kind of *binary-code* system for detecting errors.

SINGLE-PHASE. A qualifying term applied to a system or apparatus to denote one in which there is a single *alternating* voltage. Compare *Polyphase*.

SINGLE-POLE SWITCH. A switch which closes only one circuit when thrown, in contrast to a *double-pole switch*.

SINGLE-SHOT MULTIVIBRATOR. A *monostable multivibrator*.

SINGLE-SHOT TRIGGER. A *trigger circuit* in which the triggering pulse initiates one complete cycle of conditions, ending with a stable condition.

SINGLE-SIDEBAND MODULATION. *Modulation* whereby the spectrum of the modulating wave is translated in frequency by a specified amount.

SINGLE-SIDEBAND TRANSMISSION. That method of operation in which one *sideband* is transmitted and the other sideband is suppressed. The *carrier* may also be reduced or suppressed.

SINGLE-TUNED CIRCUIT. A circuit which may be represented by a single *inductance* and a single capacitance, together with associated resistances.

SINUSOIDAL. Having the characteristics of the function sin x, or of the function cos x.

SINUSOIDAL WAVE. A *wave* whose displacement is the sine (or cosine) of an angle proportional to time or distance or both.

SKIATRON. A type of *cathode-ray tube* having a screen coated with a material such as potassium chloride, which darkens under the action of the electron beam. It is the all-electronic counterpart of the *scophony system*.

SKIN EFFECT. In a *conductor* carrying an alternating *current*, an electromagnetic effect which results in the current density being greater at the surface of the conductor than in the centre. At sufficiently high frequencies the current is practically confined to the surface of the conductor. See Figure 145.

SKIP. In *computer* terminology, an instruction to proceed to the next instruction, or a blank instruction. See Figure 146.

SKIP DISTANCE. The minimum separation for which *radio* waves of a specified frequency can be transmitted at a particular time between two points on the earth's surface, by reflection from the *ionosphere*.

SKY WAVE. The portion of a radiated wave which travels outward from the earth's surface and may or may not be reflected or refracted in the *ionosphere*.

low frequency   medium frequency   very high frequency

effect of frequency on distribution of current in
cross-section of conductor

Fig. 145. Skin effect

SKYNET. The first British military telecommunications satellite, launched into an equatorial stationary orbit above the Indian Ocean in November 1969.

SLAVE SWEEP. A *time base* triggered or synchronized by a waveform from an external source, or a source common to a number of separate time bases.

SLEEPING SICKNESS. A 'disease' sometimes observed in *junction transistors,* the main symptom being the gradual development of a leakage path between emitter and collector. It appears to be due to the accumulation of moisture on the base layer and may be one of the factors in *slow death.* The term is also used to describe a slow increase in resistance between the coating and core of some *coated cathodes* in valves whose anode current is cut off for long periods.

SLIDEBACK VOLTMETER. See *Voltmeter, slideback.*

SLIDE-WIRE. A wire of uniform resistance with which a sliding contact makes connection at any desired point.

SLIP RING. A conducting ring connected to a winding on a rotating part and rotating with it, serving to make connection with an external circuit through a *brush.*

SLOPE. *Mutual conductance.*

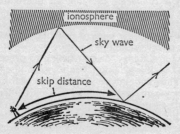

Fig. 146. Skip distance

SLOTTED LINE. A section of a *waveguide, coaxial line*, or shielded *transmission* line, in which the shield is slotted to receive a travelling probe for examination or control of *standing waves*.

SLOW DEATH. The gradual drift of *transistor* characteristics with time, especially the reverse currents of *collectors*; probably due to the effects of *ions* on the active surfaces.

SLUG. (1) A metal ring or short-circuited winding placed on the core of a *relay* to introduce a time delay in its operation. (2) A metallic core which can be moved along the long axis of a coil so as to alter its *inductance*. See *Slug tuning*.

SLUG TUNING. A method of altering the *resonant frequency* of a circuit by introducing a *slug* into the electric or magnetic fields or both. See *Permeability tuning*.

SMALL-SIGNAL TRANSISTOR PARAMETERS. See *Transistor parameters*.

SMEARER. A circuit used to cancel the *overshoot* of a *pulse*.

SMITH CHART. A convenient and practical adaptation of the *circle diagram*.

SMOOTHING CHOKE. A *choke* used to reduce the fluctuations in the output of a *rectifier* power supply.

SMOOTHING CIRCUIT. A circuit designed to reduce *ripple*.

SNOW. *Interference* resembling falling snow which may appear on a *television screen* or *radarscope* in the absence of a signal or in the presence of a weak signal, owing to *noise* in the equipment.

SODIUM (Na). Very reactive metallic element, Atomic No. 11.

SOFT VALVE (TUBE). A *vacuum valve* or tube whose performance has been affected by the presence of gas within the envelope. See *Hard valve*.

SOFTWARE. Of *computers*: a general term for all programmes and routines used to implement and extend the capabilities of the computer: e.g. assemblers, compilers, and sub-routines.

SOLAR BATTERY. A device for converting sunlight into electrical power, especially a device based on the *photovoltaic* effects which can be obtained in a *semiconductor* crystal having a *p-n junction* near an illuminated surface.

SOLDER. Material used to join metals by alloying with their surface layers. Ordinary 'soft' solder used in electronics for making permanent *ohmic contacts* is a lead-tin alloy. In addition, a variety of solders are used with flow-points ranging from 200 to 2000 degrees Centigrade and consisting of alloys of metals such as silver, copper, gold, nickel, and rhodium.

SOLENOID. A coil, usually of tubular form, for producing a *magnetic field*; sometimes including an iron bar or rod which is free to move along the axis of the coil under the influence of its magnetic field. See Figure 147.

SOLID CONDUCTOR. A conductor composed of a single wire rather than a number of strands.

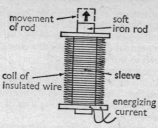

Fig. 147. Principle of solenoid

SOLID-STATE DEVICE. An electronic device composed entirely of solids.

SOLID-STATE MASER. See *Maser, solid-state.*

SOLID-STATE PHYSICS. That branch of physics which deals with the structure and properties of solids, especially semiconductors.

SOLION. A device, used to detect very low-frequency *audio* signals, which converts the movement of *ions* in an electrolyte into electrical signals.

SONAR. SOund Navigation And Ranging. Apparatus or technique for using under-water acoustic energy to obtain information about objects or events below the surface. See *Echo ranging, Echo sounder, Hydrophone, Underwater sound projector.*

SONDE. See *Radiosonde.*

SONE. A unit of loudness. A simple tone of frequency 1,000 cycles per second, 40 decibels above the listener's threshold, produces a loudness of 1 sone. See *Audibility threshold.*

SONIC ALTIMETER. An *absolute altimeter* which uses short pulses of sound to measure the distance of an aircraft above the ground.

SOPHIA. A learning machine built at Kent University in 1969 which uses twelve digital microcircuit binary cells occupying no more space than the head of a match. The machine learns to recognize simple patterns based on its 36 photocell 'eye'.

SOUND ANALYSER. A device consisting of a microphone, amplifier, and *wave analyser* used to measure the amplitude and frequency of the components of a complex sound.

SOUND LEVEL METER. An instrument for the measurement of noise and sound level in a manner approximating to the results obtained by an ear in judging *loudness level.* It includes a microphone, amplifier, output meter, and frequency *weighting* networks.

SOUND PROBE. A device, usually a very small *microphone,* for exploring a sound field without any significant disturbance to the field.

SOUND RECORDING AND SOUND REPRODUCING SYSTEM. A combination of *transducers* and associated equipment suitable for storing sound in a form which can be reproduced, and the complementary equipment for reproducing it.

SOUND TRACK. A narrow band, usually along the margin of a sound film, which carries the sound record. A number of such bands may be used on one film.

SOUND TRACK, VARIABLE AREA (WIDTH). A *sound track* divided laterally into transparent and opaque areas, the sharp line of demarcation between these areas forming an *oscillographic trace* of the wave shape of the recorded signal.

SOUND TRACK, VARIABLE DENSITY. A *sound track* of constant width and of which the average light transmission varies along the longitudinal axis in proportion to some characteristic of the applied signal.

SOUNDING ELECTRODE. An electrode used as a probe for measurements in a *gas discharge*.

SOURCE IMPEDANCE. The *impedance* presented by a source of energy to the input terminals of a device.

SPACE CHARGE. Generally, the net *electric charge* within a given volume. In a valve, a negative charge due to electrons, emitted from the *cathode* and not immediately drawn to the *anode*, remaining in the space between cathode and anode.

SPACE WAVE. The component of a *ground wave* which travels more or less directly from transmitter to receiving aerial.

SPACE-CHARGE DEBUNCHING. See *Debunching*.

SPACE-CHARGE DENSITY. The net *electric charge* per unit volume.

SPACE-CHARGE GRID. A *grid* placed near the *cathode* and positively biased, to reduce the effect of *space charge* in limiting the current through a valve.

SPACE-CHARGE REGION. (1) In a *semiconductor*, the *depletion layer* in which the net charge density is not zero. (2) In a valve, the region near the cathode containing the *space charge*.

SPACE-CHARGE-LIMITED CURRENT. Of *vacuum valves,* the current passing through an interelectrode space which contains a *virtual cathode*.

SPACISTOR®. A class of *semiconductor* devices using a single *n-p junction* which is biased oppositely to a normal *transistor* junction to avoid the frequency limitations due to *carrier transit time* imposed by conventional transistors.

SPAGHETTI. Flexible insulating tubing.

SPARK. A *disruptive discharge* through an insulating material accompanied by a flash. A spark is of very short duration compared with an *arc*.

SPARK GAP. An arrangement of two electrodes between which a *disruptive discharge* takes place at some prescribed potential difference between the electrodes, and such that the insulation is self-restoring after passage of a discharge.

345

SPARK TRANSMITTER. A transmitter using the oscillatory discharge from a *capacitor* through an *inductor* across a *spark gap*.

SPARK-GAP MODULATION. *Modulation* using pulses of energy from a controlled *spark gap*.

SPARKING VOLTAGE. The minimum voltage across a *spark gap* which causes a *spark*.

SPECIFIC INDUCTIVE CAPACITY. See *Dielectric constant*.

SPECIFIC RESISTANCE. See *Resistivity*.

SPECTRAL ANALYSER. See *Pulse height analyser, Spectrum analyser*.

SPECTRAL CHARACTERISTIC. Of a *camera tube*, a graph showing the relation between wavelength and sensitivity per unit wavelength interval of the tube. Of a *photoemissive* surface, the relation between wavelength of incident radiation and electron current emitted per unit incident energy.

SPECTRORADIOMETER. An instrument which measures spectral energy distribution of radiation, especially of the infra-red region.

SPECTRUM. Of a wave, the distribution of amplitude (and phase) of the components of the wave as a function of frequency. Also a continuous range of frequencies within which waves have some specified common characteristic. See *Electromagnetic spectrum*.

SPECTRUM ANALYSER. A test instrument which shows the distribution of energy contained in the frequencies emitted by a pulsed *magnetron*, and is also used to measure the Q of *resonant cavities* and lines.

SPECTRUM ANALYSER, PANORAMIC. A device for analysing complex waves which presents the data on a *cathode-ray screen* in the form of a spectrum pattern, the horizontal spacing indicating the frequency and frequency differences, and the vertical height indicating the amplitude.

SPEED OF LIGHT. See *Physical constants*.

SPHERE GAP. A *spark gap* in which the electrodes are spheres.

SPIKE. A *transient* of short duration, occurring during a *pulse*, whose amplitude is considerably greater than the average amplitude of the pulse. See *Pulse spike*.

SPILL. In a *storage* device, e.g. a *charge-storage tube*, the loss of information from a storage element by redistribution.

SPIN RESONANCE. See *Electron paramagnetic resonance, Nuclear magnetic resonance*.

SPIRAL SCANNING. Of a *cathode-ray tube*, a type of *scanning* in which the electron beam executes a spiral trace.

SPLIT-ANODE MAGNETRON. See *Magnetron, split-anode*.

SPOT. The area immediately affected by the impact of an *electron beam*.

SPOT SPEED. In *television*, the product of the number of spots in the *scanning line* by the number of scanning lines per second. In *facsimile*, the speed of the scanning or recording spot within the available line.

SPOT-KNOCKING. The process of removing local imperfections on the surface of an anode or other high-voltage electrode of a valve, by operating the valve at abnormally high voltages until all flashing inside the valve ceases.

SPRAT. Small Portable Radar Torch: a hand-held *radar* based on the Gunn diode and developed at the Royal Radar Establishment. It weighs less than 5·5 lbs. and can detect movement within 625 metres.

SPREADING RESISTANCE. The effective series resistance of a *point-contact rectifier* which is not due to the resistance of the *barrier layer* but is contributed by the body of the *semiconductor*, through which current passes to reach the junction.

SPURIOUS RESPONSE. Of an electric *transducer*, any response other than the desired response.

SPUTTERING. A method of plating metal films on glass, quartz, and similar materials by releasing atomic particles of the metal by bombardment with electrons or ions in a gaseous atmosphere. See also *Cathode sputtering*.

SQUARE WAVE. A *periodic wave* which alternately assumes, for equal lengths of time, one of two fixed values, the time of transition being negligible in comparison. See *Rectangular pulse*.

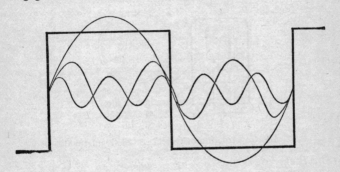

a square wave is made up of the sum of a series of sine waves: the lowest frequency component, with the same period as the square wave, and the next two odd harmonics are shown

Fig. 148. Square wave

SQUARE WAVE GENERATOR. A *signal generator* which supplies an output voltage of *square wave* form.

SQUARE WAVE RESPONSE. Of a *camera tube*, the peak-to-peak signal amplitude given by a test pattern which consists of alternate black and white bars of equal width.

SQUARE-LAW DETECTOR. A *detector* whose output voltage is sub-stantially proportional, over its useful range, to the square of the voltage of the input wave.

SQUARE-LOOP MATERIAL. A material having a *rectangular hysteresis loop*.

SQUARENESS RATIO. Of *square-loop materials*, the ratio of *retentivity* to saturation *flux density*. See *Magnetization curve*.

SQUEGGING. The condition of self *blocking* in a valve *oscillator* circuit.

SQUEGGING OSCILLATOR. A *relaxation oscillator* in which a series of periodically recurring oscillations build up to a fixed amplitude and then die away to zero.

SQUELCH CIRCUIT. A circuit which prevents a radio receiver from producing *audio-frequency* output in the absence of signals of a predetermined character.

SSB. Abbreviation for *single sideband*.

STABILIVOLT. A *gas tube*, usually a *neon lamp*, which maintains a constant voltage drop across its terminals over a relatively wide range of variation of current.

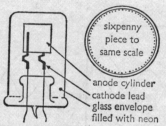

sixpenny piece to same scale

anode cylinder
cathode lead
glass envelope
filled with neon

Fig. 149. Construction of typical voltage-regulator tube

STABILIZER, VOLTAGE. See *Voltage regulator*.

STABLE OSCILLATION. *Oscillation* which tends to decrease rather than increase in amplitude.

STACKED-CERAMIC VALVE. A valve in which the electrodes are mounted on a series of conical metal supports separated by ceramic discs; it is designed to operate at high frequencies at elevated temperatures.

STAGE EFFICIENCY. Of an *amplifier*, the ratio of useful power de-livered to the load of a stage of the amplifier to the anode power input for that stage.

STAGGER-TUNED AMPLIFIER. An *amplifier* which includes two or more tuned stages, which are tuned to different frequencies.

STANDARD CELL. A *cell* which maintains a very constant voltage over a long period of time provided little current is drawn from it. It is not

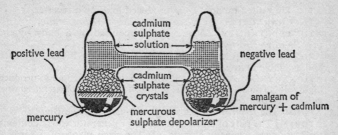

Fig. 150. Standard cell, Weston type

intended as a source of energy but as a reference standard of *electromotive force*.

STANDARD RESISTOR (RESISTANCE STANDARD). A *resistor* which is precisely adjusted to a specified value and is designed to be little affected by temperature and remain constant over long periods.

STANDING WAVE (STATIONARY WAVE). A *periodic wave* whose amplitude of displacement is a function of the distance in the direction of any line of propagation of the wave, and which results from the combination of two or more waves which have the same period.

STANDING-WAVE DETECTOR. An instrument containing a detecting device such as a *bolometer* or *thermocouple*, used to determine the ratio of maximum current or voltage of a *standing wave*.

STAR NETWORK. A set of three or more branches with one terminal of each connected to a common *node*.

STARTER (ELECTRODE). Of a *glow discharge tube*, an auxiliary electrode used to initiate conduction.

STARTER GAP. In a *cold-cathode glow discharge* tube, the conduction path between a *starter* and the other electrode to which starting voltage is applied.

STARTER VOLTAGE. Of a *cold-cathode glow discharge tube*, the voltage of the *starter* with respect to the cathode.

STARTING ANODE. An electrode of a *rectifier* which is used to establish the initial *arc*.

STARTING CURRENT OF AN OSCILLATOR. The value of electron-stream current through a valve *oscillator* at which self-sustaining oscillations will start under given load conditions.

STARTING POTENTIAL. See *Counter starting potential*.

STATAMPERE. The *electrostatic* c.g.s. unit of *current*. See *Units*.

STATCOULOMB. The *electrostatic* c.g.s. unit of *charge*. See *Units*.

STATFARAD. The *electrostatic* c.g.s. unit of *capacitance*. See *Units*.

STATHENRY. The *electrostatic* c.g.s. unit of *inductance*. See *Units*.

STATIC. Unwanted random *noise* produced in a radio receiver.

STATIC CHARACTERISTICS. See *Valve characteristics, static*.

STATIC ELECTRICITY. *Electric charge* acquired by a body by the process of *electrostatic induction*.

STATIC MEMORY. A *memory* in which information is fixed in space and available at any time.

STATIONARY GAP. A device used with a *pulse-forming* line to produce high-voltage pulses of very short duration for *modulating* a *microwave pulsed oscillator*, e.g. a *trigatron*.

STATOHM. The *electrostatic* c.g.s. unit of *resistance*. See *Units*.

STATOR. (1) The part of an electric motor or generator which contains the stationary magnetic circuits. (2) The fixed plates of a *variable capacitor*.

STEADY-STATE OSCILLATION. *Oscillation* existing in a system in which the motion at each point is a periodic quantity. This normally occurs with *forced oscillation*.

STEEL-TANK RECTIFIER. A *mercury-arc rectifier* whose anode is a steel tank.

STEINMETZ COEFFICIENT. A constant of proportionality in a formula devised by Steinmetz to express the electric *hysteresis loss* of different materials.

STEP FUNCTION. See *Unit-step function*.

STEP-DOWN TRANSFORMER. A *transformer* in which the transfer of energy is from a high- to a low-voltage winding.

STEPPING RELAY. A *relay* whose contact arm may rotate through 360° but not in one operation.

STEP-UP TRANSFORMER. A transformer in which the transfer of energy is from a low- to a' high-voltage winding.

STEREOPHONIC SYSTEM. A system of sound recording and reproduction employing groups of microphones, loudspeakers, and a corresponding number of channels, to give listeners an impression of direction and so enhance the realism of the reproduction.

STICKING POTENTIAL. The maximum attainable voltage between the screen and the cathode of a *cathode-ray tube* having a non-conducting *phosphorescent screen*. It represents the potential at which as many electrons arrive in the electron beam as *secondary electrons* leave the screen.

STOCHASTIC PROCESS. A random process.

STOPPING POTENTIAL. The *potential* necessary to bring to rest an electron emitted from a surface.

STORAGE (DEVICE). Any device or method of storing information by whatever physical or chemical means. Specifically, in a *computer*, the section used primarily for storing information.

STORAGE BATTERY. An *accumulator*. See *Storage cell*.

STORAGE CAPACITY. See *Memory capacity*.

STORAGE CELL. A *voltaic cell* which is reversible and which can be restored, after discharge, to its initial condition by passing a current through it opposite to that of discharge.

STORAGE EFFECT. In a *semiconductor junction*: temporary storage of injected excess *minority carriers* in the higher-resistivity side of the junction.

STORAGE ELEMENT. In a *charge-storage tube*, an area of the storage surface which retains information distinguishable from that of adjacent areas. Also a unit of a computer *memory*.

STORAGE REGISTER. See *Memory register*.

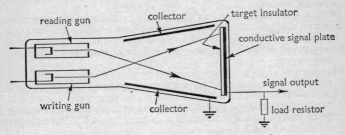

Fig. 151. Diagram of electrostatic storage tube

STORAGE TUBE. An electronic tube into which information can be introduced and later extracted. See *Charge-storage tube, Graphecon, Radechon, Electrostatic memory*.

STORE. (1) A *storage* system or device. (2) As a verb, introduce or retain information in such a device.

STRAIGHT-LINE CAPACITOR. A *variable capacitor* whose plates are shaped so that its *capacity* varies directly with the angle of rotation of the moving plates. Other forms of this component are designed so that resonant frequency or wavelength of the circuit in which it is used varies directly with the angle of rotation.

STRAIN GAUGE. A *transducer* which converts mechanical stresses and strains into electrical signals. An electric strain gauge usually consists of a *printed circuit* or wire-wound resistor, in which a change in stress produces a change in *resistance*. The electromagnetic type, used extensively in *telemetering* systems, consists of a small armature and coil, the *inductance* of which is caused to vary by changes in the armature gap which follow movements of the stressed structure being measured. The inductance variations are used to control the frequency of a transmitting oscillator.

STRAPPING. In a *multi-cavity magnetron*: connecting together resonating segments which have the same polarity, in order to suppress undesired *modes* of oscillation.

STRAY CAPACITANCE. In electric circuits: *capacitance* arising from the proximity of components, connecting wires, and metal chassis, etc. It is usually undesirable but is occasionally used as a part of the tuning capacity in the circuit.

STRETCH. A class of *digital computers* employing *transistors*, made by International Business Machines. They were claimed to be capable of more than one million logical operations per second and of performing more than one hundred thousand million (10^{11}) arithmetical operations per day.

STRIKING POTENTIAL. The *potential* necessary to start an *arc* between two electrodes.

STRING ELECTROMETER. An *electrometer* in which a fine quartz fibre with a thin metallic coating is stretched under tension between two parallel metal plates. The measured voltage is applied between fibre and plates, and the deflection of the fibre is read by a calibrated microscope.

STROBE PULSE. A *pulse* superimposed on the trace for the echo on a *radarscope* to act as a marker for measuring the range of the target.

STROBOSCOPE. An instrument for studying moving machinery (usually rotating) by flashing a light at the same repetition rate as the rotating or oscillating part. It consists of a variable frequency *multivibrator* which drives the flasher or *strobotron*.

STROBOTRON. A gas-filled *tetrode* designed to produce brilliant flashes of light at a rate which can be controlled by the frequency of a voltage applied to its control grid. It is used in the *stroboscope*.

STRONTIUM (Sr). Metallic element, Atomic No. 38.

STUB. A device used in *microwave* and *ultra-high-frequency* systems to match the *impedance* of a *transmission line* to a load. It usually consists of an adjustable section of line similar to the one it is matching, and its position and length are adjusted for maximum energy transfer.

SUBCARRIER. A *carrier* which is applied as a *modulating* wave to another carrier.

SUBHARMONIC. A *sinusoidal* quantity whose frequency is an integral submultiple of the *fundamental frequency* to which it is related. See *Harmonic*.

SUBMICRON. A particle having a diameter less than a *micron*.

SUBMINIATURE VALVE (TUBE). An extremely small valve designed for miniature equipment such as hearing aids. Typical dimensions are about $1\frac{1}{2}$ ins. long and less than $\frac{1}{4}$ in. diameter with the pins or wire leads emerging directly through the glass base.

SUB-ROUTINE. In a *computer,* a portion of a *routine* which causes the computer to carry out a well-defined logical or mathematical operation. See also *Closed sub-routine*.

SUBSONIC FREQUENCY. A frequency below the *audio* frequency range.

SUDDEN DEATH. The abrupt reduction of the current multiplication factor, alpha, in *point-contact transistors*, especially at low voltages, probably due to shock or moisture absorption. Compare *Slow death*.

SUHL EFFECT. The observed effect of the deflection of *holes* injected into a bar of *semiconductor* material by a magnetic field. This is an extension of the *Hall effect*.

SULPHUR (S). A reactive non-metallic element, Atomic No. 16.

SUMMATION CHECK. In a computer, a *redundant check* in which groups of digits are summed and the answer checked against a previously computed sum to verify accuracy.

SUPERCONDUCTIVITY. The property exhibited by some metals and metal compounds of a steady fall of *resistance* as their temperature is lowered, until, at a critical temperature within a few degrees absolute zero ($-273°C$), the resistance falls to zero or nearly to zero. In this condition, currents induced by a magnetic field in a ring of the material continue to circulate long after the magnetic field has been removed.

SUPERCONDUCTOR. A material such as tin, lead, or thallium, which exhibits *superconductivity*.

SUPER HIGH FREQUENCY. A frequency in the band 3,000 Mc/s to 30,000 Mc/s.

SUPER-EMITRON®. An *image iconoscope*.

SUPERHETERODYNE (RECEIVER). A *receiver* in which the incoming signal is mixed with a locally generated signal to produce an *intermediate frequency* which is then amplified and *demodulated*.

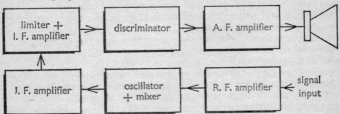

Fig. 152. Block diagram of a typical FM superheterodyne receiver

SUPERHETERODYNE CONVERTER. The section of a *superheterodyne* receiver in which the *radio-frequency* signal is converted to the *intermediate frequency* signal by the *local oscillator* and *mixer*. The last two devices are usually combined in one valve envelope.

SUPERHETERODYNE RECEPTION. See *Heterodyne*, *Superheterodyne*.

SUPERMALLOY®. A magnetic material of extremely high maximum *permeability*.

SUPERMENDUR®. A magnetic core material with a *rectangular hysteresis loop*.

SUPERREGENERATIVE RECEIVER. A *regenerative receiver* in which the *feedback* is greater than that required to produce oscillations, and a separate voltage is provided to quench the oscillations at regular intervals.

SUPERSENSITIVE RELAY. A *relay* which operates with very small currents, usually less than 250 microamperes.

SUPERSONICS. Formerly the study of high-frequency sound, but the term is now replaced by *ultrasonics*.

SUPERVISORY CONTROL. A system for selective control and automatic indication of remotely located units by electrical means, using a relatively small number of common transmission channels.

SUPPRESSED-CARRIER TRANSMISSION. A system of *transmission* in which the *carrier* component of a *modulated* wave is suppressed, only the *side frequencies* being transmitted.

SUPPRESSED-ZERO INSTRUMENT. A recording or indicating instrument which has its zero position below the limit of indication.

SUPPRESSOR GRID. Of a valve, a *grid* interposed between two positive electrodes, usually *screen-grid* and *anode*, primarily to reduce the flow of *secondary electrons* from one to the other.

SUPPRESSOR-GRID MODULATION. *Modulation* in which the modulating signal is applied to the *suppressor grid* of a *pentode* which acts as an amplifier for the *carrier* signal.

SURFACE BARRIER. A potential barrier at the surface of a *semiconductor* created by charge *carriers* trapped at the surface.

SURFACE CONDUCTIVITY. The *conductance* between two opposite sides of a unit square area on the surface of the conductor. The reciprocal of *surface resistivity*.

SURFACE LEAKAGE. The passage of current over the surfaces of a material rather than through its volume.

SURFACE NOISE. See *Noise, surface*.

SURFACE PASSIVATION. The application or growth of protective layers on the surface of a *semiconducting device*, after the formation of *p-type* and/or *n-type* layers by diffusion or alloying.

SURFACE RECOMBINATION VELOCITY. In a *semiconductor*, the velocity with which *electrons* and *holes* would have to drift to the surface of the material in order to account for the observed rate at which they tend to combine there, and are thus lost.

SURFACE RESISTIVITY. Of any material, the *resistance* between two opposite sides of a unit square of its surface. It may vary widely with the conditions of measurement. See *Surface conductivity, Resistivity*.

SURFACE WAVE. The component of a *ground wave* which travels along the surface of the earth.

SURFACE-BARRIER TRANSISTOR. See *Transistor, surface barrier*.

SURGE. A transient and abnormal rush of electricity along a conductor.

SURGE ELECTRODE CURRENT. See *Fault current*.

SURGE GENERATOR. An electrical apparatus suitable for the production of *surges*. See *Impulse generator*.

SUSCEPTANCE. The component of the current in *quadrature* with the applied voltage, divided by the voltage. The *reactance* divided by the square of the *impedance*. Symbol: *B*. Practical unit: *mho*. See also *Admittance*.

SUSCEPTIBILITY. Of a magnetic material, the ratio of the intensity of magnetization to the *magnetizing force*. See also *Dielectric susceptibility*.

SUSTAINED OSCILLATION. The *oscillation* of a system in which forces outside the system, but controlled by it, maintain the oscillation with a period that is nearly the *natural period* of the system. One example is a pendulum actuated by a clock movement.

SWEEP (CIRCUIT). A circuit which provides a *time* base for *cathode-ray tube* presentation, i.e. it causes the *spot* to move across the screen as a linear function of time and return abruptly to the starting point at the end of each sweep. See *Circular sweep*, *Expanded sweep*, *Linear sweep*, *Scanning*, *Slave sweep*.

SWEEP FREQUENCY. The rate at which the electron beam is swept to and fro across the face of a *cathode-ray tube*.

SWEEP GENERATOR. The *sweep* circuit in a *cathode-ray oscilloscope* which makes the deflection of the electron beam a known function of time so that other periodic quantities, e.g. modulated waves, may be compared with it.

SWEEP GENERATOR, MILLER. See *Miller sweep generator*.

SWEEP VOLTAGE. The voltage applied to a *cathode-ray tube* which produces the horizontal or vertical deflection, or both, of the *electron beam*.

SWING. The range of variation in amplitude or frequency of an electrical quantity. See *Grid swing*, *Frequency swing*.

SWINGING. Momentary variation in frequency of a received wave.

SWINGING CHOKE. A choke, generally a *smoothing choke,* whose effective *inductance* varies with the amount of current through it.

SWITCH. Any device for opening, closing, or directing an electric circuit. See the following kinds of switch: *Anti-capacitance*, *ATR*, *Band*, *Commutation*, *Double-pole*, *Electronic commutator*, *Function*, *Glow*, *Many-one function*, *Muting*, *One-many function*, *Single-pole*, *Time*, *TR*, *Vacuum*.

SWITCHING CONSTANT. In materials having a *rectangular hysteresis loop,* the ratio of *switching time* to *magnetizing force*.

SWITCHING TIME, FERROELECTRIC. The time required for complete reversal of the alignment of *electric dipoles* in a saturated sample of *ferroelectric* material.

SWITCHING TIME, FERROMAGNETIC. The time required for complete reversal of the *magnetic flux*, for a given *magnetizing force*, in a saturated sample of *square-loop ferromagnetic* material.

SWITCHING TUBE (VALVE). See *Transmit-receive switch*.

SWTL. Abbreviation for surface-wave transmission line. A single wire whose surface is threaded or coated with a suitable dielectric to reduce the *phase velocity* of a wave transmitted along it. The line is electrically equivalent to a *coaxial line* with its outer annular conductor removed to infinity.

SYLLABIC COMPANDING. *Companding* in which the effective gain variations are made to allow response to the syllables of speech but not to individual cycles of the signal wave.

SYMBOLIC LOGIC. See *Boolean algebra*.

SYMMETRICAL NETWORK. A *network* arranged so that a cut can be made through it to produce two mirror images.

SYMMETRICAL TRANSDUCER. See *Transducer, symmetrical*.

SYMPATHETIC VIBRATION. Vibration (oscillation) due to *resonance*.

SYNC. Abbreviation for synchronous or *synchronizing signal*.

SYNC SEPARATOR. In a *television receiver*: a circuit which separates *synchronizing signals* from *video signals*.

SYNCHRO. An electrical device for the instantaneous transmission or reception of the angular movements of rotating parts. The transmitter is a synchro generator, and the receiver a synchro motor. See *Autosyn, Selsyn, Servomechanism*.

SYNCHRO-CYCLOTRON ACCELERATOR (FM CYCLOTRON). A *cyclotron* in which the electric field is *frequency-modulated* to accelerate the particles to speeds approaching the speed of light.

SYNCHRONISM. The relation between two or more periodic quantities when the phase difference between them is zero. They are then said to be *in phase*.

SYNCHRONIZING LEVEL. In *television*, the level of the peak of the *synchronizing signal*.

SYNCHRONIZING SIGNAL. In *television*, the signal employed for the synchronization of *scanning*.

SYNCHRONOMETER. A device which counts the number of cycles in a given signal during a given time.

SYNCHRONOUS CLOCK. A clock driven by an electric motor whose speed is governed by the frequency of the alternating current supply.

SYNCHRONOUS COMPUTER. A *computer* in which the timing of all operations is controlled by equally-spaced signals from a clock.

SYNCHRONOUS CONVERTER. A *synchronous machine* which converts alternating to direct current or vice-versa.

SYNCHRONOUS GATE. A time *gate* in which the output intervals are synchronized with an incoming signal.

SYNCHRONOUS MACHINE. A machine whose average speed of normal operation is exactly proportional to the frequency of its electrical power supply. Examples are synchronous generators and motors.

SYNCHRONOUS VOLTAGE. Of *travelling-wave tubes*, the voltage required to accelerate electrons from rest to a velocity equal to the phase velocity of a wave in the absence of electron flow.

SYNCHROSCOPE. An *oscilloscope* designed for the observation of short pulses, using a fast *sweep* synchronized with the observed pulses.

SYNCHROTRON. An *accelerator* in which particles, usually electrons, are accelerated in circular orbits in an increasing magnetic field, by means of an alternating electric field synchronized with the orbital motion.

SYNCOM. Abbreviation for Synchronous Communication Satellite. The Syncom range of satellites, originally designed by the Hughes Aircraft Co. of the U.S.A., are intended to orbit at about 22,000 miles from the earth, where they synchronize with the rotation of the earth and appear to hover in a fixed location. They have facilities for relaying, amplifying, or storing communication signals which can be transmitted in this way between points separated by nearly one-third the circumference of the earth. See *Early Bird*.

T

TACAN. TACtical Air Navigation, an American ultra-high frequency radio navigation system for aircraft, in which both bearing and range are indicated directly by the airborne equipment for any ground or shipboard station to which the aircraft receiver is tuned.

TACHOGENERATOR. A rotating device used in *servo systems* which produces an output voltage proportional to the speed of rotation of its shaft.

TACHOMETER. An instrument for measuring speed of rotation.

TACITRON. A form of *thyratron* in which the flow of anode current can be interrupted by the voltage on the grid.

TAG. See *Sentinel*.

TAILING. Excessive prolongation of the decay of a signal-wave tail.

TANDEM. Two *networks* are in tandem when the two output terminals of one network are connected to the two input terminals of the other.

TANGENT GALVANOMETER. A suspension *galvanometer* whose indication is proportional to the tangent of the angle of deflection.

TANK. (1) See *Mercury tank*. (2) See *Tank circuit*.

TANK CIRCUIT. A circuit designed to store energy over a band of frequencies continuously distributed about a single frequency at which the circuit is said to be *resonant* or tuned. It is usually a *parallel resonant* circuit in the *anode* circuit of a valve oscillator, e.g. a *Hartley oscillator*, and the ratio of the energy stored to the energy dissipated is called the Q of the tank circuit.

TANTALUM (T). A metallic element, Atomic No. 73. It is very resistant to corrosion and is used in the construction of valves, lamps, capacitors, and other applications.

TANTALUM ELECTROLYTIC CAPACITOR. A very compact *electrolytic capacitor* in which the anode is made of tantalum foil or a sintered rod of tantalum. See Fig. 45.

TANTALUM RECTIFIER. An *electrolytic rectifier* in which the electrodes are tantalum and lead, and the electrolyte is dilute sulphuric acid.

TAPE. See *Magnetic tape*.

TAPE FEED. In *computing* and *date-processing* equipment, the part which supplies any kind of magnetic or paper tape to the reading or sensing components of the system.

TAPE READER. A device for translating information recorded on magnetic, punched paper, or other kind of tape into a series of electrical impulses which are usually transferred to some other *storage* medium. See Figure 153.

TAPE RECORDER. A system of recording *audio* signals which uses a *magnetic tape* magnetized along its length in accordance with the

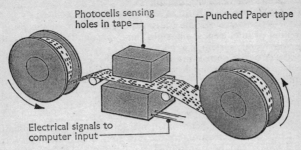

Fig. 153. Principle of punched tape reader

signals impressed on it. The term also includes the reproducing system.

TARGET. Of a *camera tube,* the structure whose surface is scanned by the electron beam to generate a signal output current corresponding to the charge-density pattern stored on the surface.

TARGET CAPACITANCE. The *capacitance* between the scanned area of a *camera-tube target* and the *backplate.*

TARGET CUT-OFF VOLTAGE. In *camera tubes,* the lowest target voltage at which any detectable electrical signal corresponding to a light image on the sensitive surface of the tube can be obtained.

TARGET VOLTAGE. In a *camera tube* with low velocity scanning, the potential difference between the thermionic cathode and the *backplate.*

TASI®. Time Assignment Speech Interpolation. A high-speed switching and transmission system for telephone circuits designed by Bell Laboratories, in which the pauses in the speech of one subscriber are used to interpolate the speech of another subscriber using the same channel.

TEARING. Of a *television* picture, the apparent breaking up of the picture which is caused by faults in the synchronizing circuits.

TECNETRON®.Trade name for a unipolar *field-effect transistor* invented in France, which uses centripetal striction to vary the conductance of the *semiconductor.*

TEFLON®. Polytetrafluoroethylene, an insulating material with extremely high *resistivity* which is little affected by heat or moisture.

TELCOMP. A conversational computing system developed in the U.S.A. which allows man and machine to interact closely with one another when problem-solving. The language used comprises only 22 instructions and can be learnt in a few hours.

TELEAUTOGRAPH. See *Telewriter.*

TELECAMERA. See *Camera, television*.

TELECHROME. An early form of *television colour tube*.

TELECOMMUNICATION. The transmission, emission, or reception of any kind of intelligence, e.g. signs, images, sounds, by any kind of electromagnetic system, e.g. wire, radio, or visual.

TELECONTROL. Control of any device by *telecommunication*. A spectacular recent example was the soft landing on the moon of Lunik 4.

TELEGRAPHY. A system of telecommunication for the transmission of graphic symbols, usually letters or numerals, by the use of a signal code. See *Carrier telegraphy, Radiotelegraphy*.

TELEMETER. An instrument which measures any physical quantity, e.g. temperature, pressure, cosmic radiation, etc., converts it into a signal which can be transmitted, e.g. voltage, pressure, etc., and transmits the measurements to a distant station.

TELEMETERING. Measurement at a distance using a *telemeter*. The distinctive feature in telemetering is the means used for translation and transmission, the actual distance being irrelevant.

TELEMETERING CHANNEL. The route required to convey the magnitude of a single measured quantity in *telemetering*.

TELEMETERING RECORD. A chart of a telemetered signal plotted against time on film, tape, or paper.

TELEPHONE CHANNEL. A channel used in *telephony* for the transmission of telephone signals.

TELEPHONE TRANSMITTER. A *microphone* for use on a telephone system.

TELEPHONY. A system of *telecommunication* for the transmission of speech or other sounds, with or without connecting wires. A complete system includes a transmitter for converting the sound vibrations into electrical signals (the telephone); transmission circuits; and a receiver for converting the electrical signals back into sounds. See *Carrier telephony, Quiescent-carrier telephony, Radiotelephone*.

TELEPRINTER. A system of telecommunication which employs a typewriting device at both transmitting and receiving ends.

TELETYPE®. A printing telegraph apparatus produced by the Teletype Corporation.

TELEVISION. The electrical transmission and reception of transient visual images; the branch of electronics which deals with this. In popular use the term includes the actual pictures or programme shown on the *television screen* and also the *television receiver*. In current technique, light waves from the subject are converted by a *television camera* into electronic impulses which are then transmitted from a *television transmitter* and received and reconverted into *electron beams* projected on to the *cathode-ray screen* to form the picture. See *Closed-circuit television, Colour television*, and the entries following.

TELEVISION APERTURE. The diameter of an individual picture

element into which the televised object is divided for *scanning* and for production of transmitted signals.

TELEVISION BROADCASTING. *Radio* transmission of synchronized picture and sound signals intended for general reception.

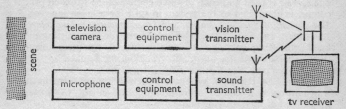

Fig. 154. Elements of a television broadcasting system

TELEVISION CAMERA. See *Camera, television.*

TELEVISION CAMERA TUBE. See *Camera tube.*

TELEVISION CHANNEL. A channel suitable for transmission of television signals. It may or may not include the associated channel for sound signals.

TELEVISION DIRECT TRANSMISSION. A television system in which the object to be televised is focused directly on to the *scanning* device.

TELEVISION FIELD. One of the two or more equal parts into which the *frame* is divided in *interlaced scanning.*

TELEVISION FRAMING. Adjustment of the picture to a desired position with respect to the field of view, usually a central position.

TELEVISION RECEIVER. A *receiver* for conversion of incoming electric signals into television pictures and associated sounds. It normally includes a *superheterodyne* radio-frequency receiver and separate *intermediate amplifiers, detectors,* and output units for the *audio* and *video* signals, together with the supply and control circuits for the *cathode-ray tube* which receives the video output.

TELEVISION SCANNING. The process of analysing successively, according to a predetermined method, the light values of picture elements constituting the whole picture area.

TELEVISION SCREEN. A cathode-ray screen used in a television receiver.

TELEVISION SYNCHRONIZING SIGNAL. See *Synchronizing signal.*

TELEVISION SYSTEM, CLOSED CIRCUIT. See *Closed-circuit television system.*

TELEVISION SYSTEM, COMPATIBLE. A *colour television* system which allows existing black and white television receivers to view the colour transmissions in black and white.

TELEVISION SYSTEM, DOT-SEQUENTIAL COLOUR. See *Dot-sequential colour television.*

TELEVISION SYSTEM, LINE-SEQUENTIAL COLOUR. See *Line-sequential colour television*.

TELEVISION SYSTEM, SEQUENTIAL COLOUR TRANSMISSION. A system of transmitting television signals in colour in which only one *carrier* is used and each of three cameras, one for each primary colour, is connected to the system in turn for a given interval. See *Dot-sequential, Field-sequential,* and *Line-sequential colour television,*

TELEVISION TRANSMITTER. The equipment for transmitting modulated radio-frequency power representing a complete television signal, including *audio, video,* and *synchronizing* signals.

TELEVISION TUBE. See *Camera tube, Picture tube*.

TELEWRITER. A telegraph system in which writing movement at the transmitting end causes corresponding movement of a writing instrument at the receiving end.

TELEX. An audio-frequency teleprinter system provided by the G.P.O. for use over telephone lines.

TELLURIUM (Te). An element similar to sulphur, Atomic No. 52.

TELLUROMETER®. A navigational device which measures distances using phase comparison.

TELSTAR. A series of active satellites designed and built by the Bell Telephone Laboratories to obtain data on communication by satellite. One of the series, a rough sphere about a yard in diameter, was launched in 1962 and provided the first live television exchange between America, Britain and France.

TEMPERATURE COEFFICIENT OF RESISTANCE. The fractional change in *resistance* or *resistivity* of a material or *resistor* per degree change in temperature. It is positive for most metals and negative for many *semiconductors* and *non-metals*.

TEMPERATURE COEFFICIENT OF VOLTAGE DROP. In *glow discharge tubes,* the ratio of the change of tube voltage drop to the change of ambient temperature.

TEMPERATURE SATURATION. Of a *valve,* that point in the operation of the valve at which, if the anode and other electrodes are held at constant potential, an increase in the cathode temperature is not followed by a corresponding increase in the valve current. The effect is due to the formation of a *space charge* between the electrodes near the cathode which repels electrons back to the cathode.

TEMPERATURE-LIMITED EMISSION. In a *thermionic valve,* the condition which exists in the valve when all of the electrons emitted by the cathode are collected by the anode and other electrodes, and any further increases of potential on any of the electrodes has negligible effect on the anode current. See *Current saturation, Schottky effect*.

TEMPORARY MEMORY. In a *computer,* internal *storage* locations reserved for intermediate or partial results.

TENSION. A synonym for *voltage* as in 'high tension'.

TERACYCLE. A million, million cycles per second, or 10^{12} c/s.

TERATRON. A generator developed at the National Physical Laboratory, capable of producing oscillations at a frequency of about 1 million, million cycles per second. It is based on stimulated emission from molecules of cyanogen which have been produced by passing a high voltage, high current, pulsed discharge through the vapour of acetonitrile in a tube.

TERMINAL IMPEDANCE. The complex *impedance* at the unloaded output or input terminals of a transmission line or equipment which is otherwise in its normal operating condition.

TERMINATING. The closing of the circuit at either end of a line or *transducer* by the connection of some device.

TESLA. The unit of magnetic flux density in the *SI* and *mksa* electromagnetic systems.

TESLA COIL. An *induction coil* used to develop a high-voltage, high-frequency, oscillatory discharge.

TEST PATTERN. In *television broadcasting,* a chart transmitted for general testing purposes, or for a prescribed time before the transmission of programmes. Test patterns are also produced by special testing equipment for bench testing of television receivers.

TEST ROUTINE. In *computers*, a *check routine* or *diagnostic routine*.

TETRODE. A valve containing four electrodes: anode, cathode, control electrode, and one additional electrode which is usually a *screen grid*.

TETRODE, TRANSISTOR. See *Transistor tetrode*.

THALLIUM (Tl). A metallic element with radioactive isotopes used in scintillation crystals. Atomic No. 81.

THEORETICAL CUT-OFF FREQUENCY. See *Cut-off frequency*.

symbol

THERMAL BATTERY. A voltage source consisting of a number of bimetal junctions connected so as to produce a voltage when heat is applied. The voltage produced is a measure of the current supplied to the heater.

THERMAL EXPANSION INSTRUMENT. See *Hot-wire instrument*.

THERMAL IMAGING DEVICE. A device which uses the infra-red radiation emitted by objects in the field of view to produce a picture of the scene in terms of thermal constants. In contrast to television this technique does not require an external source of radiation or illumination.

THERMAL INSTRUMENT. Any instrument which depends on the heating effect of a current.

THERMAL NOISE. *Random noise* due to the thermodynamic interchange of energy between a material and its surroundings.

THERMAL RESISTANCE. Of a *semiconductor device*: the temperature rise (per unit power dissipation) of a junction above the temperature of the ambient or of a stated external reference point under conditions of thermal equilibrium.

THERMAL TUNING TIME. Of a *microwave valve* or *tube*, the time required to tune through a specified frequency range when the tuning power is instantaneously changed from the specified maximum to zero (cooling time) or from zero to the specified maximum (heating time).

THERMION. A positively or negatively charged *ion* emitted by an incandescent material.

THERMIONIC CATHODE. A *cathode* which functions primarily by the process of *thermionic emission*.

THERMIONIC EMISSION. Emission of *electrons* or *ions* due to the high temperature – i.e. the high thermal energy – of the emitter. See *Edison effect, Richardson–Dushman equation*.

THERMIONIC GRID EMISSION. In a *thermionic valve*, current produced by thermionic emission of electrons from a *grid*.

THERMIONIC VALVE (TUBE). A *vacuum valve* (tube) in which one of the electrodes, normally the *cathode*, is heated for the purpose of causing *electron* or *ion emission* from that electrode.

THERMIONIC WORK FUNCTION. See *Work function, thermionic*.

THERMISTOR. A *resistor* of special material whose value decreases with temperature in a definite desired manner. Thermistors have a number of applications: compensation for temperature variations of other components; use as non-linear circuit elements; measurement of temperature and power. See *Barretter, Bead thermistor, Bolometer, Thermistor bridge*.

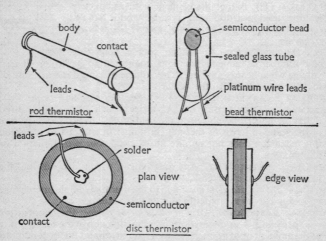

Fig. 155. Construction of types of thermistor

THERMISTOR BRIDGE. A *bridge* network used for measuring power by the change in the resistance of *thermistors* in the bridge arms.

THERMOAMMETER. See *Thermocouple ammeter*.

THERMOCOUPLE. A pair of different conductors joined so as to produce a *thermoelectric effect*.

symbol

THERMOCOUPLE, VACUUM. A *thermocouple* used to measure very small currents. The couple is usually between platinum and a platinum–rhodium alloy of which one junction is attached to a fine platinum wire to carry the current to be measured. The assembly is sealed in an evacuated container.

THERMOCOUPLE AMMETER. An *ammeter* employing a thermocouple as the measuring instrument.

THERMOCOUPLE INSTRUMENT. A *thermal instrument* in which one or more thermocouples are heated directly or indirectly by the current to be measured and supply a direct current to operate a suitable direct-current meter.

THERMOCOUPLE VOLTMETER. See *Voltmeter, thermocouple*.

THERMOCOUPLE WATTMETER. See *Wattmeter, thermocouple*.

THERMOELECTRIC EFFECT. An *electromotive force* resulting from a difference in temperature between two junctions of different metals in contact in the same circuit. See *Peltier effect*.

THERMOELECTRIC MODULE. An assembly using the *Peltier effect* to provide built-in cooling or automatic temperature control for compact equipment and devices energized with direct current.

THERMOELECTRON. An electron liberated by heat. Also called negative *thermion*.

THERMOELEMENT. A device consisting of a *thermocouple* and a heating element arranged for measuring small currents.

THERMOGRAPHY. The use of a *thermal imaging device* to produce pictures of the pattern of infra-red radiation emitted by an object or scene.

THERMOJUNCTION. One of the surfaces of contact between the two different conductors of a *thermocouple*.

THERMOMAGNETIC. Descriptive of the mutual effects of heat and magnetic fields in certain bodies.

THERMOPHONE. An electroacoustic *transducer* which produces sound waves of calculable magnitude from the expansion and contraction of the air near a conductor whose temperature varies in response to a current input.

THERMOPILE. A device used either to measure radiant power or as a source of electric energy, consisting of a number of *thermocouples* connected in series so that their outputs are added together.

THERMOREGULATOR. A high accuracy or high sensitivity *thermostat*.

THERMOSTAT. A device consisting of a *bimetal strip* arranged to open and close the contacts of a heating (or cooling) circuit so that the temperature of a medium or equipment can be maintained constant or within two set limits.

THEVENIN'S THEOREM. This states that an electrical *network* may be replaced by an *electromotive force*, equal to the open circuit voltage of the network, in series with an *impedance*. This series impedance is the impedance measured back into the network with all the internal *e.m.f.s.* reduced to zero but with their internal impedances remaining in the network.

THICK-FILM RESISTOR. Films of resistive glazes and *cermets* up to about 1 mil thick, fired on to glass or ceramic substrates, are used to provide stable *resistors*.

THIN-FILM CIRCUIT. A miniature electronic circuit which dispenses with separate containers for passive elements, and which is constructed by depositing those elements, together with their interconnections, in the form of a thin film on a glass or ceramic substrate. Active components – transistors, valves, etc. – are added afterwards by bonding, welding, or soldering the contact tabs of diodes and transistors to suitable contact areas of the circuit. See *Integrated circuits, Micro-electronics*.

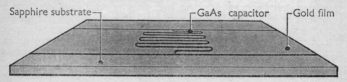

Fig. 156. Thin-film circuit capacitor formed by laser

THIN-FILM MEMORY. A computer *memory* made by evaporating a thin film of magnetic material on to a heated glass base in the presence of a dc magnetic field parallel to the surface of the base. Thousands of such elements can be formed in one operation to produce a large-capacity magnetic memory.

THIN-FILM TRANSISTOR (TFT). A *field effect transistor* in which a semiconductor film forms one plate and a metal electrode the other plate of a parallel-plate capacitor. The capacitor is filled by a suitable solid dielectric layer. See Figure 157.

THIN-WALL (THIN-WINDOW) COUNTER TUBE. A *counter tube* in which a defined part of the enclosure has low absorption for the radiation to be measured.

THOMAS RESISTOR. A standard *resistor* made of *manganin* which is annealed in an inert atmosphere and sealed into a container.

THOMSON EFFECT. When current flows between portions of a conductor which are at different temperatures, heat is liberated or absorbed depending on the direction of the current and the conducting material.

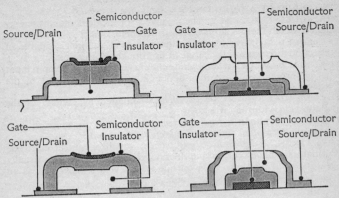

Fig. 157. Thin-film transistors

THOMSON SCATTERING. Scattering of electromagnetic radiation by electrons.

THORIATED-TUNGSTEN FILAMENT. A tungsten filament of a valve to which a small amount of thorium has been added to improve the *emission*.

THORIUM (Th). Radioactive element, Atomic No. 90.

THREE-GUN COLOUR PICTURE TUBE. See *Colourtron.*

THREE-LEVEL MASER. A *solid-state maser* in which three energy levels are used.

THREE-PHASE CIRCUIT. A circuit in which the alternating voltages differ in *phase* by one third of a cycle or 120°.

THRESHOLD CURRENT. Of a *gas discharge,* the current at which the discharge becomes self-sustaining.

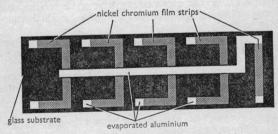

Fig. 158. Thin-film resistor networks interconnected by evaporated aluminium

THRESHOLD OF AUDIBILITY. See *Audibility threshold*.

THRESHOLD FREQUENCY. The frequency of incident radiant energy below which there is no *photo-emissive effect*.

THRESHOLD SIGNAL. In navigation, the smallest signal capable of effecting a recognizable change in positional information.

THRESHOLD VALUE. In *automatic control systems*, the minimum input which produces a corrective action in the power element of the system.

THROAT MICROPHONE. A *microphone* actuated by contact of the diaphragm with the throat.

THROUGH PATH. In a *feedback control loop*, the transmission path from the *loop input signal* to the *loop output signal*.

THYRATRON. A *hot-cathode gas tube* in which a control electrode initiates the anode current but does not normally limit it. It is usually a triode but may also be a tetrode and is very widely used as an electronic switch in control circuits.

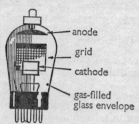

anode

grid

cathode

gas-filled glass envelope

Fig. 159. Construction of thyratron

THYRATRON, HIGH-VOLTAGE HYDROGEN. A specially constructed *thyratron* filled with hydrogen used as a fast switch in *pulse modulators*.

THYRATRON CONTROL CHARACTERISTIC. The relation between the anode voltage and the critical grid voltage which causes the *thyratron* to fire.

THYRATRON EXTINCTION. The cutting off of the anode current of a *thyratron* which must be accomplished by an external circuit if the thyratron is used in dc applications.

THYRATRON FIRING ANGLE. The electrical angle between the time a sinusoidal voltage applied to the *thyratron* anode passes through zero, rising in the positive direction, and the time at which the thyratron fires.

THYRATRON INVERTER. An *inverter* circuit in which *thyratrons* convert dc power to ac power.

THYRISTOR. A *transistor* having a thyratron-like characteristic. See *Silicon controlled rectifier*.

THYRITE. Trade name for silicon carbide with a high negative *temperature coefficient* of *resistivity*.

TIME BASE. A circuit, usually including a *sawtooth oscillator*, which provides a periodic rise and fall in voltage or current for the *sweep* of an electron beam along prescribed paths on a cathode-ray screen so that the resultant trace is a function of time. In a *radar* display this trace appears as a bright line which is synchronized with transmitted pulses, so that the range of an object can be measured as the distance of the echo signal along the line. In *television*, time bases are employed for line and frame *scanning* and are synchronized by pulses in the television waveform, or by the properties of the scanning circuits as in the *flywheel time base*. See also *Miller sweep generator*.

TIME CONSTANT. The time required for an electrical quantity to rise to 63·2 per cent of its final value or to fall to 36·8 per cent of its initial value.

TIME DISCRIMINATOR. A circuit in which the sense and magnitude of the output are functions of the time difference and relative time sequence of two *pulses*.

TIME GATE. A transducer which gives an output only during chosen time intervals.

TIME MODULATION. *Modulation* in which a time interval associated with a transmission varies in accordance with the signal. The interval chosen may be the length of a *pulse* or the interval between two pulses.

TIME SHARING. A technique employed in high speed *digital computers*, based on the principle of *time-division multiplexing*, to enable independent programmes to be processed concurrently in the computer.

TIME SWITCH, AUTOMATIC. A device for opening or closing, or changing the connections in, an electric circuit at predetermined times.

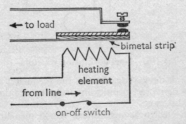

Fig. 160. Principle of a thermal time-delay relay

TIME-DELAY RELAY. A *relay* which incorporates a time-delay mechanism which functions between the instant that the relay is energized and the time when the relay contacts open or close.

TIME-DELAY SWITCH, AUTOMATIC. A switch for carrying out a predetermined sequence of operations with definite time intervals between each operation.

TIME-DIVISION MULTIPLEXING. The transmission of two or more signals over a common path using different time intervals for the different signals. See *Commutation switch.*

TIME-INTERVAL METER. An instrument capable of measuring the time interval between two events or the average time interval between a series of events.

TIME-SHARING. In *computers*: a form of concurrent working in which equipment is used to deal with two or more sequences concurrently by sharing the time of parts of the equipment among the sequences.

TIN (Sn). Metallic element with good resistance to corrosion. Atomic No. 50. It alloys readily with copper and other good conductors and is extensively used in electronic equipment for making permanent *ohmic contact.* See *Solder.*

TIROS. A series of Television and Infra-Red Observation Satellites. Tiros I, designed and built by R.C.A., was launched in April 1960 and transmitted some 14,000 usable cloud pictures to ground stations during 78 days of operating life.

TITANIUM (Ti). Metallic element, Atomic No. 22.

T-JUNCTION. In *waveguide* technique, a junction between a main waveguide and a perpendicular branch waveguide.

T-NETWORK. A *network* of three *impedance* branches connected in the form of a 'T'.

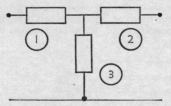

Fig. 161. T-network using resistors

TONE CONTROL. A device for altering the frequency response of an *audio-frequency amplifier*, usually in order to obtain a result more pleasing to the listener.

TONOMETER. An electronic instrument which measures hydrostatic pressure within the eye.

TOROIDAL COIL. A coil wound in the form of a toroidal helix.

TORQUE AMPLIFIER. An *analogue computer* device with input and output shafts which supplies the work required to rotate the output shaft in positional correspondence with the input shaft, but without exerting significant torque on the input shaft.

TORQUE MOTOR. A motor designed to exert torque while stalled or rotating slowly.

TOTAL CAPACITANCE. Of any conductor in a system containing other conductors, the *capacitance* between that conductor and all the other conductors connected together.

TOTAL ELECTRON BINDING ENERGY. The energy required to remove all the electrons surrounding the nucleus of an atom to infinite distance from the nucleus and from each other leaving only the bare nucleus.

TOTAL POWER LOSS. Of a *semiconductor device*: the sum of losses due to forward and reverse currents under specified conditions.

TOTAL WORK FUNCTION. The total work done by an electron in freeing itself from the binding forces of a solid. See *Work function, thermionic*.

TOUCH CONTROL. A circuit which closes a relay when two metallic areas are bridged by a finger or hand.

TOWNSEND AVALANCHE. A breakdown process, which is not self-sustaining, occurring in *gas-discharge tubes*, *semiconductors*, and *Geiger counters*. It is promoted by high-energy *ions* and electrons and it is supposed that collisions between ions lead to the productions of additional ions which in turn lead to further collisions, the process building up like an avalanche.

TOWNSEND DISCHARGE. A *discharge* in a gas due to ionization which is not a result of the applied voltage. See *Townsend avalanche*.

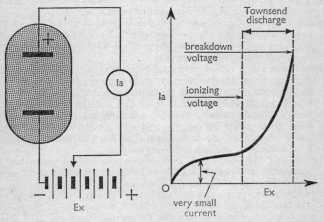

Fig. 162. Townsend discharge – circuit and graph showing typical behaviour in gas-discharge tube

TR SWITCH. See *Transmit-recieve switch*.

TRACE INTERVAL. In a *cathode-ray tube*, the interval corresponding to the direction of the *sweep* used for delineation.

TRACK. In *computers*, that portion of a moving *storage* medium, e.g. film, drum, or discs, which is accessible to a given reading station.

TRACKING. (1) The maintenance of the correct frequency relations in circuits which are designed to be varied together by ganged operation. (2) The process of keeping a radio beam set on a target while determining the range of the target. (3) The accuracy with which the stylus of a gramophone pick-up follows a prescribed path.

TRAILING EDGE. See *Pulse trailing edge*.

TRANSADMITTANCE. Of a valve, the ratio of the alternating component of the current in the second electrode (of a pair of electrodes) to the alternating component of the voltage in the first electrode, all other electrode voltages being maintained constant.

TRANSADMITTANCE, FORWARD. Of any two gaps between *electrodes*, the complex ratio of the fundamental component of the short-circuit current induced in the second of the gaps, to the fundamental component of the voltage across the first.

TRANSCONDUCTANCE. Between two electrodes of a vacuum valve: the ratio of the in-phase component of the alternating current of a second electrode to the alternating voltage on the first electrode, all electrode voltages other than that of the first remaining constant.

TRANSCONDUCTANCE METER. An instrument for indicating the *transconductance* of a grid-controlled valve.

TRANSCRIBER. Equipment associated with a *computer* for the purpose of transferring input or output data from a record of information in a given *language*, to the medium and language used by a digital computer, or vice versa.

TRANSDUCER. A device which can be actuated by *waves* from one or more media or transmission systems, and supply related waves to one or more other media or transmission systems. The term 'wave' is used in the widest possible sense and includes such concepts as energy, signal, current, voltage, pressure, motion, etc., whether these are constant or varying, and whether the input and output waves are of the same or different type. See *Conversion transducer*, *Ideal transducer*, and the following entries.

TRANSDUCER, ACTIVE. A *transducer* whose output waves are dependent on and control sources of power independent of the power supplied by any of the actuating waves.

TRANSDUCER, BILATERAL (BIDIRECTIONAL). Any *transducer* which is not a *unilateral* (unidirectional) *transducer*.

TRANSDUCER, CONVERSION. See *Conversion transducer*.

TRANSDUCER, DISSYMMETRICAL. A *transducer* is dissymmetrical with respect to a pair of terminations when the interchange of that pair will affect the transmission. Compare *Transducer, symmetrical.*

TRANSDUCER, ELECTRIC. A *transducer* in which all the waves concerned are electric.

TRANSDUCER, ELECTROACOUSTIC. A *transducer* for receiving waves from an electric system and delivering waves to an acoustic system, e.g. a *loudspeaker*, or vice versa, e.g. a *microphone.*

TRANSDUCER, ELECTROMECHANICAL. A transducer for receiving waves from an electric system and delivering waves to a mechanical system, e.g. an electric motor, or vice versa, e.g. a *piezolectric crystal.*

TRANSDUCER, HARMONIC CONVERSION. See *Frequency divider, Frequency multiplier.*

TRANSDUCER, HETERODYNE CONVERSION. See *Converter.*

TRANSDUCER, IDEAL. See *Ideal transducer.*

TRANSDUCER, LINEAR. A *transducer* for which all the waves concerned are related by linear functions.

TRANSDUCER, MODE. See *Mode transducer.*

TRANSDUCER, PASSIVE. A *transducer* whose output waves are independent of any sources of power controlled by the actuating waves.

TRANSDUCER, REVERSIBLE. A *transducer* whose *loss* is independent of the direction of transmission.

TRANSDUCER, SYMMETRICAL. A *transducer* is symmetrical with respect to a specified pair of terminations, or *port*, when the interchange of that pair will not affect the transmission.

TRANSDUCER, TWO-PORT LINEAR. A *transducer* with one input *port* and one output port.

TRANSDUCER, UNILATERAL (UNIDIRECTIONAL). A *transducer* which cannot be actuated at its output by waves in such a manner as to supply related waves at its input.

TRANSDUCER GAIN. Of a *transducer* connecting a specified source to a specified load under specified operating conditions, the ratio of the power delivered to the load, to the *available power* of the source. Compare *Available power gain.*

TRANSDUCER INSERTION GAIN. See *Insertion gain.*

TRANSDUCER INTERACTION FACTOR. See *Interaction factor.*

TRANSDUCER LOSS. Of a *transducer* connecting a specified source and load under specified conditions, the transmission loss expressed by the ratio of the power available from the source to the power delivered to the load. Usually expressed in *decibels.*

TRANSDUCTOR®. A *saturable reactor.*

TRANSFER. In *computers*, transmit or copy information from one device to another. Also, in programming, *jump.*

TRANSFER ADMITTANCE, TRANSDUCER. From one pair of terminals of an electrical *transducer* to another pair, the complex ratio of the current at the second pair of terminals to the electromotive force applied between the first pair, all pairs of terminals being terminated in any specified manner.

TRANSFER CHARACTERISTIC. (1) A relation between the voltage of one electrode and the current to another electrode, all other electrode voltages being maintained constant. (2) Of *camera tubes*, a relation between the illumination of the tube and the corresponding output signal current, under specified conditions of illumination.

TRANSFER CURRENT. In a *glow discharge cold-cathode tube*, the *starter-gap* current required to cause conduction across the *main gap*. In *gas tubes*, the current to one electrode required to initiate *breakdown* to another electrode.

TRANSFER FUNCTION. The function relating the output of a *closed-loop servo* system to its input. See *Actuating transfer function, Difference transfer function, Forward transfer function, Loop transfer function*.

TRANSFER IMPEDANCE. Between any two pairs of terminals of a *network*, the ratio of a potential difference applied at one pair of terminals to the resultant current at the other pair, all terminals being terminated in any specified manner.

TRANSFERRED ELECTRON EFFECT. If an electric field is applied and increased across a semiconductor the velocity of the charge carriers at first increases linearly with the field, then decreases, and finally increases again at a reduced rate. This central region of the carrier velocity curve, which has a negative slope, occurs in gallium arsenide at about 3000 V/cm and is observed as the *Gunn effect*.

TRANSFLUXOR®. A magnetic core having two or more apertures and three or more limbs carrying flux, used as a computer memory element, switch, or control element.

TRANSFORM. In *digital computers*: to change the form of information without significantly altering its meaning.

TRANSFORMER. A device without moving parts, which by *electromagnetic induction* transforms electrical alternating or intermittent energy from one or more circuits to one or more other circuits at the same frequency, usually with changed values of voltage and current. The following types of transformer are described in this book: *Autotransformer, Constant-current, Constant-potential, Coupling, Current, Doorknob, Filament, Instrument, Intermediate-frequency, Modulation, Output, Peaking, Power, Primary winding, Pulse, Radio-frequency, Rectifier, Repeating coil, Step-down, Step-up, Tesla coil, Transformer, Ideal, Tuned, Variac, Variocoupler, Wave guide*.

symbol

TRANSFORMER, IDEAL. A hypothetical device with the following characteristics: *coupling coefficient* equal to unity; *self-inductances* of the windings so large that their *reactances* can be considered infinite; ratio of the self-inductances exactly equal to the square of the turns ratio; no *resistive, hysteresis, eddy-current*, or any other kind of loss; no *capacitance*. An ideal transformer is a useful device in the analysis of networks and electronic circuits.

TRANSFORMER, THREE-PHASE. A *transformer* for *three-phase* ac circuits with windings placed on separate legs of a common core.

TRANSFORMER COUPLING. *Inductive coupling.*

TRANSFORMER PRIMARY. The input winding of a transformer.

TRANSFORMER RATIO. The ratio between the numbers of turns on primary and secondary windings.

TRANSFORMER SECONDARY. The output winding of a transformer, usually connected to the load.

TRANSFORMER TAP. A connection brought out of a transformer winding at some point between its extremities, usually to enable the voltage ratio to be changed.

TRANSIENT. A term applied to phenomena, usually *damped oscillatory* motion, voltage, current, etc., which take place in a system owing to a sudden change of conditions, and which persist for a relatively short time after the change has occurred.

TRANSIENT ANALYSER. A device for producing a repeated succession of equal electric *transients* in a test circuit, and displaying their waveform, which is usually adjustable, on an *oscilloscope.*

TRANSIENT RESPONSE. Of an *amplifier* or other device, the response to a *transient* of the input signal, or to an input voltage suddenly applied.

TRANSISTOR. A *semiconductor device* capable of providing amplification and having three or more electrodes. See *Field-effect transistor, Junction transistor, Phototransistor, Point-contact transistor,* and transistor entries following.

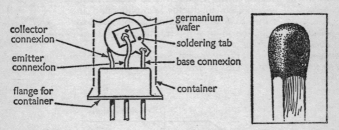

Fig. 163. A typical germanium transistor and container compared with a match

TRANSISTOR, BI-DIRECTIONAL. A *transistor* which has substantially the same electrical characteristics when the emitter and collector terminals are interchanged.

TRANSISTOR, BIPOLAR. A *transistor* which makes use of both positive and negative *charge carriers.* Contrast *Transistor, unipolar.*

TRANSISTOR, CONDUCTIVITY MODULATION. A *transistor* in

which the active properties are derived from *modulation* by *minority carriers* of the bulk *resistivity* of a *semiconductor*.

TRANSISTOR, FILAMENTARY. A *conductivity-modulation transistor* whose length is much greater than its transverse dimensions.

TRANSISTOR, GRADED-BASE. A transistor in which the resistivity of the material of the base increases smoothly between emitter and collector junctions.

TRANSISTOR, HOOK. A *junction transistor* which uses an extra *p-n junction* to act as a trap for *holes* and so increase the amount of current amplification.

TRANSISTOR, JUNCTION. A *transistor* having a *base electrode* and two or more electrodes connected to a *semiconductor junction*. See *Transistor, hook, Transistor, n-p-n, Transistor, p-n-p*.

TRANSISTOR, N-P-N. A *transistor* which consists essentially of a thin slice of *p-type semiconductor* between two pieces of *n-type semiconductor*. The middle slice forms the *base* and the n-type pieces form the *collector* and *emitter*.

TRANSISTOR, P-N-I-P. A *p-n-p transistor* with a layer of *intrinsic semiconductor* (high-purity germanium) between the *base* and *collector*, designed to extend the high-frequency range. In *n-p-i-n* transistors this construction is applied to an *n-p-n transistor*.

symbol
(with collector connected to envelope)

TRANSISTOR, P-N-P. A *transistor* in which a thin slice of *n-type semi-conductor* is sandwiched between two pieces of *p-type* semiconductor. Amplification in this type is due to *hole* conduction, as opposed to *electron* conduction in an n-p-n transistor. See *Semiconductor junction, p-n-p* and *n-p-n*.

symbol

TRANSISTOR, POINT-CONTACT. See *Point-contact transistor*.

TRANSISTOR, POWER. A *transistor* designed specially to handle

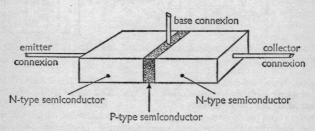

Fig. 164. Basic structure of an n-p-n transistor

relatively large power, i.e. between 1 and 100 watts. Apart from the increased size of such transistors compared to models which handle a fraction of a watt, the design takes into account the necessity for efficient cooling to prevent damage due to excessive temperature rise. See Fig. 112.

TRANSISTOR, SURFACE BARRIER. A *transistor* designed for operation at frequencies up to and above 100 mc/s, produced by accurate electrochemical etching and plating techniques which allow extremely thin barriers to be used in the *semiconductor*.

TRANSISTOR, UNIPOLAR. A *transistor* whose operation depends on *charge carriers* of one polarity only as opposed to *bipolar* transistors. See *Tecnetron, Transistor, bipolar*.

TRANSISTOR AMPLIFIER. An *amplifier* using *transistors* to control the power from a local source.

TRANSISTOR AMPLIFIER, COMPLEMENTARY. A *transistor amplifier* which makes use of the complementary symmetry of *n-p-n* and *p-n-p* transistors which have substantially identical characteristics but the polarities of voltages and currents reversed. *Push-pull amplifiers* using only transistors can readily be constructed from such components. See *Transistor, n-p-n*; *Transistor, p-n-p*.

TRANSISTOR CHARACTERISTICS. Sets of curves showing the relations between input and output currents and voltages, according to the type of connection of the transistor. The three main configurations are: *common-base connection, common-collector connection, common-emitter connection*.

TRANSISTOR CURRENT GAIN (CURRENT-AMPLIFICATION FACTOR). For a *junction transistor* with *common-base connection* the current gain, alpha, can be defined as the ratio of the small current change in the *collector* to the small current change in the *emitter* which produces it. Alpha is always less than 1. For the *common-emitter connection*, the current gain, beta, is the ratio of the change in the collector current to the small change in the base current required to produce it. Beta is always greater than 1.

TRANSISTOR EQUIVALENT CIRCUITS. In circuit analysis it is useful to consider the *transistor* as represented by a *four-terminal network* of which two terminals are common. This in effect enables the transistor performance to be described in terms of only four independent variables which correspond to input and output currents and voltages in the equivalent network. Several types of equivalent circuit are used for each of the three possible transistor configurations: *common-base, common-collector*, and *common-emitter*. See *Transistor parameters*.

TRANSISTOR INPUT RESISTANCE. The resistance presented by the input terminals of a *transistor* stage in any one of the three possible configurations. See *Transistor parameters*.

TRANSISTOR OUTPUT RESISTANCE. The resistance across the

output terminals of a *transistor* stage in any of the three possible configurations. See *Transistor parameters*.

TRANSISTOR PARAMETERS. If a *transistor* is represented by a *four-terminal network* its circuit performance can be completely defined in terms of the instantaneous currents and voltages at the input and output of this network. Various sets of transistor *parameters* are used to express the circuit equations derived from the network under conditions of low signal input. These sets of parameters are then generally called the small-signal parameters. The z-parameters represent the open-circuit *impedance* parameters. The y-parameters represent the short-circuit *admittance* parameters. The h-parameters or hybrid parameters combine input impedance and output admittance.

TRANSISTOR PENTODE. A *point-contact transistor* with three *emitters* and one *collector* designed for switching, mixing, or modulation.

TRANSISTOR TETRODE. A *transistor* with two *base* connections, an *emitter* and a *collector*, which operates at higher frequencies than the three-electrode design.

TRANSISTORIZED. Descriptive of an equipment which uses *transistors* as opposed to *thermionic valves*.

TRANSISTORS, COMPLEMENTARY. See *Complementary transistors*.

TRANSIT ANGLE. Of a valve or tube, the product of the *transit time* and the angular frequency.

TRANSIT TIME, ELECTRON. The time required for an *electron* to travel between the electrodes in a *vacuum valve* or tube. It becomes increasingly important as the frequency of operation is increased and may form the limiting factor to the speed of operation of the valve. In *klystrons* and *travelling-wave tubes* it becomes an integral part of the design.

TRANSIT TIME, TRANSISTOR. The time required for injected *charge carriers* to diffuse across the *barrier* region.

TRANSITION REGION. The region between two homogeneous *semiconductor* regions where the impurity concentration changes.

TRANSITRON OSCILLATOR. A *negative-resistance oscillator* which employs a *pentode* with a negative *transconductance* characteristic due to the action of a retarding field set up between the *suppressor* and *screen grid* of the valve.

TRANSLATOR. (1) Generally, a device capable of converting information from one form into another form. In *computers*, a network or system having a number of inputs and outputs, connected so that signals representing information expressed in a certain code, when applied to the inputs, cause output signals to appear which are a representation of the input information (with no significant alteration to the meaning) in a different code. (2) In *television* transmitters: a satellite transmitter which supplements the service area of the transmitter.

TRANSMISSION. In this dictionary the term is used to mean the act of transmitting a signal by an electromagnetic, audio, or pressure wave over a wire, waveguide, radio channel, or any other medium or media.

TRANSMISSION LEVEL. Of the signal power at any point in a *transmission system*, the ratio of the power at that point to the power at some other point in the system chosen as a reference point. Usually expressed in *decibels*.

TRANSMISSION LINE. A material structure, e.g. telephone line, *waveguide*, power line, etc., which forms a continuous path from one place to another, along which electric or electromagnetic energy can be directed and transferred.

TRANSMISSION LINE, BALANCED. A *transmission line* which is symmetrical with respect to *ground*.

TRANSMISSION LINE, MATCHED. See *Matched termination*.

TRANSMISSION LINE, SHIELDED. A *transmission line* whose elements confine the propagated electrical energy to a finite space inside a conducting sheath.

TRANSMISSION LOSS. A general term denoting a decrease in power in transmission from one point to another, usually expressed in *decibels*. Frequently abbreviated to 'loss'.

TRANSMISSION MEASURING SET. An instrument comprising a signal source and a signal receiver with known *impedances*, used to measure the *insertion gain* or *loss* of a network or path between those impedances.

TRANSMISSION MODE. A *mode* of propagation along a *transmission line*.

TRANSMISSION PRIMARYS. In *colour television*, the set of three primarys chosen so that each corresponds in amount to one of the three independent signals contained in the colour signal.

TRANSMISSION SYSTEM. Any assembly of elements which are capable of functioning together to transmit signal waves.

TRANSMISSIVITY. Of radiation incident on the boundary between two media, the ratio of the radiation transmitted through the boundary to the component of radiation perpendicular to the boundary.

TRANSMIT–RECEIVE SWITCH (TR SWITCH). An electronic switch, frequently of the gas-discharge type, used in a *radar* system which has a common transmitting and receiving aerial, which automatically decouples the receiver from the aerial during the transmitting period. It usually combines a *gas discharge tube* with a *resonant cavity*, the resonance being destroyed each time the tube fires. Also called a TR cell, or TR box. An *anti-transmit–receive switch* (ATR switch) is often used in conjunction with a TR switch as additional protection for the transmitter from signals returned from the aerial.

TRANSMITTANCE. The ratio of the radiant power transmitted through a material to the incident radiant power.

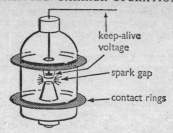

Fig. 165. TR gas-discharge tube

TRANSMITTED-CARRIER OPERATION. That form of AM carrier transmission in which the *carrier wave* is transmitted. Compare *Suppressed carrier transmission.*

TRANSMITTER. A device or equipment which converts *audio, video,* or coded signals into *modulated radio-frequency* signals which can be propagated by *electromagnetic waves.* See the following types of transmitter: *Crystal-controlled, Double-sideband, Facsimile, Frequency-shift, Jim Creek, Radar, Radio, Single-sideband, Spark, Telephone, Television, Vestigial-sideband.*

TRANSMITTER, ALTERNATOR. A radio transmitter which uses power generated by a *radio-frequency alternator.*

TRANSMITTER, AMPLITUDE-MODULATED. A *transmitter* which transmits an *amplitude-modulated* wave. See *Amplitude modulation.*

TRANSMITTER, DOUBLE-SIDEBAND. See *Double-sideband transmission.*

TRANSMITTER, FREQUENCY-MODULATED. A *radio transmitter* which transmits a *frequency-modulated* wave. See *Frequency modulation.*

TRANSMITTER, MULTI-CHANNEL. A *radio transmitter* capable of operating on different frequencies either individually or simultaneously.

TRANSMITTER, MULTI-FREQUENCY. A *radio transmitter* capable of operating on two or more selected frequencies, one at a time, using pre-set adjustments of a single radio-frequency portion.

TRANSMITTER, PHASE-MODULATED. A transmitter which transmits a phase-modulated wave. See *Phase modulation.*

TRANSMITTER, PULSE. A transmitter whose output envelope is in the form of *pulses.*

TRANSMITTER, SINGLE-SIDEBAND. See *Single-sideband transmission.*

TRANSMITTER, VESTIGIAL-SIDEBAND. See *Vestigial-sideband transmission.*

TRANSMITTER FREQUENCY TOLERANCE. Of a *radio transmitter,* the extent to which the *carrier frequency* of the transmitter may be permitted to depart from the assigned frequency.

TRANSPONDER. A transmitter–receiver system whose function is to transmit signals automatically when the proper interrogation is received.

TRANSPORT NUMBER. Of a given type of *ion* in an *electrolyte*, the fraction of the total current carried by that type.

TRANSPOSITION. In open-wire circuits, an interchange of positions of the conductors of a circuit.

TRANSVERSE MAGNETIZATION. In *magnetic recording*, magnetization of the recording medium in a direction perpendicular to the line of travel, and parallel to the greatest cross-section.

TRANSVERSE WAVE. A wave in which the direction of displacement at each point of the medium is perpendicular to the direction of propagation. Transverse electric and magnetic waves play an important part in *waveguide* propagation.

TRANSVERSE-BEAM TRAVELLING WAVE TUBE. A *travelling wave tube* in which the direction of motion of the electron beam is transverse to the average direction in which the signal wave moves.

TRANSVERSE-FIELD TRAVELLING WAVE TUBE. A *travelling wave tube* in which the travelling electric fields which interact with the electrons are essentially transverse to the average motion of the electrons.

TRAP. See *Ion trap, Semiconductor trap, Wave trap.*

TRAVELLING WAVE MAGNETRON. A *travelling wave tube* in which electrons move in crossed electric and magnetic fields which are substantially at right-angles to the direction of propagation.

TRAVELLING WAVE MASER. A *maser* in which the interaction between paramagnetic material and radiation occurs in a non-resonant travelling wave structure.

TRAVELLING WAVE TUBE. An electron tube (valve) used to amplify *ultra-high* and *microwave frequencies*, in which the distributed interaction between an electron beam and an electromagnetic wave travelling along the axis of the tube is used to abstract energy from the beam and to obtain amplification of the wave.

simplified symbol of forward travelling wave amplifier tube.

TREE. A set of connected circuit branches which includes no meshes.

TRIBOELECTRICITY. *Electricity* generated by friction.

TRIBOLUMINESCENCE. *Luminescence* generated by friction.

TRICKLE CHARGE. Of a storage battery, a continuous charge at a low rate approximately equal to the internal losses and suitable to maintain the battery in a fully-charged condition.

TRICOLOUR TUBE. See *Colourtron, Colour television.*

TRIGATRON

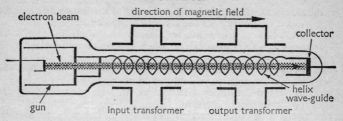

Fig. 166. Diagram of axial field travelling-wave amplifier

TRIGATRON. A *gas-filled spark-gap* switch, e.g. filled with a mixture of argon and oxygen, used in line *pulse modulators*. It includes cathode, anode, and *trigger electrode*.

TRIGGER. Start an action in a system by using a circuit or element which is controlled externally, but which continues to function for a time under its own control once it has started.

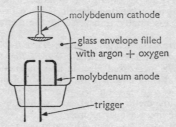

Fig. 167. Trigatron

TRIGGER CIRCUIT. A circuit with two conditions of stability and means for passing from one to the other when certain conditions are satisfied, either spontaneously or by the application of an external stimulus. See *Multivibrator*, *Flip-flop*.

TRIGGER ELECTRODE. A *starter electrode*.

TRIGGER LEVEL. Of a *transponder*, the minimum input to the receiver which is capable of causing a transmitter to emit a reply.

TRIGGER PULSE. A *pulse* which starts a train of events.

TRIMMER CAPACITOR. A small variable *capacitor* associated with a large capacitor and used to adjust the total capacitance.

TRINISTOR®. A range of commercially available *silicon controlled rectifiers*.

TRINITRON®. A television colour tube developed by the **SONY**

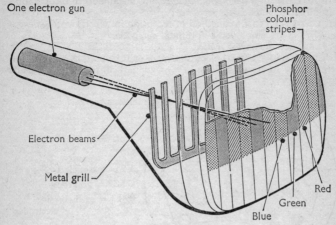

Fig. 168. Trinitron

Corporation which has 3 cathodes but only a single electron gun. It employs a grid like structure, which is much cheaper to produce than a *shadow mask*, and a simple system of alternating stripes of red, green, and blue phosphors, thus eliminating the need for expensive *convergence magnets*. See Fig. 168.

TRIODE. A three-electrode valve containing anode, cathode, and control grid electrode.

TRIODE AMPLIFIER. A valve *amplifier* which uses a *triode*.

TRIODE DETECTOR. A *detector* circuit which uses a *triode*.

symbol

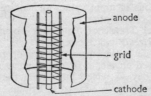

Fig. 169. Construction of a triode

TRIODE-HEXODE CONVERTER. A combination of a triode oscillator and a hexode mixer in the same valve envelope, principally used in a *frequency changer*.

TRIP ACTION. In a *magnetic amplifier*, an unstable form of operation usually due to incorrect *feedback*.

TRIPLE DETECTION. A method of reception in which two *frequency-converters* are employed before final *detection*.

TRIPLER. See *Frequency tripler*.

TROCHOTRON®. A multi-electrode valve or tube used as a *scaler* and employing an electron beam in a magnetic field to charge elements in sequence.

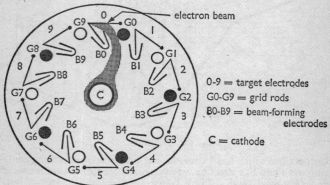

Fig. 170. Construction of trochotron switching and scaling tube

TROPOSPHERE. A region of the earth's atmosphere extending from the earth's surface to a height varying from 18 km. over the Equator to 6 km. over the Poles, which is characterized by convective air movements and a pronounced vertical temperature gradient.

TROPOTRON. A type of *magnetron*.

TRUTH TABLE. A table which describes a logic function by listing all possible combinations of input values and indicating the true output values for each combination.

TTL. Abbreviation for Transistor–Transistor Logic: descriptive of a class of *integrated circuit* logic elements.

TUBE. An electronic device in which *electrons* operate in a gas or in vacuo inside a closed envelope. In American usage this term also includes all those devices which fall within the above definition, but for which the term '*valve*' is used in Britain. There seems to be no hard and fast rule for the use of 'tube' in electronics but it is becoming more popular.

TUBE COUNT. A terminated discharge produced by an ionizing event in a radiation *counter tube*.

TUBE HEATING TIME. The time required for all essential parts of a tube or valve to attain such a temperature that it will operate satisfactorily.

TUBULAR CAPACITOR. A *fixed capacitor* having a tubular form.

TUNABLE MAGNETRON. A *magnetron* with provision for tuning by electrical or mechanical means over a range of frequencies.

TUNED AMPLIFIER. An amplifier in which the anode load impedance is supplied by a *resonant circuit*.

TUNED ANODE. See *Tank circuit*.

TUNED CIRCUIT. A circuit in *resonance*.

TUNED RADIO-FREQUENCY (TRF) RECEIVER. An *amplitude-modulated* receiver with one or more stages of radio-frequency amplification preceding the *detector*. Compare: *Superheterodyne*.

TUNED RELAY. A *relay* having mechanical or electrical resonating arrangements which limit its response to currents at one predetermined frequency.

TUNED TRANSFORMER. A *transformer* whose associated circuit elements are adjusted to be resonant at the frequency of the alternating current supplied to the *primary*, in order that the *secondary* voltage may build up to higher values than can otherwise be obtained.

TUNED-BASE OSCILLATOR. A *transistor oscillator* in which the resonant circuit determining the frequency of oscillation is in the *base* circuit of the transistor. It is the transistor equivalent of the *tuned-grid oscillator*.

TUNED-GRID OSCILLATOR. An oscillator circuit, e.g. *Hartley oscillator*, in which the *grid* is separately tuned.

TUNER. A device for *tuning*.

TUNGSTEN (W). A heavy metallic element, Atomic No. 74. It has an extremely high melting point and is extensively used in *thermionic cathodes* and lamp filaments.

TUNGSTEN CARBIDE. A very hard compound of tungsten and carbon.

TUNING. Adjustment of a circuit or system in relation to frequency; usually, adjusting a circuit or circuits to *resonance*.

TUNING, ELECTRONIC. The process of changing the operating frequency of a system by changing the characteristics – e.g. velocity or density – of an electron stream.

TUNING, SLUG. See *Slug tuning*.

TUNING, THERMAL. In *microwave tubes*, the process of changing the operating frequency of a microwave system by using a controlled thermal expansion to alter the geometry of the system.

TUNING CAPACITANCE. A variable capacitance used for tuning.

TUNING COIL. A *variable inductance* used for tuning.

TUNING INDICATOR. See *Electron-ray tube*.

TUNING RANGE, SWITCHING TUBES. The frequency range over which the resonance frequency of the tube may be adjusted by the mechanical means provided on the tube or associated cavity. See *Transmit–receive switch*.

TUNING-FORK DRIVE. The use of a tuning fork to control the frequency of an oscillator.

TUNNEL DIODE. A *semiconductor* device based on the *tunnelling effect*, consisting essentially of a heavily *doped* junction in a variety of semiconductor materials and exhibiting a negative resistance in the forward direction over a portion of its operating range. It is capable of low noise amplification at frequencies up to and above 1,000 megacycles/second.

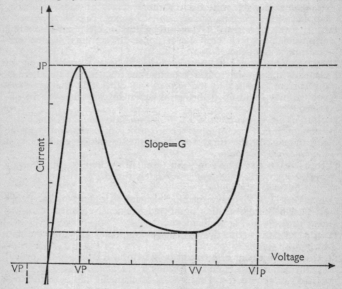

Fig. 171. Tunnel Diode Characteristic

TUNNEL(LING) EFFECT. The observed effect, explained by quantum mechanics, of the piercing of a rectangular potential barrier in a semiconductor by a particle which does not possess sufficient energy to surmount the barrier. The wave associated with the particle suffers almost total reflection at the initial slope of the barrier but a small fraction emerges on the other side.

TURING MACHINE. A mathematical model of a device which will read from, write on, and move an infinitely long tape, thereby constituting a model for computer-like behaviour.

TURRET TUNER. A tuning device for a multi-channel television receiver which allows the contacts for each set of pretuned circuits, corresponding to the various channels, to be connected in turn to the receiver aerial circuit, rf amplifier, and rf oscillator.

TWEETER. A loudspeaker used in *high fidelity* sound systems to reproduce the higher *audio frequencies*, e.g. from 5,000 c/s upwards.

TWIN CHECK. In *digital computers*: a continuous check of the computer operation achieved by duplication of equipment and continuous comparison of results.

TWIN T-NETWORK. A *network* composed of separate T-networks with their terminals connected in parallel.

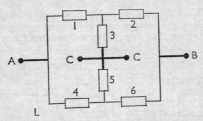

Fig. 172. Twin T-network using resistors

TWIN TRIODE. Two triode vacuum valves in the same envelope.

TWISTED PAIR. A cable made up of two small insulated conductors twisted together without a common covering. This arrangement may be used to provide a small capacity whose value is a function of the length of the pair.

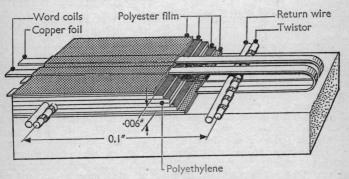

Fig. 173. Construction of twistor memory circuit

TWISTOR. An experimental *memory* device for *computers*, based on the *Wiedemann effect*. See Figure 173.

TWO-ADDRESS PROGRAMMING. *Digital computer programming* in which each complete instruction includes the operation and specific location of two registers, one of which usually contains the operand and the other the result of the operation.

TWO-POSITION ACTION. Automatic control action in which a control element is moved from one fixed position to another fixed position with no intermediate positions.

TWO-WAY COMMUNICATION. Communication between radio stations having both transmitting and receiving equipment.

TWT. Abbreviation for Travelling Wave Tube.

TYPOTRON. A *charge-storage tube* in which a character matrix directs the electron beam to write letters and words on the viewing screen in type form. See Fig. 174 and *Charactron*.

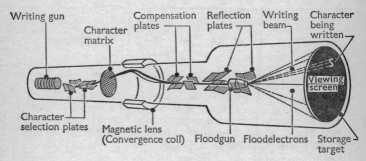

Fig. 174. Typotron

U

UBITRON. A high-power *travelling wave tube*.

UHF. Abbreviation for *ultra-high frequency*.

ULTRA-AUDION OSCILLATOR. A *Colpitts oscillator* in which a section of transmission line is used for the resonant circuit.

ULTRAFAX®. Trademark of a system which combines radio, television, facsimile and film recording for very high speed transmission of printed information.

ULTRASONIC. Descriptive of a device or system in which *ultrasonic frequencies* are used.

ULTRA-HIGH FREQUENCY. Radio frequencies between 300 and 3,000 Mc/s. See *Frequency band*.

ULTRASONIC CLEANING. A general method of separating dirt or any undesired particles from a substance by vibrating it at speeds higher than sound in a suitable stationary medium, or vice versa.

ULTRASONIC COAGULATION. The bonding of small particles into larger aggregates by the action of *ultrasonic waves*.

ULTRASONIC COMMUNICATION. Communication through water using suitably modified *sonar* equipment.

ULTRASONIC DELAY LINE. See *Delay line, ultrasonic*.

ULTRASONIC DETECTOR. A device for the detection and measurement of *ultrasonic waves*.

ULTRASONIC DISPERSION. The production of suspensions of one material in another due to high-intensity *ultrasonic waves*.

ULTRASONIC FREQUENCY. A frequency lying above the *audio frequency* range, usually taken to be above 20 kc/s. The term is mainly used for elastic waves propagated in solids, liquids, or gases.

ULTRASONIC GENERATOR. A device for the production of sound waves of *ultrasonic frequency*.

ULTRASONIC LIGHT VALVE. Essentially a quartz crystal immersed

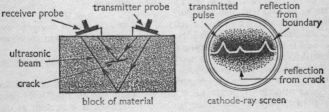

Fig. 175. Ultrasonic crack detection

389

in a transparent fluid and excited at *ultrasonic* frequencies to produce compressive waves therein. This system acts as a liquid *diffraction grating* and, by shining a light through it while modulating the vibrating crystal with the *video* frequencies of a picture signal, it can be used to transmit picture elements.

ULTRASONIC MACHINING. The use of an *ultrasonic generator* and *transducer* with certain kinds of machine tool to facilitate drilling and cutting operations on hard, brittle, or crystalline materials.

ULTRASONIC SOLDERING. A method of tinning metals, such as aluminium, which form a tenacious oxide film on their surface, by a specially designed soldering tool with a bit driven against the work at *ultrasonic* frequency as the solder is applied; or by vibrating the work ultrasonically in a bath of molten tin.

ULTRASONIC STROBOSCOPE. A light interrupter whose action is based on the modulation of a light beam by an *ultrasonic* field.

ULTRASONIC WAVES. Waves whose frequency lies in the *ultrasonic* range.

ULTRASONICS. That branch of technology dealing with sonic waves in a material medium, the waves being similar to sound waves but of higher frequency.

ULTRAVIOLET RAYS (SPECTRUM). That part of the *electromagnetic spectrum* lying between visible light and gamma rays extending from a wavelength of 4000 angstroms (4×10^{-5} cm.) to 200 angstroms (2×10^{-6} cm.). Ultraviolet light plays an important part in *fluorescence* and *ionization*.

UMBILICAL CORD. A cable, which can be disconnected or severed quickly, used to test a missile or to feed in last minute information up to the instant of launching.

UNBALANCED CIRCUIT. A circuit whose two sides are inherently unlike.

UNCONDITIONAL JUMP. See *Unconditional transfer of control.*

UNCONDITIONAL TRANSFER OF CONTROL. In a *digital computer* which obtains its instructions serially from an ordered sequence of *addresses*, an instruction which causes the following instruction to be taken from an address which starts a new sequence.

UNDERBUNCHING. In *velocity-modulated* electron streams, a condition representing less than *optimum bunching.*

UNDERDAMPING. See *Damping, periodic.*

UNDERSHOOT. The initial electrical *transient* change in response to a unidirectional change in input, which precedes the main transition and is in the opposite sense. Compare *Overshoot.*

UNDERWATER SOUND PROJECTOR. A *transducer*, e.g. a *crystal loudspeaker*, used to produce sound in water.

UNDISTORTED WAVE. A *period wave* in which both attenuation and velocity of propagation are the same for all sinusoidal components, and in which no sinusoidal component is present at one point which is not present at all points.

Quantity	Name of practical (c.g.s.) system unit	Name of mksa (rationalized) system unit	For 1 mksa (rationalized) unit equivalent number of		
			practical c.g.s. units	e.m.u.	e.s.u.
Length	Centimetre (cm.)	Metre (m.)	10^2	10^2	10^2
Mass	Gramme (g.)	Kilogramme (Kg.)	10^3	10^3	10^3
Time	Second (t)	Second (t)	1	1	1
Force	Dyne (f)	Newton (f)	10^5	10^5	10^5
Energy, work	Joule	Joule	1	10^7	10^7
Power	Watt	Watt	1	10^7	10^7
Electric charge	Coulomb	Coulomb	1	10^{-1}	3×10^9
Electric potential (e.m.f.)	Volt	Volt	1	10^8	$\frac{1}{3} \times 10^{-2}$
Current	Ampere	Ampere	1	10^{-1}	3×10^9
Surface charge density	Coulomb/cm.^{2}	Coulomb/m.^{2}	10^{-4}	10^{-5}	3×10^5
Volume charge density	Coulomb/cm.^{3}	Coulomb/m.^{3}	10^{-6}	10^{-7}	3×10^3
Electric field intensity	Volt/cm.	Volt/m.	10^{-2}	10^6	$\frac{1}{3} \times 10^{-4}$
Electric dipole moment	Coulomb/cm.	Coulomb/metre	10^2	10	3×10^{11}
Polarization	Coulomb/cm.^{2}	Coulomb/m.^{2}	1	10^{-5}	3×10^5
Electric displacement	—	Coulomb/m.^{2}	$4\pi \times 10^{-4}$	$4\pi \times 10^{-5}$	$12\pi \times 10^5$
Displacement flux	—	Coulomb	4π	$4\pi \times 10^{-1}$	$12\pi \times 10^9$
Surface current density	Ampere/cm.^{2}	Ampere/m.^{2}	10^{-2}	10^{-3}	3×10^7
Volume current density	Ampere/cm.^{3}	Ampere/m.^{3}	10^{-4}	10^{-5}	3×10^5
Resistance	Ohm	Ohm	1	10^9	$\frac{1}{3} \times 10^{-11}$
Conductance	Mho	Mho	1	10^{-9}	9×10^{11}
Resistivity	Ohm/cm.	Ohm/m.	10^2	10^{11}	$\frac{1}{3} \times 10^{-9}$
Conductivity	Mho/cm.	Mho/m.	10^{-2}	10^{-11}	9×10^9
Capacitance	Farad	Farad	1	1	9×10^{11}
Permittivity	—	Farad/m.	$4\pi \times 10^{-9}$	$4\pi \times 10^{-11}$	$36\pi \times 10^9$
Magnetic charge	—	Weber	$\frac{1}{4}\pi \times 10^8$	$\frac{1}{4}\pi \times 10^8$	$\frac{1}{12}\pi \times 10^{-2}$
Magnetic field intensity	Ampere-turn/cm. or Oersted	Ampere-turn/m.	$4\pi \times 10^{-3}$	$4\pi \times 10^{-3}$	$12\pi \times 10^7$
Magnetic induction	Gauss	Weber/m.^{2}	10^4	10^4	$\frac{1}{3} \times 10^{-6}$
Permeability	Gauss/Oersted	Henry/m.	$10^7/4\pi$	$10^7/4\pi$	$\frac{1}{36}\pi \times 10^{-13}$
Magnetic flux	Maxwell (line)	Weber	10^8	10^8	$\frac{1}{3} \times 10^{-2}$
Magnetic potential (m.m.f.)	Gilbert	Ampere-turn	$4\pi \times 10^{-1}$	$4\pi \times 10^{-1}$	$12\pi \times 10^9$
Reluctance	Gilbert/Maxwell	Amp-turn/Weber	$4\pi \times 10^{-9}$	$4\pi \times 10^{-9}$	$36\pi \times 10^{11}$
Inductance	Henry	Henry	1	10^9	$\frac{1}{3} \times 10^{-11}$

Fig. 177. Table of relations between c.g.s. system of practical units and mksa (rationalized), electromagnetic, and electrostatic unit systems

UNFIRED TUBE (GAS SWITCH). Of *TR* and *ATR switches*, the condition during which there is no radio-frequency glow discharge at either the resonant gap or the resonant window.

UNICONDUCTOR WAVEGUIDE. A *waveguide* consisting of a cylindrical metallic surface surrounding a uniform *dielectric* medium.

UNIFORM LINE. A line which has substantially identical electrical properties throughout its length.

UNIFORM WAVEGUIDE. A *waveguide* in which the physical and electrical characteristics do not change along the axis of the guide.

UNIJUNCTION TRANSISTOR. A *transistor* made from a bar of *n-type semiconductor* with a *p-type* alloy region on one side. The connections are made to base contacts at either end of the bar and to the p-region. Also called a filamentary transistor.

UNILATERAL TRANSDUCER. See *Transducer, unilateral*.

UNIPOLAR TRANSISTOR. See *Transistor, unipolar*.

UNIPOTENTIAL CATHODE. See *Indirectly heated cathode*.

UNITS. A chart showing the relation between units used in electronics is provided in Figure 177. See also *SI units*, Fig. 142.

UNIT-STEP FUNCTION. A function which is zero for all values of time prior to a certain instant and unity for all values of time following.

UNIVAC. A series of large *digital computers* and *data-processing systems* built by the Sperry Rand Corporation of the U.S.A.

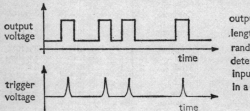

Fig. 176. Univibrator

UNIVIBRATOR. See *Monostable* and *Multivibrator*.

URANIUM (U). A metallic element with a large number of radioactive and unstable isotopes, Atomic No. 92.

V

v.a. Abbreviation for Volt-Ampere.

VACANCY. An unoccupied atomic site in the structure of a crystal from which an atom is missing. Not to be confused with '*hole*'.

VACUUM PHOTOTUBE. A *phototube* which is evacuated to the extent that its electrical characteristics are unaffected by *ionization*.

VACUUM SWITCH. A switch whose contacts are enclosed in an evacuated container, usually to minimize sparking.

VACUUM THERMOCOUPLE. See *Thermocouple, vacuum*.

VACUUM TUBE (VALVE). Any tube (or valve) evacuated to such a degree that its electrical characteristics are unaffected by the presence of residual gas or vapour. Compare *Gas tube*.

VACUUM-TUBE ELECTROMETER. See *Electrometer valve*.

VALENCE. The relative ability of an atom of an element to combine with other atoms.

VALENCE BAND. The range of energy states in the spectrum of a solid crystal which includes the energies of all the electrons binding the crystal together.

VALENCE ELECTRONS. Electrons of an atom lying farthest from the *nucleus*. Valence electrons are shared when atoms combine to form molecules.

VALVE. An electronic device in which electrons operate in a gas or in vacuo inside a closed envelope. The term is tending to be replaced by '*tube*'. See entries following.

VALVE BASE. The term is used to denote the arrangement and the construction of the part of the valve envelope through which the electrode connections are brought out (usually as pins). See Figure 178.

VALVE CHARACTERISTICS. Graphs showing the relation between the various voltages and currents present at the electrodes of the valve, used to determine the performance of the valve in a circuit.

VALVE CHARACTERISTICS, DYNAMIC. The characteristics of the valve under specified operating conditions of a circuit, applied electrode voltages, and frequency; usually represented by a graph showing the relation between a pair of variables, e.g. electrode voltage and current, when all direct electrode supply voltages are maintained constant. They may be determined graphically from the *static characteristic* if the period of the operating frequency is large compared with the electron *transit time*.

VALVE CHARACTERISTICS, STATIC. The relations between pairs of variables such as electrode voltage and electrode current with all other voltages maintained constant.

VALVE ELECTRODE. See *Electrode, Anode, Grid, Cathode*.

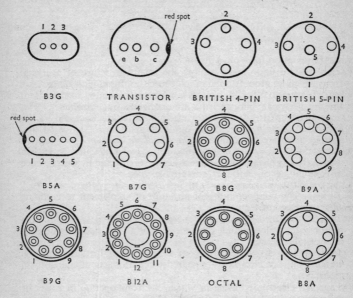

Fig. 178. Valve bases

VALVE ELEMENT. A constituent part of the *valve* which contributes directly to its operation.

VALVE NOISE. *Noise* in *valves* due to vibration, poor circulation, faulty connections, charges on the envelope. *Shot noise, flicker noise, collision ionization, thermal noise,* etc.

VALVE RECTIFIER. A *rectifier* which makes use of the unidirectional flow of current in a valve.

VALVE SOCKETS. Keyed insulating receptacles of various types, usually made of plastic or ceramic, used to fix valves to a chassis and provide a reliable means of connection to the valve pins. In many high-frequency or high-voltage applications the valve sockets are a very important part of the design.

VALVE SYMBOLS. See *Graphical symbols.*

VALVE VOLTMETER. See *Voltmeter, valve.*

VALVEHOLDER. See *Valve sockets.*

VAN DER POL OSCILLATOR. A *relaxation oscillator* employing a single *pentode.*

VANADIUM (V). Rare metallic element resembling titanium, Atomic No. 23.

VARACTOR. A *semiconductor* device characterized by a variation of *capacitance* with voltage.

VARIABLE AREA SOUND TRACK. See *Sound track, variable area*.

VARIABLE CAPACITOR. A continuously adjustable *capacitor* which can have every possible value of capacity within its range. It usually consists of one set of plates or electrodes moving in relation to another set.

VARIABLE CYCLE OPERATION. Computer operation so that cycles may be of different lengths as in an *asynchronous computer*.

VARIABLE INDUCTOR. One whose *inductance* can be varied continuously within its range.

VARIABLE RESISTOR. An adjustable *resistor* which can have every possible value within its range. See Figure 179.

symbol

VARIABLE-MU VALVE. A *valve* in which the *amplification factor* varies in a predetermined way with the voltage on the *control grid*. This is achieved by winding the control grid with variable spacing.

VARIAC®. A rotary *autotransformer* consisting of a single layer winding on a toroidal iron core, the voltage being picked up from the turns by a rotating carbon brush. Continuously variable voltage from zero to about 15 per cent above the supply voltage can be obtained.

VARIOCOUPLER. A *transformer*, used in radio circuits, the mutual impedance between whose windings is adjustable while the self-impedance of those windings remains substantially constant.

VARIOMETER. An adjustable *inductance* consisting of two coils connected in series and mounted so that one can rotate inside the other.

Variometer.

VARIOPLEX. In *time-division multiplexing*: a device which enables the multiplexed channels to be distributed among users according to the number of users.

VARISTOR. A two-terminal *resistor*, made of *semiconductor* material with suitable contacts, which has a markedly non-linear volt-ampere characteristic.

VELOCITY MICROPHONE. A *microphone* whose electrical output corresponds to the particle velocity of the impressed sound waves.

VELOCITY MODULATION. The modification of the velocity of an electron stream by the process of alternately accelerating and decelerating the electrons with a period comparable to the *transit time* in the space concerned.

VELOCITY OF LIGHT. See *Physical constants*.

VELOCITY SORTING. Any process of selecting electrons or other particles according to their velocities.

VELOCITY-MODULATED TUBE (VALVE)

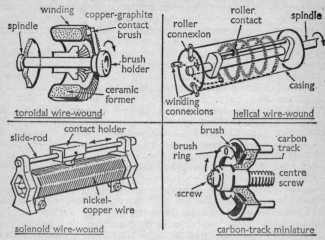

Fig. 179. Construction of types of variable resistor

VELOCITY-MODULATED TUBE (VALVE). A tube or valve for producing *velocity modulation* of a beam of electrons. See *Klystron*.

VERIFIER. A device used in *data processing* to verify the results of key punching.

VERTICAL BLANKING. Of a *cathode-ray tube*, the *blanking* applied to the tube to eliminate the trace of the vertical *sweep* during the *flyback* period.

VERTICAL CENTRING CONTROL. Of a *cathode-ray tube*, a control to shift the starting point of the vertical *sweep* with respect to the face of the tube, in a vertical direction.

VERTICAL RETRACE. In *television*, the path of the electron beam in returning from the bottom of the picture, after a *scanning* sequence, to the top, where a new sequence starts.

VERTICALLY POLARIZED WAVE. A linearly polarized wave whose direction of polarization is vertical.

VERY HIGH FREQUENCY (VHF). Radio frequencies between 30 and 300 megacycles/second. See *Frequency band*.

VERY LOW FREQUENCY (VLF). Radio frequencies below 30 kilocycles/second. See *Frequency band*.

VESTIGIAL SIDEBAND. The remaining transmitted portion of one *sideband* which has been nearly suppressed by a *transducer* with a gradual cut-off in the neighbourhood of the *carrier* frequency. The other sideband is transmitted without much suppression.

396

VESTIGIAL-SIDEBAND TRANSMISSION. A method of signal transmission in which one normal sideband and the corresponding *vestigial sideband* are used.

VHF. Abbreviation for *very high frequency*.

VIBRATING CAPACITOR. One whose *capacitance* is varied cyclically so that an alternating e.m.f. is developed proportional to the charge of the insulated electrode.

VIBRATING CONTACTOR. See *Vibrator*.

VIBRATING-REED ELECTROMETER. An instrument using a *vibration capacitor* to measure a small change.

VIBRATION. See *Oscillation*.

VIBRATION GALVANOMETER. An ac *galvanometer*, in which the moving element vibrates, used to detect the presence of low-frequency alternating currents, i.e. less than 1,000 c/s.

VIBRATOR. An electromagnetic device, usually consisting of a double-acting relay working as an *interrupter*, used to change a continuous steady current into a pulsating current. See *Chopper*.

VIBROTRON. A *triode* valve having an anode which can be moved or vibrated with a force applied external to the envelope.

VIDEO. The conversion and/or display on a *radar* or *television* screen of electronic impulses which can be seen when converted. Also, pertaining to frequencies similar to those resulting from television *scanning*. See *Video signal*.

VIDEO (FREQUENCY) AMPLIFIER. A *wide-band* amplifier capable of amplifying *video frequencies* and used to amplify periodic visual signals in *television* and pulses in *radar*.

VIDEO FREQUENCY. The frequency of the television *video signal*.

VIDEO INTEGRATION. A method of using the redundancy of repetitive signals to improve the output signal-to-noise ratio by summing the successive *video signals*.

VIDEO MAPPING. A procedure in which a chart of an area is superimposed by electronic methods on a radar display.

VIDEO SIGNAL. In a *television* system, the signal which conveys the whole of the intelligence present in the image, together with the necessary synchronizing and equalizing pulses.

VIDEO STRETCHING. A method of increasing the duration of a *video* pulse, sometimes used in navigation systems.

VIDEOPHONE®. A telephone receiver which is also capable of receiving and displaying visible images.

VIDEOTRON. A *monoscope*.

VIDICON. A *camera tube* in which a charge-density pattern is formed by *photoconduction* and stored on that surface of the photoconductor which is scanned by an electron beam, usually of low-velocity electrons. See Figure 180.

VIEW-FINDER. On a *television camera*, the electronic equivalent of an optical view-finder.

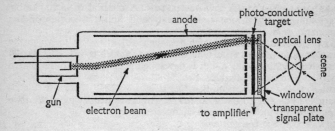

Fig. 180. Diagram of vidicon

VIRTUAL CATHODE. A region between electrodes where the potential of the field is a minimum, and which can be considered to behave as a source of electrons.

VISIBLE SPEECH. A representation on the screen of a *cathode-ray tube* of visual patterns characteristic of spoken words, used as an aid for the deaf.

VISIBILITY FACTOR. In *television* and *radar*, the ratio of the minimum signal-input power detectable by ideal instruments connected to the output of a receiver, to the minimum signal power detectable by a human operator through a display connected to the same receiver.

VISUAL CARRIER FREQUENCY. The frequency of the *carrier* which is modulated by the picture information.

VISUAL TRANSMITTING POWER. The peak power output when transmitting a standard television signal.

VLF. Abbreviation for *very low frequency*.

VOCODER. An electronic device, actuated by recorded voice signals, which will produce synthetic speech.

VODAS. Voice Operated Device Anti-Singing: used in transoceanic *radiotelephone* for automatic suppression of echoes.

VODER. An electronic device, actuated by a mechanical keyboard, which will produce synthetic speech, Compare *Vocoder*.

VOGAD. A voice-operated device used in telephone circuits to give a substantially constant volume output for a wide range of inputs.

VOICE COIL. Of a *loudspeaker*, the coil attached to the cone, which, in conjunction with the magnetic field, drives the cone. Of a *microphone*, the coil attached to the diaphragm, which, in conjunction with the magnetic field, generates the signal.

VOICE FREQUENCY. A frequency within that part of the *audio frequency* range used for the transmission of speech; usually taken to be within the range 200 to 3,500 cycles/second.

VOICE FREQUENCY TELEPHONY. That form of *telephony* in which the frequencies of the components of the transmitted electric waves are substantially the same as the frequencies of corresponding components of the actuating acoustical waves.

VOLATILE MEMORY. In a *computer*, the type of *memory* in which the stored information is lost if the power supply is cut off, e.g. an *acoustic delay line*.

VOLT. The practical unit of *electromotive force* and *potential difference*. It is that electromotive force or potential difference which, applied steadily to a conductor whose resistance is one *ohm*, produces a current of one *ampere*. See *Units*.

VOLT BOX. A series of *resistors* arranged so that a definite fraction of a given voltage may be measured in order that the given voltage may be calculated.

VOLT EFFICIENCY. Of a storage battery, the ratio of the average voltage during the discharge to the average voltage during the recharge.

VOLT, INTERNATIONAL. See *International Electrical System of Units*.

VOLTAGE. Strictly, a difference of *electric potential* expressed in *volts*; but the term is used more generally as a synonym for difference of electric potential.

VOLTAGE AMPLIFICATION. Of a *transducer*, the ratio of the magnitude of the voltage across a specified load impedance connected to a transducer, to the magnitude of the voltage across its input.

VOLTAGE AMPLIFIER. An *amplifier* designed to amplify voltage waveforms in applications where very little power is taken from the load.

VOLTAGE ATTENUATION. Of a device: the ratio of the magnitude of the voltage across the input of the device to the magnitude of the voltage delivered to a specified load impedance connected to the output.

VOLTAGE DIVIDER. A *resistor* or other *impedor* connected across a *voltage* and tapped to make available a variable fraction of the applied voltage.

VOLTAGE DOUBLER (RECTIFIER). An arrangement of *rectifiers* which separately rectifies each half-cycle of an applied alternating voltage, and adds the two rectified voltages to produce a direct voltage whose amplitude is approximately twice the peak amplitude of the applied alternating voltage. See Figure 181.

VOLTAGE DROP. The voltage between the terminals of a circuit element or component due to the flow of current through the *resistance* or *impedance* of that element.

VOLTAGE FEEDBACK. *Feedback* in which the returned voltage is dependent upon the voltage across the output load.

VOLTAGE GAIN. See *Voltage amplification*.

VOLTAGE GENERATOR. An ideal two-terminal circuit element with

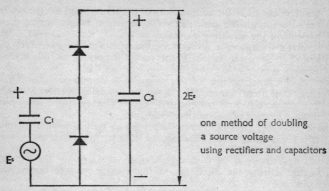

one method of doubling
a source voltage
using rectifiers and capacitors

Fig. 181. Voltage doubler

no internal *impedance* and with terminal voltage independent of the current through it.

VOLTAGE JUMP. In a *glow discharge tube*, an abrupt change or discontinuity in the tube voltage drop during operation.

VOLTAGE LEVEL. At any point in a transmission system, the ratio of the voltage existing at that point to an arbitrary reference voltage. Usually expressed in *dbv*, i.e. *decibels* referred to one volt peak to peak.

VOLTAGE MULTIPLIER. A rectifying circuit which produces a direct voltage whose amplitude is approximately equal to an integral multiple of the peak amplitude of the applied alternating voltage. See *Voltage doubler*.

VOLTAGE REGULATOR. An electronic device which functions so that it (*a*) regulates the voltage across a load for variations in either the load or the supply voltage, or (*b*) maintains the voltage of a capacitor, generator, or motor at a predetermined value or varies it in a predetermined way.

VOLTAGE RELAY. See *Relay, voltage*.

VOLTAGE SELECTOR. See *Clipper*.

VOLTAGE STABILIZER. See *Voltage regulator*.

VOLTAGE TRANSFORMER. An *instrument transformer* designed to have its primary winding connected in parallel with a circuit whose voltage is to be measured or controlled.

VOLTAGE-MULTIPLIER RECTIFIER (COCKCROFT-WALTON). A *rectifier* developed by Cockcroft and Walton to develop a high-potential direct voltage for accelerating charged particles.

VOLTAGE-REFERENCE DIODE. A *semiconductor diode* which develops across its terminals a reference voltage of specified accuracy, when biased within a specified current range.

VOLTAGE-REFERENCE TUBE. A *gas tube* in which the *voltage drop* is approximately constant over the operating range of current.

VOLTAGE-REGULATOR DIODE. A *semiconductor diode* which develops across its terminals an essentially constant voltage throughout a specified current range.

VOLTAGE-REGULATOR TUBE. A *cold-cathode glow discharge tube* which may be used as a *voltage regulator* because of its substantially constant voltage drop over its *plateau*. See *Stabilivolt* and Fig. 149.

VOLTAIC CELL. An electric *cell*, invented by Volta, consisting of two electrodes of different metals in a solution, so that the resulting chemical action produces an *electromotive force*.

VOLTAMETER. See *Coulometer*.

VOLT-AMPERE (VA). A unit in terms of which the product of the *R.M.S.* current in *amperes* and the R.M.S. *volts* is expressed.

VOLT-AMPERE METER. An instrument for measuring the *volt-amperes* in a circuit, provided with a scale graduated in volt-amperes.

VOLTMETER. An instrument for measuring the magnitude of electric *potential difference*, whose scale is calibrated in *volts*, *millivolts* (see *Millivoltmeter*), or *kilovolts*. A voltmeter usually employs the movement of a *D'Arsonval galvanometer*. See the following voltmeter entries.

VOLTMETER, AVERAGE. An ac voltmeter which indicates the average value of the alternating voltage across its terminals.

VOLTMETER, CREST. A *voltmeter* whose indications depend on the crest or *peak value* of the voltage applied to its terminals.

VOLTMETER, ELECTRONIC. A *voltmeter* which is based on the rectifying and amplifying properties of transistors, or valves.

VOLTMETER, ELECTROSTATIC. A *voltmeter* whose action depends on *electrostatic* forces, usually provided with a scale graduated in *kilovolts*.

VOLTMETER, KELVIN. An *electrostatic voltmeter* used to measure very high voltages.

VOLTMETER, MOVING-COIL. An instrument for measuring direct voltages in which a current proportional to the voltage is passed through a coil in a magnetic field producing rotation of the coil. See *D'Arsonval galvanometer*.

VOLTMETER, MOVING-IRON. An instrument for measuring alternating voltages, in which a current proportional to the voltage is passed through a thin vane of soft iron moving in a magnetic field.

VOLTMETER, PEAK. See Voltmeter, Crest.

VOLTMETER, RECTIFIER. An instrument for measuring alternating voltages in which a current proportional to the voltage is rectified in a *bridge rectifier* and read on a sensitive dc *milliammeter*.

VOLTMETER, SLIDEBACK. A *valve voltmeter* in which the alteration in value – slideback – of *grid bias* of a valve, to re-establish a condition of zero anode current after application of a signal voltage to the grid, is used as a measure of the voltage.

VOLTMETER, THERMOCOUPLE. A *voltmeter* which employs *thermocouple*, heating element, and sensitive *millivoltmeter*.

VOLTMETER, VALVE. A *voltmeter* which uses the rectifying and/or amplifying properties of a thermionic valve to measure either direct or alternating voltages. An important feature of valve voltmeters is the high input impedance, as the current used to actuate the meter movement is not taken from the circuit being measured.

VOLTMETER-AMMETER. The combination in a single container, but with separate circuits, of a *voltmeter* and an *ammeter*.

VOLUME. In an electric circuit, the magnitude of a complex *audio-frequency* wave measured on a standard *volume indicator*.

VOLUME CONTROL. See *Automatic volume control, Gain control*.

VOLUME INDICATOR. A standardized instrument for indicating the *volume* of a complex electric wave, such as that corresponding to speech or music, in *volume units*.

VOLUME LIFETIME. The average time interval between the generation and recombination of *minority carriers* in a homogeneous *semiconductor*.

VOLUME LIMITER. A device which automatically limits the output volume of speech or music to a predetermined maximum value.

VOLUME RANGE. Of a transmission system, the difference, expressed in *decibels*, between the maximum and minimum volumes which can be handled by the system.

VOLUME RESISTIVITY. See *Resistivity*.

VOLUME UNIT (VU). A transmission unit for measuring the level of non-steady-state current, for which the zero level (vu = 0) is a steady-state reference power of 1 milliwatt in a circuit of 600 ohms characteristic impedance. The volume in vu is numerically equal to the number of *decibels* expressing the ratio of the magnitude of the waves being measured to the magnitude of the above reference level.

VSWR. Abbreviation for Voltage Standing Wave Ratio.

VTVM. Abbreviation for vacuum-tube valve voltmeter. See *Voltmeter, valve*.

W

WAGNER EARTH BRIDGE. A *bridge* which eliminates the effect of *admittance* to earth.

WALKIE-TALKIE. A two-way radio communications set designed to be carried by one person and capable of operation while in motion.

WARBLE (TONE) GENERATOR. A *voice-frequency* oscillator whose frequency is varied cyclically at a sub-audible rate over a fixed range.

WATT. The practical unit of *power* and the unit of power in the *m.k.s.a.* and *SI* systems. It is the amount of energy expended per second by an unvarying current of one *ampere* under a voltage of one *volt*; or the power required to do work at the rate of one *joule* per second. See *Units*.

WATTHOUR. The product of the power in *watts* and the time in hours; equal to 3,600 *joules*.

WATTHOURMETER. An electricity meter which measures and registers the integral, with respect to time, of the active power of the circuit in which it is connected.

WATTMETER. An instrument for measuring the active power in an electric circuit, provided with a scale graduated in watts or kilo-watts. With the addition of a device for indicating the time inte-gration of the power, a wattmeter becomes a *watthourmeter*.

WATTMETER, DYNAMOMETER (ELECTRODYNAMIC). A *wattmeter* in which low-frequency power is measured by the torque exerted between fixed and moving coils connected to the load current and supply current circuits respectively.

WATTMETER, ELECTRONIC. A *wattmeter* employing valves or transistors, used to measure average ac power.

WATTMETER, ELECTROSTATIC. A *wattmeter* employing *electro-static* forces to measure ac power at high voltage.

WATTMETER, THERMOCOUPLE. A *wattmeter* employing *thermo-couples* in a *bridge* circuit to measure average ac power.

WAVE. A vibrational disturbance in a medium, or in quanta of energy, propagated through space, in which the displacement may be a function of time, or space, or both. Also a graphical representation of such a disturbance or any periodic variation.

WAVE, CENTIMETER. See *Microwaves*.

WAVE, DELTA. See *Delta wave*.

WAVE AMPLITUDE. The maximum level of disturbance of the wave source from its equilibrium state or from some arbitrary reference level (usually zero).

WAVE ANALYSER. An instrument for determining experimentally the frequency components of a complex wave.

WAVE ANALYSER, HETERODYNE. A *wave analyser* using the *heterodyne* principle to measure the frequency components of a complex audio wave.

WAVE ANGLE. Of a *wave* from one point to another, the angle of the line of propagation.

WAVE ATTENUATION. The reduction in the power intensity, or in the amplitude, of a wave with increasing distance from its source.

WAVE CLUTTER. Radar *clutter* caused by echoes from sea-waves.

WAVE DUCT. A *waveguide* with tubular boundaries capable of concentrating the propagation of waves within its boundaries.

WAVE FRONT. Of a signal wave envelope, that part between the initial point of the envelope and the point at which the wave reaches its crest.

WAVE INTERFERENCE. The phenomenon resulting when waves of the same or nearly the same type and frequency are superimposed.

WAVE TRAP. A device for rejecting an unwanted signal, consisting of a circuit tuned to *parallel resonance* connected in *series* with the signal source. When connected, e.g. between the aerial and the input of a receiver, it will impede the unwanted signal to which it is tuned and pass freely signals of other frequencies.

WAVEFORM. The shape of a wave shown graphically in amplitude and time.

WAVEGUIDE. Generally, a system of material boundaries capable of guiding *electromagnetic waves*. Specifically, a *transmission line* consisting of a hollow conducting tube within which electromagnetic waves may be propagated; or a solid dielectric or dielectric-filled conductor for the same purpose.

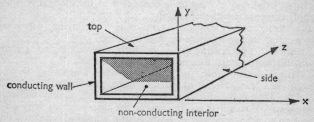

Fig. 182. Rectangular waveguide showing reference axes

WAVEGUIDE, SLOTTED. See *Slotted line.*

WAVEGUIDE BEND, EDGEWISE. See *E-bend.*

WAVEGUIDE BEND, FLATWISE. See *H-bend.*

WAVEGUIDE CONNECTOR. A mechanical device for joining electrically separable parts of a waveguide system.

WAVEGUIDE CONVERTER. A device used in a *waveguide* for converting waves of one *mode* of transmission to waves of another mode.

WAVEGUIDE CORNER. A device used to change the direction of a *waveguide* without any reflections caused by sharp corner bends.

WAVEGUIDE FILTER. Various obstacles and cavities are used in *waveguides* to eliminate undesired frequencies or modes.

WAVEGUIDE IRIS. A metallic diaphragm of a particular shape, material, and design inserted in a *waveguide* to act as a reactance, *susceptance*, or *filter*.

WAVEGUIDE JUNCTION. See *Hybrid-ring junction*.

WAVEGUIDE LENS. An assembly of short, hollow lengths of *waveguide* used to direct radio-waves by refraction.

WAVEGUIDE PLUNGER. A plunger or piston used in *waveguides* and *cavity resonators* to change their effective electrical length.

WAVEGUIDE RESONATOR. A *waveguide* device primarily intended for storing oscillating electromagnetic energy.

WAVEGUIDE STUB. An auxiliary section of *waveguide* joined at some angle with the main section of waveguide.

WAVEGUIDE TERMINATION. A device used to terminate a *waveguide* or *coaxial line* so that it absorbs all of the incident wave power. Also known as a *matched load*.

WAVEGUIDE TRANSFORMER. A device added to a *waveguide* which allows it to be coupled to another waveguide of a different *impedance*.

WAVEGUIDE TUNER. An adjustable device added to, or inserted in, a *waveguide* for the purpose of *impedance* transformation.

WAVE-HEATING. The heating of a material by energy absorption from a travelling *electromagnetic wave*.

WAVELENGTH. Of a periodic wave, the perpendicular distance between two *wave fronts* whose displacements have a difference in *phase* of one *period*. It is also the distance travelled by the wave in one period and is equal to the ratio of *phase velocity* to *frequency*.

WAVEMETER. An instrument for measuring the *wavelength* (or frequency) of a *radio-frequency* wave.

WEATHER SONDE. See *Radiosonde*.

WEBER. The unit of *magnetic flux* in the *mksa* system. See *Units*.

WEIGHTING. The artificial adjustment of measurements, or measuring networks, to compensate for factors which, in normal use of a device, would otherwise be different from those under the conditions of measurement.

WEIGHTING NETWORK. A *network* used specifically for *weighting*.

WENNER WINDING. A technique for winding *resistor* with low *reactance* by looping alternate turns of wire along the winding form.

WESTON NORMAL CELL. A *primary cell* employing mercury as a positive electrode and cadmium in a mercury amalgam as a negative electrode, with a very constant terminal voltage, used as a reference standard of electromotive force. See Fig. 150.

WET CELL. A *cell* in which the electrolyte is in liquid form which can move and flow freely.

WHEATSTONE BRIDGE. A four-arm *bridge*, in which all the arms are predominantly resistive, used for the measurement of *resistance*.

WHIRLWIND. A large *digital computer* built by the Massachusetts Institute of Technology.

WHISTLING ATMOSPHERICS. *Audio-frequency* noises sometimes heard as a background on ocean cables, long wire lines, and the usual *radio interference*; thought to be due to electrical reverberation in the ionosphere.

WHITE COMPRESSION. The reduction in gain applied to a *television* picture signal at those levels corresponding to light areas in a picture, with respect to the gain at that level corresponding to the mid-range light-value in the picture.

WHITE NOISE. See *Noise, white*.

WHITE OBJECT. An object which reflects all wavelengths of light with equal efficiency and with considerable diffusion.

WHITE PEAK. The maximum excursion of a television picture signal in the white direction at the time of observation.

WIDE-BAND AMPLIFIER. An *amplifier* designed to have a flat response over a wide frequency band.

WIEDEMANN EFFECT. The tendency of a twisted magnetostrictive rod carrying an electric current to straighten under the action of a longitudinal magnetic field. See *Magnetostriction, Twistor*.

WIEN BRIDGE OSCILLATOR. A type of *phase-shift oscillator* employing resistance and capacitance in a *bridge* circuit to control the frequency.

WIEN CAPACITANCE BRIDGE. A *bridge* used for the measurement of *capacitance* in the terms of *resistance* and frequency.

WIEN INDUCTANCE BRIDGE. A *bridge* used for the measurement of *inductance* in terms of *resistance* and frequency.

WILLIAMS TUBE. A type of *cathode-ray tube*, designed by F. C. Williams and used as an *electrostatic memory*.

WILLIAMSON AMPLIFIER. A very high quality *audio-amplifier* which uses triodes in push-pull, with negative and positive *feedback*.

WINDOW. In *radar*, long strips of metal foil intended for use as confusion reflectors.

WIRE BROADCASTING. The distribution or relaying of programmes over wire circuits to a large number of receivers, using either *voice frequencies* or *modulated carrier* frequencies. In the case of *television*, the wire circuits are designed to transmit the *video* frequencies.

WIRE GAUGE. Wire gauges normally used in electronics are the British Standard Wire Gauge (SWG) and the Brown and Sharpe Wire Gauge (B & SWG).

WIRE RECORDER. An audio recording system which employs a magnetic wire, magnetized along its length in proportion to the signal.

WIRELESS. *Radio* communication or broadcasting.

WIRELESS TELEGRAPHY. *Radio telegraphy*.

WIRE-WOUND RESISTOR. A *resistor* made from many turns of insulated wire of small cross-section, usually selected to have a low *temperature coefficient* of *resistance*. See *Resistance wire*.

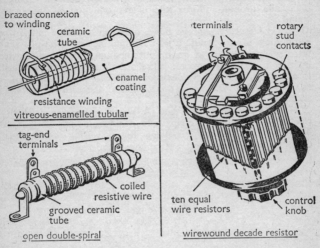

brazed connexion to winding
ceramic tube
enamel coating
resistance winding
vitreous-enamelled tubular

tag-end terminals
coiled resistive wire
grooved ceramic tube
open double-spiral

terminals
rotary stud contacts
ten equal wire resistors
control knob
wirewound decade resistor

Fig. 183. Construction of three types of wire-wound resistor (not to scale)

WIRING DIAGRAM. A diagram of an electronic equipment or circuit showing particularly the arrangement and paths of the wires connecting the circuit components and sub-assemblies.

WOBBULATOR. A *signal generator* whose frequency is automatically varied periodically over a definite range. It is normally used in conjunction with an *oscilloscope* to test the frequency response of a circuit.

WOLFRAM. See *Tungsten*.

WOOFER. A *loudspeaker* used for the low-frequency section of a high-fidelity sound reproduction system. Compare *Tweeter*.

WORD. In a *computer*, an ordered set of characters which is the normal unit in which information may be stored, transmitted, or operated upon in the computer. See *Machine word*.

WORD-TIME. In a *computer*, the time required to transport one *word* from one *storage* device to another.

WORK FUNCTION, THERMIONIC. The thermionic energy required to eject an *electron* from a heated metal. It is the difference between the *total work function* and the energy of the highest-energy electron when the metal is at absolute zero temperature.

WORKING MEMORY. A part of the internal *memory* of a *computer* on which operations are being performed.

WOW. In recorded sound reproduction, a slow periodic change in pitch due to non-uniform rate of reproduction of the original sound.

WRITE. (1) Of *charge-storage tubes*, establish a charge pattern corresponding to the input. (2) In computing and data-processing, transfer information to an output medium; copy, usually from internal to external storage; or record information in a register, location, or other *memory* device. See *Read*.

WRITING SPEED. Of *charge-storage tubes*, the rate of writing on successive storage elements.

WULF-STRING ELECTROMETER. A *string electrometer* which is relatively insensitive to changes in position, and is used in conjunction with a *ionization chamber* for measuring short pulses of current due to ionization.

X

X-AXIS DEFLECTION. Horizontal deflection on the screen of a *cathode-ray tube*.

X-BAND. *Radar* frequencies from 5,200 to 10,900 Mc/s.

XENON (Xe). An inert gaseous element, Atomic No. 54.

XEROGRAPHY. A type of photography in which a selenium-coated surface forms an electrostatic image when exposed to an optical image. A dark powder carrying opposite charge to the electrostatic image adheres to the charged regions and is fixed by heating.

X-GUIDE. A surface wave transmission line with a dielectric structure whose section is X-shaped.

X-RAY CRYSTALLOGRAPHY. The study of the arrangement of the atoms in a crystal by the use of *X-rays*.

X-RAY FOCAL SPOT. The part of the target of an *X-ray tube* which is struck by the main electron stream.

X-RAY MACHINE. An assembly of electric devices for the control and operation of an *X-ray tube*.

X-RAY SPECTROGRAM. A recorded spectrum or diffraction pattern produced by *X-rays*.

X-RAY SPECTROGRAPH. An instrument for producing an *X-ray spectrogram*.

X-RAY SPECTROMETER. An instrument for producing an X-ray spectrum and measuring the wavelengths of its components, using a crystal or radiation counter.

X-RAY TUBE. A vacuum tube designed for the production of *X-rays* by accelerating electrons to a high velocity through an electrostatic field and stopping them suddenly by collision with an anode which forms the target.

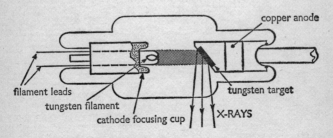

Fig. 184. Construction of typical X-ray tube

X-RAY TUBE, HOT-CATHODE. A high-vacuum *X-ray tube* in which the cathode is heated to produce copious emissions of electrons.

X-RAY TUBE TARGET. An electrode, or part of an electrode, on which *cathode rays* are focused and from which *X-rays* are emitted.

X-RAYS. *Electromagnetic* radiation of wavelength less than about 100 ångströms (10^{-6} cms.) produced when high-energy electrons lose kinetic energy in striking a target; or by the transitions of electrons from higher to lower energy states in an atom. In the latter case, a spectrum typical of each element is produced and the radiation is called characteristic X-rays. When the energy is distributed continuously over a wide spectrum, as is usually the case in an *X-ray tube*, the radiation is known as continuous X-rays. See *Electromagnetic spectrum*.

X-RAYS, SECONDARY. The X-rays emitted by any matter subjected to radiation from primary X-rays, due to the production of high-velocity electrons by the primary radiation and the subsequent production of secondary radiation by these.

XY PLOTTER. A device used in conjunction with a *computer* to plot coordinate points in the form of a graph.

Y

YAGI AERIAL. An aerial array which is the prototype of most television receiving aerials.

Y-AXIS DEFLECTION. Vertical deflection on the screen of a *cathode-ray tube*.

YIG. Yttrium Iron Garnet: one of the most widely used *ferrites* for *microwave devices*.

Y NETWORK. A star network having three branches.

YOKE. A piece of *ferromagnetic* material which permanently connects two or more magnet cores.

YOKE, SCANNING. See *Deflecting yoke*.

Y-PARAMETERS (TRANSISTORS). See *Transistor parameters*.

Y-SIGNAL. In *colour television*, the monochromatic signal which conveys the intelligence of *brightness* or of *luminance*.

Z

ZEEMAN EFFECT. The increase in the number of spectrum lines produced by a light source when subjected to a strong magnetic field.

ZENER BREAKDOWN. Of a *semiconductor diode*: a breakdown caused by the transition of electrons from the valence band to the conduction band due to the *tunnel effect* occurring in a sufficiently strong electric field.

ZENER CURRENT. In a *semiconductor* or *insulator*, the current made up of *electrons* which have escaped from the *valence band* into the *conduction band* under the influence of a strong electric field.

ZENER DIODE. A silicon diode in which the *Zener voltage* is used for voltage stabilizing or as a voltage reference.

ZENER (BREAKDOWN) VOLTAGE. In a *semiconductor*, the voltage in the reverse direction at which the insulating properties of the material break down. Under these conditions the voltage across a *semiconductor junction* remains substantially constant and the current is limited only by the circuit external to the junction.

ZERO LEVEL. A reference level used for the comparison of the relative intensities of sounds or signals. For *audio-frequency* work it is usually taken to be about a power level of 6 milliwatts.

ZERO POTENTIAL. See *Earth*.

ZERO SUPPRESSION. In computing and *data-processing*, elimination of non-significant zeros to the left of the integral part of a quantity especially before the start of printing operations.

ZERO-ACCESS MEMORY. In a *computer*, *memory capacity* for which the waiting time is very small compared to the time for operations other than access to the memory.

ZERO-ADDRESS INSTRUCTION. An instruction in computing or data-processing which does not require an explicit *address*.

ZETA. Zero Energy Thermonuclear Assembly, used for the study of controlled thermonuclear reactions at Harwell.

ZINC (Zn). Metallic element, Atomic No. 30, used as a negative electrode in some electric cells.

ZINC TELLURIDE. A *semiconductor* compound with a *forbidden band* gap of 2·2 electron-volts, and a maximum operating temperature as a transistor of 780° C.

ZIRCONIUM (Zr). Metallic element, Atomic No. 40, used as a *getter* in valves.

ZONE LEVELLING. In the processing of *semiconductors*, the passage of one or more zones, by melting along the body of the material, so as to distribute impurities uniformly throughout the material.

ZONE PURIFICATION (REFINING). The passage of one or more molten zones along a *semiconductor* for the purpose of reducing the concentration of impurity in a part of the material.

ZOOM LENS. A lens system employed in conjunction with a television camera which allows the focal length to be varied continuously over a considerable range, and so avoids the interruptions which occur with a lens turret.

Z-PARAMETERS. See *Transistor parameters.*

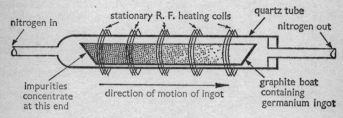

Fig. 185. Zone purification of germanium

Natural Causes

Michael Palmer is the international bestselling author of eleven previous novels, including, most recently, *The Society*. His novels have been translated into twenty-six languages and have been adapted for film and television. He trained in internal medicine at Boston City and Massachusetts General Hospitals, spent twenty years as a full-time practitioner of internal and emergency medicine, and is now involved in the treatment of alcoholism and chemical dependence. He lives in Massachusetts.